Astral Sciences in Early Imperial China

Challenging monolithic modern narratives about 'Chinese science', Daniel Patrick Morgan examines the astral sciences in China *c.*221 BCE–750 CE as a study in the disunities of scientific cultures and the narratives by which ancients and moderns alike have fought to instil them with a sense of unity. The book focuses on four unifying 'legends' recounted by contemporary subjects: the first two, redolent of antiquity, are the 'observing of signs' and 'granting of seasons' by ancient sage kings; and the other two, redolent of modernity, involve the pursuit of 'accuracy' and historical 'accumulation' to this end. Juxtaposing legend with the messy realities of practice, Morgan reveals how such narratives were told, imagined, and re-imagined in response to evolving tensions. He argues that, whether or not 'empiricism' and 'progress' are real, we must consider the real effects of such narratives as believed in and acted upon in the history of astronomy in China.

DANIEL PATRICK MORGAN graduated from the University of Chicago in 2013, and is a researcher at the Centre national de la recherche scientifique, Laboratoire SPHERE (Sciences, Philosophie, Histoire), Université Paris Diderot, where he spent 2013–16 as a member of the European Research Council project Mathematical Sciences in the Ancient World (SAW). Trained as a sinologist, and now working in an interdisciplinary team of historians of science, the author is dedicated to building bridges through the history of science, scholarship, religion, sport and manuscript studies to follow the ancient polymath as he collapses, compartmentalises and cross-pollinates the categories defining his world.

Astral Sciences in Early Imperial China

Observation, Sagehood and the Individual

Daniel Patrick Morgan

Centre national de la recherche scientifique (CNRS), Paris

Shaftesbury Road, Cambridge CB2 8EA, United Kingdom

One Liberty Plaza, 20th Floor, New York, NY 10006, USA

477 Williamstown Road, Port Melbourne, VIC 3207, Australia

314–321, 3rd Floor, Plot 3, Splendor Forum, Jasola District Centre, New Delhi – 110025, India

103 Penang Road, #05–06/07, Visioncrest Commercial, Singapore 238467

Cambridge University Press is part of Cambridge University Press & Assessment, a department of the University of Cambridge.

We share the University's mission to contribute to society through the pursuit of education, learning and research at the highest international levels of excellence.

www.cambridge.org
Information on this title: www.cambridge.org/9781107139022

DOI: 10.1017/9781316488270

First published 2017

A catalogue record for this publication is available from the British Library

Library of Congress Cataloging-in-Publication data
Names: Morgan, Daniel (Daniel Patrick)
Title: Astral science in early China : observation, sagehood, and society / Daniel Morgan, Centre National de la Recherche Scientifique (CNRS), Paris.
Description: Cambridge : Cambridge University Press, 2017. | Includes bibliographical references and index.
Identifiers: LCCN 2017012397 | ISBN 9781107139022 (alk. paper)
Subjects: LCSH: Astronomy – China – History. | Astronomy, Ancient – China. | Astronomical instruments – China – History.
Classification: LCC QB17 .M67 2017 | DDC 520.931–dc23
LC record available at https://lccn.loc.gov/2017012397

ISBN 978-1-107-13902-2 Hardback

To my father,
who taught me how to shoot

Contents

Figures

Acknowledgements

The research behind the present monograph was funded in reverse chronological order by the European Research Council, under the European Union's Seventh Framework Programme (FP7/2007–2013)/ERC Grant agreement no 269804; by the Andrew W. Mellon Foundation; and by the University of Chicago's Center for East Asian Studies. If the ideas in this book prove to be any good, it is because they were incubated in a warm room, on a full stomach, at some of the best research institutions in the world, then flown around to cavort with others. First thanks go to those who provided the means.

The present work grew out of a hurried dissertation at the University of Chicago, for which I have Donald Harper, Edward Shaughnessy and Paul Copp to thank for the opportunity, criticism and guidance that they devoted to me as their student. Equally valuable in the formative stages of this project were the friends and colleagues passing through Chicago *circa* 2010–13 who answered my questions, gave me new ones and helped me where I was weak: Kate Agnew, Chen Jian, Chen Wei, Feng Shengjun, Howard Goodman, Ethan Harkness, He Youzu, Marc Kalinowski, Kim Taeju, Esther Klein, David Lebovitz, Alan Levinovitz, Lu Jialiang, Song Huaqiang, Yin Shoufu, Jeffrey Tharsen, Andy Yamazaki, Zhou Boqun and many others. In the middle, in 2011–12, the Mellon Foundation afforded me the opportunity to conduct the early stages of my research at the Needham Research Institute, in Cambridge, where I spent many a late night reading and walking home beneath the stars. It is there that I first felt like a historian of science, and I am grateful to Christopher Cullen, Dong Qiaosheng, Jiří Hudeček, Amy Li, Michael Loewe, Sir Geoffrey Lloyd, John Moffett and the others there who welcomed and nurtured me in this capacity.

The first time I had ever been to Paris was on 15 December 2011, at the invitation of Karine Chemla and the ERC project SAW (Mathematical Sciences in the Ancient World). They asked for a three-hour talk. One talk led to another, which snowballed in three short years into a special volume, a postdoc and my recruitment into the CNRS, all of which have left this project, and this life, forever changed. I have foremost Karine to thank for leading the way and for kicking down doors for me to be here. I also have Hirose Shō, Matthieu

Husson, Agathe Keller, Christine Proust and the other members of SAW to thank for teaching me, in their capacity as Assyriologists, Europeanists and Sanskritists, just as much as any sinologist about how to approach my sources (and how to reference theirs). No less important are the many visitors and outside members with which SAW has put me in regular contact, conversations with Alexander Jones, Rich Kremer, Li Liang, Clemency Montelle, Qu Anjing, Thies Staack and John Steele, in particular, having directly shaped the questions that I have thought here to pose.

I used to giggle at the use of 'we' in academic writing, but what the last few years have taught me is just how collaborative a single-author monograph really is. This book would be a different beast, or nothing, were it not for my two anonymous referees; both were, if anything, far too kind with the original manuscript. As friends, with an intimate knowledge of my strengths and weaknesses, Karine, Howard, Alan and Garret Olberding helped to scour the finished manuscript, providing the exact encouragement and restraint that they know my writing to need. Particularly special thanks go, of course, to my editor, Lucy Rhymer, who has made what was in my imagination an impossible ordeal come together in a matter of months with the grace and concision of a Cook Ding. Would that everything in my life were as effortless as Lucy makes it look.

Reading a manuscript, of course, is nothing like living through its writing. I was silent, I ate poorly, I forgot things and I left several messes, but there with me throughout was Laure, to whom I am grateful for her support and for so enriching my here and now with her kindness, her silliness and her wisdom. I am likewise grateful to Maryse, Antoine, Natacha and Pierre-Louis for giving me a sense of family here, in my new home, while this project came together, and to Maryse and Antoine in particular for the generous use of their château, where I was able to commit my scattered thoughts to writing in peace and country air.

Daniel Patrick Morgan
Pierreclos, 25 February 2016

Introduction

There is something special about astronomy. For Mencius (fl. 320 BCE), writing in Confucius' (551–479 BCE) wake, it was the perfect illustration of knowledge done right.

> Mencius said, 'All discussion of the nature [of things] in the world-below-heaven comes down to precedence/reason (*gu*), and nothing else. Fundamental to precedent-reasoning (*gu*) is the sharp/smoothness [with which it cuts]. What is detestable about philosophers is their chiselling their way through (to their preferred conclusions). If philosophers could simply act as Yu did in guiding the [flood] waters, then there would be nothing to detest of them: Yu guided the water by simply conducting it where it was wont to go without imposition; if philosophers too could act without imposition, then great indeed would their philosophy be. Whatever the heights of the heavens and the distance of the stars, if one seeks out former instances (*gu*), then one can render the solstices of a thousand years without stirring from one's seat' (*Mencius* IVB.26).[1]

Call it *zhi* or *scientia*, the danger of philosophy, one might say, is the way that it devolves into self-referential systems of meaning, its practitioners twining precedence and reason into beguiling yarns when they should be plumbing the order that is. The stakes were high, and no less so for Mencius than for Yu the Great. In Yu's day, humanity clung to high ground, drowned, displaced and ravaged by its rivers, and so too, like the failure of a dyke, did the fall of Zhou in 771 BCE unleash a torrent across the land – a torrent of blood and battle, in the wash of which Mencius mourned as civilisation slipped away. It was by using the nature of water against it that Yu managed to turn the tide, dredging, damming and dyking the river's inexorable course to the sea; and so too, from his *moral* high ground, did Mencius labour to harness the nature of man that it may follow its course to goodness. In the end, the difference between the philosopher and the world-mover, between the crackpot and the prophet, Mencius tells us, is their respective grasp of nature.

Writing a decade after the Great War, amid the rise of the National Socialist movement, the German-born scientist Moritz Schlick could probably sympathise. Prior to his assassination in 1936, Schlick led weekly gatherings at

[1] Tr. modified from Lau (1970, 133) and Pankenier (2013, 420–1).

the University of Vienna to discuss issues of the philosophy of science. The 'Vienna circle', as they were later called, were dedicated in the words of their manifesto to an 'empiricist and positivist' world-conception: that 'there is knowledge only from experience', and that legitimate science depends upon the 'application of a certain method, namely logical analysis'.[2] One of the aims of this 'logical positivist' or 'logical empiricist' movement was to apply the scientific method to philosophical problems so as to bring matters of truth to resolution and relegate matters of meaning to 'metaphysics'. An appeal to Enlightenment values, theirs was a losing battle fought against a culture of growing romanticism, mysticism, nationalism and the Hegelian and Heideggerian 'idealism' that lent it intellectual force. Little had changed, one might say, between Mencius' day and Schlick's, but one detects an echo of Mencian optimism in the latter's words at Stockton, California, in 1931. Speaking on 'The Future of Philosophy', Schlick prophesies how 'imposition' will give way to reason, and how the 'chiselling' will one day stop:

> Of course, the mere fact that thus far the great systems of philosophy have not been successful . . . is no sufficient reason why there should not be some philosophical system discovered in the future that would universally be regarded as the ultimate solution of the great problems. This might indeed be expected to happen if philosophy were a 'science.' . . . There is not the slightest doubt that science has advanced and continues to advance, although some people speak skeptically about science. It cannot be seriously doubted for an instant that we know very much more about nature, for example, than people living in former centuries knew (Schlick 1932, 48).

Now, ironically, we do know better. The value for the solar year in Mencius' time (365¼ days) would, over 'a thousand years', produce an error of some eight days and fourteen degrees in right ascension. Schlick's 'error' is less readily quantified. Logical positivism/empiricism quickly found itself embroiled in the problem of confirmation – the inference of generalisations through particular observations – leading to Rudolf Carnap and Karl Popper's respective retreats to probability and falsification. 'Science' proved difficult demarcate, the floodwalls leaving things like unified field theory out, whilst letting chiselling charlatans in.[3] The end, however, was spelled by Thomas Kuhn's 1962 *Structure of Scientific Revolutions* and by the Edinburgh school's 'strong programme' in the sociology of scientific knowledge. Arguing that science functions by long periods of puzzle solving punctuated by crisis-induced gestalt switches, Kuhn turned the ladder of progress into a carousel, turning us in circles, and the magnifying glass into a mirror reflecting our questions back. The strong programme, on the other hand, demanded that 'good' science is just as deserving of

[2] Neurath (1973, 309).

[3] For criticism of 'positivist' and Popperian philosophies of science, see Putnam (1974), Salmon (1981) and Newton-Smith (1981, 44–76).

explanation as the 'bad', focusing on impartiality, symmetry and reflexivity as concerns why scientists and sociologists believe the things they do.[4] The effect, on both ends, was to desacralise the subject, opening it to sociological analysis and deconstruction like any other element of culture. Shapin and Schaffer (1985) have oriented us towards the objects binding the people and practices into specific scientific cultures; Latour and Woolgar (1986) have invited us into the laboratory to watch these tribes at work; and what has emerged, in the decades since, is something a lot less like Schlick's 'ultimate solution' than a cat's cradle of objects, agencies and cultures.

History remembers Schlick less kindly than it does Mencius. 'Confucianism' is a philosophy; 'P/positivism', on the other hand, is now invoked by many as a curse. 'Positivism' is the name they give the demon possessing men to write in a celebration of modern powers; 'positivism' is the name they bark to command it from its host, but 'positivism', let us not forget, was an actual philosophy, and an exigency of other -isms. 'Naive positivist history of science' is the spectre that Sivin (2009, 551–7) sees in the 'pageant of progress' that modern scholars have made of the history of astronomy. In China, he complains, 'positivism' leads to a focus on decontextualised aims 'that enabled the continuity of China's brilliant civilization' and, in Europe, to the presumption that 'the right aims are those of immaculate European astronomy', the common thread being that 'the teleological force of objective modern knowledge, like an immense magnetic field, pulled all the ancient sciences hesitatingly, against the drag of the past, toward that goal'. But is this *positivism*? Kitcher (1993) gives the object of our struggles a different name:

> Once, in those dear dead days, almost, but not quite beyond recall, there was a view of science that commanded widespread popular and academic assent. That view deserves a name. I shall call it 'Legend'.
>
> Legend celebrated science. Depicting the sciences as directed at noble goals, it maintained that those goals have been ever more successfully realized. For explanations of the successes, we need look no further than the exemplary intellectual and moral qualities of the heroes of Legend, the great contributors to the great advances ...
>
> The noble goals of science have something to do with the attainment of truth ... According to Legend, science has been very successful in attaining these goals. Successive generations of scientists have filled in more and more parts of the COMPLETE TRUE STORY OF THE WORLD ... Champions of Legend acknowledged that there have been mistakes and false steps here and there, but they saw an overall trend toward accumulation of truth, or, at the very least, of better and better approximations to truth. Moreover, they offered an explanation both for the occasional mistakes and for the dominant progressive trend: scientists have achieved so much through the use of SCIENTIFIC METHOD (Kitcher 1993, 3, emphasis in original).

[4] Bloor (1976, 4–5). For a more recent synthesis of the strong programme, see Barnes, Bloor and Henry (1996).

The goal of this book is to come to terms with 'Legend', or something like it, in early imperial China. There too, in the first few centuries of the Common Era, scholars told grandiose stories about the past, present and future of the astral sciences, and we will speak in this book to four of their common themes. Two of these smack of antiquity: the 'observation of signs' and 'granting of seasons' by hallowed sage kings in ancient scripture. The other two smack of modernity, though they are right there in the ancient past: the legend, for the purposes of the introduction, of 'empiricism' and 'progress'. 'Science' is clearly an anachronism when discussing Mencius – which is why I use the term hesitatingly in the plural, denoting 'technical knowledge' – but that is no reason, I must insist, to ignore his 'positivist' faith in solstice reckoning; it is not, from a modern perspective, the strangest thing that he believed. In fitting with the 'strong programme', my approach to such legend will be symmetric, making no distinction between convictions that seem reasonable – reasonable, indeed, for the ancient Chinese mind to have – and those that do not. We will question the reality of these beliefs as social constructs, but what interests us is *how* they were constructed and the realness of their effects upon the practices of their adherents.

What is at stake is precisely 'the ancient Chinese mind', because there is, I should like to emphasise, no one such thing. China remains the home, in many quarters, to a timeless and monolithic scientific tradition defined by its 'practicality', its 'official nature', its deference to authority and its holism. Writing on three neglected divination-board manuals from the eleventh century, Ho Peng Yoke offers a global presentation of 'Chinese science' framed in terms of J.G. Frazer's (1854–1941) laws of sympathetic magic. The divination board, Ho states, 'introduces [a] dimension of matter without form by referring to something that reminds us of what we now call the sixth sense and telepathy. Hence, what the traditional Chinese person viewed as science embodied the non-materialistic world as well as the tangible' (Ho Peng Yoke 2003, 9). Frazer, to be clear, identifies sympathetic magic with 'the savage', calling it 'a spurious system of natural law', 'a fallacious guide of conduct', 'a false science', and 'an abortive art',[5] but Ho appeals to relativism to turn this into a source of modern pride: 'The traditional East Asian view was far more universal than Newton's . . .

[5] Frazer (1911, vol. 1, 53). Citing 'J.G. Frazer, the Cambridge anthropologist of *Golden Bough* fame' (Ho 2003, 8, 23), the essence of Ho Peng Yoke's (ibid., 8–10) headings 'What science was to the traditional Chinese' and 'A basic difference between East and West' comes down to this, in the words of Frazer's first edition: 'A savage hardly conceives the distinction commonly drawn by more advanced peoples between the natural and the supernatural . . . Side by side with [this view], primitive man has another conception in which we may detect a germ of the modern notion of natural law or the view of nature as a series of events occurring in an invariable order without the intervention of personal agency. The germ of which I speak is involved in that sympathetic magic, as it may be called, which plays a large part in most systems of superstition' (Frazer 1890, vol. 1, 8–9).

Today some people advocate "fuzzy logic" and the "chaos theory" as a departure from the rigid Greek rationalism. Perhaps this is a step forward in the direction of the East Asian tradition' (ibid., 10). We have stopped asking why 'the Chinese' did not develop (the real) science (of 'the Europeans'),[6] but whatever our justification for reducing our subject's mental world to a singular metaphysics of correlation – for stopping at Frazer's 'Practical Magic' – it leaves them in a dragon-haunted world absent the possibility of distinction or disenchantment. It leaves 'the Chinese', via post-modern sleight of hand, the self-same 'savage' invented by the nineteenth-century anthropologist.[7]

The problem is that our sources are actually quite explicit about distinctions and disenchantments, and 'Chinese science' gets in the way of us seeing that. It is to this end that we will focus on the astral sciences, and not just because there are more than one. The astral sciences (*tianwen* and *li*) span a fertile range of cultural activities – data collection, theorizing, testing, time control, politics and ritual – they combine what is often presumed to be the most abstract and universal of sciences – mathematics – with practical and philosophical questions like observation, proof and instrument building; furthermore, they connect the world of rarefied intellectual pursuits with that of the clock-punching, omen-fearing Everyman. Heaven is vast, and it was experienced in different ways. Legend is to blame for reducing this variety into unity; my goal, however, is not to overthrow legends, as such, but to return us to those through which our subjects saw the world. What this means is that we need to redraw distinctions of terminology, genres, institutions and exemplars as appropriate to our subjects' usages – the sort of existential boundaries that they argued with the intensity with which we now do 'science' and 'religion'. What it also means is recognising how disagreement shaped and reshaped those distinctions – disagreement between voices, disagreement between centuries, and disagreement, most importantly, between philosophical ideals and everyday practice. 'Legend', I would like to show, provides a solution to disagreement by furnishing it a venue.

As to methodology, I intended for the present work to be theoretical and comparative and, conversely, for theory and comparison not to stand in the way of primary sources. To me, the best way that I can think of to apply recent insights in the history and philosophy of science to my sources is as an aide to *forgetting*, the cornerstone of our trade. Forgetting is essential to the historian

[6] For the closing of the 'why not' question, see Sivin (1982), Hart (1999) and Kim (2004).

[7] Now referred to as 'Chinese cosmology', Henderson (1984) provides an excellent introduction to the topic of correlative thinking in China, and the modern anthropological and sinological approach thereto, as well as how such thinking came under criticism and, ultimately, scepticism over the course of Chinese history. Harper (1998, 10–11; 1999) likewise reminds us of the limited and piecemeal nature of five-agents correlative metaphysics within early 'numbers and procedures' technical knowledge. On the perseverance of nineteenth- and early twentieth-century theories of the 'primitive' in Chinese studies, see Brown (2006).

of the ancient world, because it is only when we let go of modern assumptions about ethics, power, gender, reason and daily tasks that the grunt work of textual and terrestrial archaeology can reveal something of the other beyond the self. I would not want to read Daston and Galison's (2007) conclusions about modern scientific atlases onto premodern omen compendia, for example, but what they tell us to forget about a trans-historical conception of 'objectivity' is as priceless as what we must forget about writing in the age before print. My attempts to reconstruct a coherent discourse from original objects and authors' words often lead in odd directions, and where I feel myself lost is where I turn to comparison. The question for me is not how China and Greece were *different* – a question better entrusted to G.E.R. Lloyd (2006; 2007; 1996) – but whether I can find where Ptolemy, Āryabhaṭīya or a cuneiform tablet might say the same thing in different words. Where they do, I readjust my sinologist's assumptions about what is plausible. Where they do, moreover, the scholars in these respective fields often pose questions that I, as a sinologist, have never thought to ask. At the end of the day, however, I prefer to quote a historian of the seventh century over one of the twenty-first, and to highlight parallels within my subjects' cultures over those without. This is a history of China, so 'China for essence, West for use'.

As to presentation, we begin in Chapter 1 by setting the conceptual, historical and sociological stage for the rest of the book. We begin with an analysis of what the actors' categories *tianwen* and *li* entail as words, practices, textual genres and knowledge of legendarily divine origin, asking, at every step, why actors thought to juxtapose the two. From there, I offer a historical overview of the one – *li* mathematical astronomy – over the course of the Han (206 BCE–220 CE), recounting the same history twice so that we may separate public policy from private practice. We move from there to an overview of the cast of characters and the values, motivations, education, career paths and epistemic contentions at play in that history. With an idea of the players, playing field and rules, we will proceed from there to take up each of the legends holding this world together.

In the following two chapters, we will take an object-oriented approach to the founding legend of *tianwen* and *li* – that the sage kings at the beginning of time realised paradise on earth by 'observing the signs' and 'granting the seasons', and that paradise lost may be regained by much the same. In Chapter 2, we take up 'observing' as mediated by the material instruments through which our early imperial subjects saw. Establishing what was available to whom, when and what they thought to do with it, we shall ask why experts for so many centuries waxed ecstatic on the armillary sphere – an instrument whose history was one of want, waste, confusion, foreign production and approximation. The answer, I offer, is that the philosophical potential of the sphere loomed larger than its material reality in the literati's mind, for which 'observation' (*guan*) was something different than it is today. In Chapter 3, we then take up the question of 'granting the seasons' as mediated by the material instruments of time control – calendars. Using calendars excavated from contemporary tombs and administrative dumps, we will examine the sort of material practicalities that went into the production, distribution and use of calendars in a manuscript culture. Juxtaposing how things *worked* with how things *ought to have worked* according to the classicist's ideology of time control and the astronomer's ideology of accuracy, we will identify contradictions in theory and administrative practice and explore how those contradictions were negotiated. In both cases, we find that what it means to 'observe the signs' and 'grant the seasons' changes over centuries of debate.

In the following chapters, we then turn to the legend of 'empiricism' and 'progress', which I will swap for actors' categories, substantiate with thick description, critically evaluate, and attempt to ground in cultural phenomena external to the astral sciences. In Chapter 4, we will take up the question of 'tightness' and how it is 'verified' via live trial and public debate in the context of calendro-astronomical policy reform. Making a case study of a failed reform at the Cao Wei (220–65 CE) court, I lay out the context of the debate in court and interpersonal politics and offer an analysis of the rhetoric and data mounted by each side of the aisle. The numbers, in this case, speak louder than the words, but the numbers, I argue, are arrived at by something of a game. In Chapter 5, we then turn to the theme of 'accumulation' as the historical trajectory of the astral sciences. Regardless of whether there *is* progress, we will examine how actors recount the history of the field to their own day, spotlighting the passing comments of the practitioner and the fastidious narrative of the historian. 'Accumulation' is so recurrent a theme in the astral sciences, I attempt to show, that it proves a point of contention as to *how*, and not *if*, human knowledge advances, the question being one that we can likewise trace through participants' writings on religious salvation.

In Chapter 6, finally, we will turn to comparison, using the struggle between mathematics and divinity as it is negotiated in early imperial historiography as a lens through which to consider how Greek, Mesopotamian, Indian and European writers thought to construct their respective histories of science and civilisation. Faced with a common dilemma – how to reconcile the infallibility of ancient gods with the advancement of human knowledge – I will attempt to argue that the place for the East in the Western past is a product of the same forces that turned the Chinese mathematician against his gods.

Conventions

Before we begin, I owe the reader an explanation of my conventions. As concerns the Chinese language, I shall be using the modern, international academic standard of Hanyu Pinyin for romanisation, which, for the non-sinologist, is different from the Wade–Giles (Nathan Sivin) and Needham systems. The Pinyin and Chinese characters for people, e.g. Li Chunfeng 李淳風 (602–70 CE), and works, e.g. the Quarter-remainder *li* (*Sifen li* 四分曆) (85/86 CE), are supplied in the Index. In translation, I shall stick as closely to the original language of my sources as possible, preferring to translate *chi-ji* 遲疾 'slow–fast' rather than 'equation of centre', for example, so as to avoid that we think about such things in Ptolemaic terms. As to dates, note that I shall render dates in Y-M-D format, giving 'Martial Establishment 19-IV^2-5, *wuzi*.$_{25}$' instead of 'Martial Establishment era, year nineteen, intercalary month four, day five, day *wuzi* (25) in the sexagenary cycle' (1 June 43 CE). I translate official and noble titles as per Hucker (1985) with slight modifications for astronomical offices. As to co-ordinates in the sexagenary cycle (e.g. *wuzi*.$_{25}$), heavenly stems (e.g. *wu*.$_{S05}$), earthly branches (e.g. *zi*.$_{B01}$), twenty-eight lodges (e.g. Wings.$_{L27}$), and twenty-four *qi* (e.g. spring equinox.$_{Q04}$), diagrams await you in the Appendix.

1 The World Below

The science of heaven is conducted here, in the world below, so it is here that we must set our stage. If the stage takes several pages to set, it is because the story that we are about to hear is not that of *astronomy* but of her cousin thrice removed and otherwise conceived. 'Astronomy', one might say, is the modern study of space and/as time; it is *exact knowledge* grounded in methodology, precision instruments and institutions, the purpose of which is to resolve questions about the origins and workings of our world otherwise left to charlatans and speculation. That too is what the ancients said about what they were doing. The questions, methods and tools were different, of course, but so too were the charlatans. There are *always* charlatans, even in fables, and this is as much their story as anyone else's, for though it may not have been 'science', 'peer review' or 'superstition', rest assured that the 'expert' defined himself with labels meant to set the two apart. There were fables too, for that matter, as it is by storytelling that man substantiates the labels defining his community; where we might invoke a Copernicus (1473–1543), however, our subjects might reach instead for Liu Hong (fl. 167–206 CE), there being heroes elsewhere and prior to our own.

If we want to understand what our subjects are doing we must start from what they *think* they are doing. With what field do they identify a given practice? How do they define that field in relation to others? Whence did it arise, and to what end? To what examples did they aspire? What terms do they actually use, and what do these terms imply in context? Early imperial China presents us with a world of distinctions every bit as complex and contentious as our own, and since we cannot rely upon ours alone to navigate that world, we therefore begin in Section 1.1 with an outline of textual genres, actors' categories and legend intended to reveal the conceptual frameworks at play.

The history of the astral sciences in China begins in earnest at the inception of empire, in the third century BCE. It is from this point on, that is to say, that the textual record begins and the subject is picked up in the 'monographs' (*zhi*) of the 'standard histories' (*zheng shi*). It is from this point on, moreover, that the narrative transitions from sage kings and culture heroes to the sort of men and women who leave us with biographies, authored works and records of their careers. In Section 1.2, we move to the Qin (221–207 BCE) and Han (206 BCE–220 CE) and into the history, properly speaking, of astronomy and calendrics. The history of this period is one that is often told through a sequence of government reforms – a frame that supports an impression of contiguity, practicality and bureaucratic restraint. As a corrective, I shall tell this history twice, once from the perspective of policy, and once again from that of practice. The goal of this exercise, beyond familiarising the reader with the subject, is to disentangle these historical threads and to provide a framework by which to consider the dynamics between them.

Whether our focus is practice or policy, the history of the astral sciences remains foremost a history of *people*, which brings us finally to the cast of characters. The state, we know, involved itself in astronomy for reasons of ideology and legitimation, but what about the individual? What was his motivation? How did he come to learn and practise *li*? How was he employed, and in what relation did he stand to the state astronomical office? What methods and values, finally, do we see each party express, and how were these negotiated? To address these questions, Section 1.3 offers an overview of the sociology of knowledge and practice in the Qin and Han, covering matters of institutions, individuals, education, employment, methodology, motivations, rhetoric and epistemology. The point here, needless to say, is to reveal (and revel in) the diversity of this community rather than distil a single archetype from its members.

1.1 Intellectual Framework

1.1.1 Genre

Throughout this book, I shall borrow the term 'astral sciences' from Assyriology to refer collectively to what our historical subjects call *tianwen* and *li*. Practices handily identifiable as such are central to the earliest myths and written records in China, but we shall be focusing in this book on the period in which *tianwen* and *li* are well in place as self-identified fields of scholastic discourse.[1] There were *two* astral sciences, so if we are to understand either, in itself, we must begin at the line that our subjects thought to draw between them. Where such distinctions are considered, sinologists tend to trace that line along one of three divides:[2]

[1] On pre-imperial practices pertaining to *tianwen* and *li*, see Keightley (2000, 17–53), Smith (2011) and Pankenier (2013).

[2] The position that ① *tianwen* and *li* are complementary aspects of a single 'astronomy' is implicit in the way in which most studies deploy primary sources; for an explicit statement, see Zhang Wenyu (2008, 5–7). For the position that ② *tianwen* and *li* are sciences of divination and

① *tianwen* : observation :: *li* : computation
② *tianwen* : qualitative : 'astrology' :: *li* : quantitative : 'astronomy'
③ *tianwen* : space : 'astronomy' :: *li* : time : 'calendrics'
The disconnect of these options hints at the difficulty of rendering *tianwen* and *li* into modern terms. Is *li* 'astronomy', or is *tianwen*? And how could either function independent of space or time? For now, let us stick to premodern terms and appeal to ancient parallels.

Most of *tianwen* and *li* comes down to us through the eponymous 'monographs' of the standard histories, of which we possess thirty in 131 fascicles spanning 2,116 years. What compilers thought to put in these monographs tells us something about the scope, organisation and evolution of these categories in their day. The 'Tianwen zhi' inevitably comprises one to three sections. At its core is an annals of observed phenomena, the focus of which is the *zhan* ('reading') and *yan* ('verification') of omens relating to national security – what the Assyriologist might label an 'astronomical diary'.[3] The annals are usually accompanied by a catalogue of heavenly bodies and anomalies listing physical descriptions, cultural associations, and 'when/if'–'then' *zhan* formulae – what the Assyriologist would place with *Enūma Anu Enlil* under 'astral omen literature'. Starting in the fifth century CE, some *tianwen* monographs also open with a history of instrumentation (e.g. armillary spheres) and cosmology (i.e. the shape and workings of 'heaven').[4] Turning to *li*, one notes that *li* often features in a joint monograph with *lü* ('pitch-pipes', 'tono-metrics' or 'standards'); be it a 'Li zhi' or a 'Lü-li zhi', however, *lü* and *li* are segregated such that we may speak of *li* separately. Organised chronologically (rather than topically), the *li* monograph (or monograph section) chronicles noteworthy events up to the end of the dynasty in question in the history of what the Assyriologist would call 'mathematical astronomy' and 'the civil calendar'.[5]

mathematics, respectively, looking incommensurately at the same sky, see Chen Zungui (2006, 1002), Nakayama (1965; 1966), Jiang Xiaoyuan (1991, esp. 1–6, 109–15) and Chen Meidong (2007, 1–16). For the position that ③ the one deals exclusively with space and the other with time, see Needham (1959, 390–408), Nakayama (1965), Kalinowski (1996, 71–2) and Henderson (2006, 97). For an attempt to go beyond these definitions, see Lloyd (2008).

[3] I prefer to render *zhan* as 'omen-reading' rather than 'prognostication', because, as modern and pre-modern scholarship alike affirm, *zhan* are just as often diagnostic as they are prognostic; see for example Chen Meidong (2007, 696–702) and *Yisi zhan*, 1.2a–3a.

[4] On the 'Tianwen zhi' genre, see the translation of the *Book of Jin* monograph in Ho (1966) and the author's forthcoming chapter, 'Heavenly patterns', in Chaussende, Morgan and Chemla (forthcoming). Note that I use the observer's category 'cosmology' as a stand-in for the actor's category *tian ti* ('heaven's form') and 'metaphysics' for *yin-yang wuxing* ('*yin-yang* and five agents' correlative thought) following the distinction made by Sivin (1969, esp. 9 n2), Cullen (1996, xi n2), Kawahara (1996, esp. 3, 288) and pre-1980s sinological usages.

[5] On the 'Lü-li zhi' genre, see the author's forthcoming chapter with Howard L. Goodman in Chaussende, Morgan & Chemla (forthcoming). On the early *lü-li* synthesis, see Kawahara (1991) and Vogel (1994).

Bibliographic monographs from the self-same standard histories shed further light on these categories. The earliest such monograph, the *Book of Han* 'Yi wen zhi' (92 CE), places *tianwen* and *li* as the first of six subcategories of 'Numbers and Procedures' (*shu shu*):[6]

1. '*Tianwen*'
2. '*Li* and Genealogies' (*li pu*)
3. 'Five Agents [Hemerology]' (*wu xing*)
4. 'Milfoil and Tortoise[-shell Divination]' (*shi gui*)
5. 'Miscellaneous [Terrestrial] Omen-Reading' (*za zhan*)
6. 'Morphomancy' (*xingfa*)

All but one of the 190 works that the *Book of Han* lists under 'Numbers and Procedures' are now lost, but their *titles* nonetheless reveal something about their contents. The '*Tianwen*' titles fall into two groups: *zhan* omen series and *zhan-yan* verificatory records. The '*Li* and Genealogies' titles fall into four: *li*, gnomonics (*rigui*), genealogies (*pu*) and mathematics (*suan*). This was not the last word of bibliographic classification. By the time of the *Book of Sui* (656 CE), bibliographers had rolled headings 4–6 into 'Five Agents' and moved 1–3 under 'Masters' (*zi*), alongside philosophy, agriculture, warfare and medicine. '*Tianwen*' and '*Li and* Genealogies' (now '*Li* and Numbers') remained basically unchanged except for their absorption of new genres: under '*Tianwen*' appear titles on cosmology, armillary spheres and 'Brahman' *tianwen*; under '*Li* and Numbers' appear 'Brahman' *li* and mathematics, as well as archaeo-astronomical studies of ancient methods and records. Later bibliographies expand upon the *Book of Sui* framework.[7]

It is interesting to note what sky- and time-related knowledge bibliographers *exclude* from *tianwen* and *li*. First, one finds ritual and festival calendars (e.g. the 'monthly ordinance' genre of *parapegmata*) under the 'Ritual' (*li*) subheading of 'Classics' (*jing*). Second, hemerology (calendar divination) is consistently placed under 'Five Agents'. Third, techniques for absorbing celestial *qi* and travelling to the stars are found alongside works on bodily cultivation and alchemy under 'Recipes and Skills' (*fang ji*).[8] When it came to organising the imperial library, these were different things, and they were shelved accordingly.[9]

[6] *HS* 30.1763–75. [7] *Sui shu*, 34.1018–26; *JTS* 47.2036–9; *XTS* 59.1543–9.

[8] On calendar divination and ritual, see Chapter 3 below. On magico-religious practices involving space and time travel, see Schafer (1977, 234–69), Schipper & Wang (1986), Andersen (1990) and Raz (2005).

[9] On the bibliographic monographs and early imperial libraries, see Drège (1991). Over the last two decades, the *Book of Han* category 'Shu shu' (Numbers and Procedures) has emerged as a cause célèbre among scholars of excavated divination literature as suggestive of the parity (if not indistinction) of the mathematical and divinatory sciences; see Li Ling (2006, 1–24), Song Huiqun (1999), Liu Lexian (2002, 3–52) and Kalinowski (2004). For a word of caution about over-reading and over-extending this actor's category, see Harper (1999) and Xia De'an (2007). On *shu shu* from the perspective of the history of astronomy, see Jiang Xiaoyuan (1991, 47–55) and Kawahara (1996, 54–79).

What survives of *tianwen* and *li* – i.e. what survives of texts self-identified therewith and/or later bibliographically filed thereunder – basically comports with the scope of the respective fields as circumscribed in the standard history 'monographs'. *Li* texts come down to us in two forms: 'procedure texts', to borrow another term from Assyriology, and calendar-tables. Procedure texts, as preserved/excerpted in the monographs, present us with strings of algorithms for computing the position and time of calendro-astronomical phenomena via 'numbers' (*shu*), 'procedures' (*shu*), 'sequence-tables' (*li*) and/or '*du* travelled' (*xingdu*) planetary models (see Section 3.3 below); calendars and planetary tables, mostly recovered archaeologically, provide us with tangible products of such calculation. *Tianwen* literature, on the other hand, comes down to us in the form of compendia – compendia dedicated to *zhan* omen series and astral lore bearing a close relationship with the 'Tianwen zhi' catalogues.[10] By their nature and size, these compendia do, however, tend towards the miscellaneous as concerns bibliographic classification: the ten-fasicle *Yisi zhan* (645 CE), for example, contains whole sections on cosmology, *li*, terrestrial omens and hemerology, while the 120-fascicle *Kaiyuan zhanjing* (729 CE) preserves the earliest sine table, zero and 'Brahman' procedure text in the Chinese language.[11]

Tianwen and *li* form the two sides of a theoretically perfect dichotomy – *that* first-millennium historians, bibliographers and authors could agree on, if not, in practice, where to draw the line. Some topics come to traverse *tianwen* and *li*. The emergence of the instrument-cosmos as a historical topic, for example, shifts some of what once was discussed in the *lü-li* monograph (and what remained '*li*' for the sake of bibliography) into the *tianwen* monograph. Some sub-genres defy easy classification. Omen compendia, as already mentioned, might include fascicles on *li* and mathematics, and some treatises on instruments and cosmology are thoroughly quantitative. Though monograph compilers exclude it from their histories, 'Five Agents' hemerology likewise crept into the *li* procedure text and *tianwen* compendium alike.[12] More importantly, *li*

[10] Chinese omen compendia are collected in *DT-TW* vols. 4 and 5 and *Xijian Tangdai tianwen shiliao san zhong*. Important studies on the topic include Nakayama (1964), Jiang Xiaoyuan (1992b; 2009), Lu Yang (2007) and Chen Meidong (2007, 669–756). I identify these works with *tianwen* on the basis of their titles (many include '*tianwen*' and '*zhan*') and their classification in pre-modern bibliographies.

[11] There are two noteworthy exceptions to the *tianwen–li* distinction: the 'Tianwen xun' chapter of the *Huainanzi* (139 BCE), which is devoted primarily to 'Five Agents' hemerology (see Major 1993, 55–139; Tao Lei 2003), and the *Gnomon of Zhou* (first century BCE/CE), which combines *li* calendrics, *suan* mathematics and cosmology (see Cullen 1996). Tellingly, both date to the beginning of the emergence of the conceptual vocabulary of *tianwen* and *li*, and bibliographers disagree on whether the latter belongs to '*Tianwen*' or 'Li and Numbers' (e.g. *Sui shu*, 34.1018; *JTS* 47.2036; *XTS* 59.1534, 1544, 1547; *Song shi*, 207.5271). On the compilation and hemerological contents of the *Yisi zhan* and *Kaiyuan zhanjing*, see Harper (2010).

[12] See, for example, Martzloff (2009, 82–99).

is almost necessarily paired with some other thing in bibliography and monograph writing – with tono-metrics, genealogy or mathematics – and as much as each pairing makes its own historical sense, one suspects that the issue is more aesthetic than academic: '*tian-wen li-something*' is just more syllabographically balanced.

However nebulous their boundaries, there is nonetheless something to be said for actors' insistence upon dichotomy. The logic to this distinction is not one that is easily captured in modern terms. ① *Observation versus computation* is somewhat apt, but *tianwen* cosmology and observational data rely on calculation (Chapter 2 below), and so too does *li* rely upon observation (see Chapter 4 below). ② *Qualitative versus quantitative* is less apt, given the presence of star catalogues and precision instruments in *tianwen*, and (some) hemerology in *li*. ③ *Space versus time*, lastly, is fatuous, because space and time are of the same fabric in any astronomy. Judging from the contents of their respective literary genres, we might say that the distinction was more one of venue (outdoors versus indoors), hours (night versus day), tools (armillaries versus counting rods), object (sky versus data), assumptions (anomaly versus regularity) and, of course, literatures. It is also worth noting that the corpora embody different cultures of attestation: *tianwen* is a morass of anonymity and pseudepigraphy supposedly transmitted from high antiquity, while *li* seeks to credit and date 'creations' (*zao*) to plausible historical figures.[13] More than anything, the distinction probably reflects the sort of institutional division of labour that we will see in Section 1.3.1 below.

1.1.2 Legend

With a sense of what ancient man was doing under the name of *tianwen* and *li*, let us inquire about the gods. No question is more crucial to our history of science than this, for man told stories about the origins of his activities, and these stories, we shall see, provided him exemplars through which to think and speak about what he was doing. When it comes to myth and the pre-imperial classics, early China studies tend to focus on issues of provenance and authenticity, but it is important that we not fester on what is and isn't 'real'.[14] Our interest is not how these stories were *written* but how they were *read* in later times, and, in this regard, the sage was a historical agent every bit as real as any person or institution in early imperial times.

The story of *tianwen* appears famously in the *Book of Changes* 'Appended Statements' commentary (< 221 BCE). The commentary pairs *tianwen*

[13] Liu Lexian (2002, 341–51; 2007) provide an excellent example of the fluidity of *tianwen* knowledge in comparing parallels between the *Wu xing zhan* from Mawangdui tomb 3 (sealed 168 BCE) and the later River Chart weft *Di lan xi*.

[14] On the provenance and authenticity of the classical sources cited throughout, see Loewe (1993).

('heavenly patterns') with *dili* ('terrestrial forms') as manifestations of an occult order apprehended by the sages and distilled thereby into symbols and human language:

In antiquity, when [Fu]xi reigned as king over the subcelestial realm, [he] looked up and observed (*guan*) the signs (*xiang*) in heaven, [he] looked down and observed the principles (*fa*) of earth, and [he] observed the patterns (*wen*) of birds and beasts and [their] suitabilities with the earth. [He] took what was near at hand from himself, [he] took what was far away from things, and from this [he] invented (*shizuo*) the eight trigrams to enter into connection with the virtue of the light of the spirits and to categorise the circumstances of the myriad things (*Zhouyi zhushu*, 8.166b).[15]

The 'Appended Statements' goes on to narrate how human civilisation unfolded from Fuxi's act of *tianwen*, with later sages like the Divine Husbandman and Yellow Emperor extrapolating key technologies such as fishing, agriculture, markets, boats, carts, city defences, archery, shelter and writing. In so doing, the *Changes* runs through the semantic range of *wen* 文: the 'patterns' manifest in nature, the 'cultural patterns' distilled therefrom, the sages' project of 'civilisation' and the linchpin thereof – 'writing'.[16] 'Heavenly patterns' are more than just patterns in space (constellations) and time (periodicities); they are the very germ and blueprint of civilisation.

The legend of *li* is one whose tendrils thread through pre-imperial literature, appearing in the *Book of Documents*, the *Analects*, the *Discourses of the States*, and the *Classic of Mountains and Seas*.[17] Let us turn, however, to Ban Gu's (32–92 CE) *Book of Han lü-li* monograph preface, which offers a typical early imperial synthesis:

The origin of *li* numbers lies in high [antiquity]. Tradition states that [Thearch] Zhuanxu commanded Chong, the Rector of the South, to administer Heaven, and Li, the Rector of Fire, to administer Earth. Afterwards, the Sanmiao [tribes] corrupted their virtue, and both offices were abandoned. [As a result,] the intercalary remainder got out of order, the *meng-zou* [correspondence] was quashed into extinction, and the Regulators went in the wrong direction. [Sage King] Yao re-educated Chong and Li's descendants that they might inherit their patrimony. Thus do the *Documents* say:

'And so [Yao] charged Xi and He: "In reverent accordance with prodigious heaven, [thou shalt] *li* and *xiang* [defined in Section 1.1.3 below] the sun, moon and stars and respectfully grant human seasons/time . . . The agricultural year (*sui*) has three hundred, sixty and six days, [so] fix the four seasons and complete the agricultural year by means of intercalary month. Earnestly regulate [all the] officials, and all [thy] many achievements shall be resplendent".'[18]

15 Tr. modified from Wilhelm (1967, 328–9). 16 On *wen*, see Schaberg (2001, esp. 57–95).
17 On the mythology of *li*, see Birrell (1993, 91–5) and Jiang Xiaoyuan (1991, 9–108).
18 *Shangshu zhushu*, 2.21a.

> Afterwards, [Yao] transferred [the throne] to Shun in kind, saying 'Oh, thou, Shun! The *li* numbers of heaven rest in thy person', and 'so too did Shun command Yu [the Great] in kind'.[19]
>
> When King Wu of Zhou (r. 1049/45–1043 BCE) called on Jizi, Jizi spoke of the great method in nine chapters and illuminated *li* methods [via] the five regulators (year, moon, day, stars and numbers).[20] Thus, of all those since the Yin (?–1045 BCE) and [Western] Zhou (1045–771 BCE) who have founded patrimonies and initiated reform – all of them rectified the *li* regulators (*ji*) and conformed the colour of court costume to [the *li*], obeying the *qi* of the times in response to the *dao* of heaven.
>
> After the Three Dynasties (Xia, Yin-Shang and Zhou) had disappeared, and at the end of the Five Earls (770–481 BCE), the Clerk's Office (*shiguan*) lost its regulation/discipline (*ji*), and the disciples of its hereditary practitioners were dispersed, some to the Yi and Di [tribes], which is why their records [leave us with] the Yellow Emperor, Zhuanxu, Xia, Yin, Zhou and Lu *li* (the 'six ancient *li*').[21] In the tumult of the Warring States (480–221 BCE), the Qin [managed to] consolidate the word-under-heaven but not pacify it. [The Qin Empire,] for its part, rather pushed the five conquests [theory], believing themselves to have obtained the virtue of water, and so took month X as the first [month of the civil calendar] and promoted the colour black (*HS* 21A.973).

What does all this mean? Well, the sage's first matter of business is intercalation. The civil calendar comprised three parallel cycles: (1) a ≈ 354½-day lunar year (*nian*) of twelve ≈ 29½-day months (*yue*) beginning at new moon (*shuo*); (2) a ≈ 365¼-day solar/agricultural year (*sui*) comprising four seasons (*shi*), twelve medial *qi* (*zhong qi*) and/or twenty-four *qi* ('medial' + 'nodal'); (3) a sexagenary day-count (see Appendix). There is a ≈ 11-day gap between the lunar and solar years – the 'intercalary remainder' (*runyu*). Allowed to accumulate (11 + 11 + 11 . . .), the seasons would fall progressively later in the civil year, driving apart the *zou* (first month of the civil year) and *meng* (first

[19] *Analects* XX.1; tr. modified from Lau (2000, 201). Here the *Book of Documents* 'Dayu mo' chapter reads: 'We luxuriate in thy virtue and esteem thy great achievements; the *li* numbers of heaven rest in thy person, and thou shalt eventually ascend [to the throne] of the great sovereign' (*Shangshu zhushu*, 4.55b).

[20] Yan Shigu (581–645 CE) explains that 'the "great method in nine chapters" is the nine divisions of the "Great Plan" (Hong fan), the fourth of which is the co-ordination of the "five regulators" ' (*HS* 21A.973 (comm.)). Preserved in the *Book of Documents*, the 'Hong fan' is the manifesto for good governance that Jizi, a virtuous minister of the Yin-Shang, supposedly passed to King Wu upon his founding of the Zhou. The 'Hong fan' identifies the 'five regulators' thus: 'the first is the agricultural year (*sui*); the second is the moon; the third is the sun; the fourth is the stars; and the fifth is the *li* numbers' (*Shangshu zhushu*, 12.171b). On the term *ji* 'regulator', see Section 1.1.3 below.

[21] The 'six ancient *li*' are those supposedly implemented by the ruling clans of the Three Dynasties and their bloodlines in the dukedoms, marquisates and kingdoms of the subsequent period of disunion (770–221 BCE). The Yellow Emperor's *li* activities are clearly legendary, but the other 'ancient *li*' are serious historical questions explored in Hirase (1996) and Gassmann (2002). The 'six ancient *li*' became a topic of scholarly interest as early as the Han, by which time what has survived of them was already considered suspect; see Zhang Peiyu et al. (2008, 251–390).

month of the season).[22] The solution is the timely insertion of intercalary months to offset this 'intercalary remainder' (11 + 11 + 11 – 29 . . .). Such a lunisolar calendar, properly regulated, allows you to 'fix the four seasons and complete the agricultural year'. But how do you know? Seasonal and solar progress can be observed indirectly via fixed stars, e.g. the Northern Dipper (Beidou; UMa) or the Regulators (Sheti; Boo) lying on a straight line off the Dipper's handle. Each night at dusk, the Dipper–Regulator line shifts slightly 'leftward' (clockwise) in relation to the horizon, which is divided into twelve branch-directions. It is by these directions that the twelve seasonal months – the first (*meng*), middle (*zhong*), and final (*ji*) months of each season – are 'established' (*jian*) by that line (see Table 1.1).[23] The Regulators, in other words, are the seasonal hour hand of a grand celestial clock.

So sage kings harmonise lunar and solar/agricultural time – a common preoccupation of among premodern states – but what has this to do with court clothing, 'five conquests' and the beginning of the year? In an attempt to reconcile the idea of a 'mandate of heaven' (*tianming*) with the realities of dynastic change, early thinkers argued that kingship passed from one bloodline to another in a natural cycle of 'three concordances' (*san tong*) or 'five virtues' (*wu de*), each of which was associated with parallel series of colours, qualities, months and so on (see Table 1.1). To legitimise their military conquest of the beleaguered Eastern Zhou (770–256 BCE) and its nominal vassal states, therefore, the Qin called 'water' on the Zhou's 'fire', and it symbolically reinforced this by changing its colours from red to black, and New Year's Day from XI-1 to X-1. In modern terms, imagine that there was a failed state whose national flag was a pair of scissors. What the conquerors did was declare themselves 'rock', issue their officers appropriate insignia, and mandate that the year now begin in 'Rocktober'. Such are the subtleties of political ideology to which one must be sensitive in Chinese history.[24]

'*Li* numbers' arose in the context of sagely rule, to get back to the *Book of Han*, and so central were they to this project that it was their transfer alone by which Yao and Shun conveyed their abdication. The hallmark of sagehood, after all, was assuring the 'observation of signs and granting of seasons/time' (*guan xiang shou shi*) – to perceive the cosmic order of spirits and nature and choreograph the world of man to its rhythms. I say 'assure' because sagehood is not about *doing* but about *delegating*, and this, their most solemn of responsibilities, the sages delegated to professionals. The classics record not only the names and deeds of these experts – Chong and Li, Xi and He – but also the very

[22] *Zou* is the first of the twelve months of the civil year as enumerated in the *Erya* and the Chu Silk Manuscript from Zidanku, Changsha; see Wang Zhiping (1998).
[23] On the elements of the Chinese civil calendar, see Martzloff (2009, esp. 23–106). On the Northern Dipper as seasonal indicator, see Chen Jiujin (1994).
[24] On the reform of regalia and the first month, see Kawahara (1996, 80–113).

Table 1.1 *The Dipper directions, first months, three concordances and five virtues*[*]

Date	Medial *qi*	Solar station		Dipper direction		Month	Three Concordances			Five Virtues		
		lodge	*du*				(conc.)	(dyn.)	(colour)	(vir.)	(dyn.)	(colour)
Dec. 24	winter solstice.$_{\text{Q22}}$	Dipper.$_{\text{L08}}$	21 $\frac{8}{32}$	*zi*.$_{\text{B01}}$	N	win II	man	Zhou	red	fire	Zhou	red
Jan. 24	great cold.$_{\text{Q24}}$	Tumulus.$_{\text{L10}}$	5 $\frac{14}{32}$	*chou*.$_{\text{B02}}$	NNE	win III	heaven	Yin	white	metal	Yin	white
Feb. 23	rain water.$_{\text{Q02}}$	Hall.$_{\text{L13}}$	8 $\frac{28}{32}$	*yin*.$_{\text{B03}}$	ENE	spr I	earth	Xia	black	wood	Xia	green
Mar. 26	spring equinox.$_{\text{Q04}}$	Straddler.$_{\text{L15}}$	14 $\frac{10}{32}$	*mao*.$_{\text{B04}}$	E	spr II				earth	Y. Emp	yellow
Apr. 25	grain rains.$_{\text{Q06}}$	Mane.$_{\text{L18}}$	2 $\frac{24}{32}$	*chen*.$_{\text{B05}}$	ESE	spr III						
May 25	small but full.$_{\text{Q08}}$	Triad.$_{\text{L21}}$	4 $\frac{6}{32}$	*si*.$_{\text{B06}}$	SSE	sum I						
Jun. 25	summer solstice.$_{\text{Q10}}$	Well.$_{\text{L22}}$	25 $\frac{20}{32}$	*wu*.$_{\text{B07}}$	S	sum II						
Jul. 25	greater heat.$_{\text{Q12}}$	Stars.$_{\text{L25}}$	4 $\frac{2}{32}$	*wei*.$_{\text{B08}}$	SSW	sum III						
Aug. 25	abiding heat.$_{\text{Q14}}$	Wings.$_{\text{L27}}$	9 $\frac{16}{32}$	*shen*.$_{\text{B09}}$	WSW	aut I						
Sep. 24	autumn equinox.$_{\text{Q16}}$	Horn.$_{\text{L01}}$	4 $\frac{30}{32}$	*you*.$_{\text{B10}}$	W	aut II						
Oct. 25	frost settles.$_{\text{Q18}}$	Root.$_{\text{L03}}$	14 $\frac{12}{32}$	*xu*.$_{\text{B11}}$	WNW	aut III						
Nov. 24	lesser snow.$_{\text{Q20}}$	Basket.$_{\text{L07}}$	1 $\frac{26}{32}$	*hai*.$_{\text{B12}}$	NNW	win I				water	Qin	

[*] *Note:* The 'Date' column, meant only for reference, is calculated for 111 CE (the *Book of Han*'s date of completion) according to the Han Quarter-remainder *li* (85/86 CE). The 'Solar station' column is likewise taken from the Quarter-remainder *li*.

minutiae of their office. In its fantastical blueprint of the Zhou bureaucratic machine, the *Offices of Zhou* assigns the astral sciences to the 'Great Clerk' and his subordinates, the 'Minor Clerk', the 'Reckoner of the Signs', the 'Guardian of the Rule', etc. Together, the Clerk's Office (*shiguan*) served to warn the throne of observed anomalies (*tianwen*), maintain a lunisolar civil calendar (*li*) and select auspicious days for ceremonies of state (hemerology).[25]

One finds versions of this legend repeated throughout *li* discourse – in prefaces,[26] memorials,[27] debates[28], and contemporary philosophy.[29] One finds it also in the very social and institutional order that sustained it. Emperors wanted to play the role of sage king, as did the Qin in the *Book of Han*'s story. Far more interesting than the throne's investment in this legend, however, was the expert's. *Li*, as extolled in Liu Hong and Cai Yong's (133–92 CE) monograph in the *Book of Later Han*, implicated the practitioner himself in sagehood and its most holy and transcendental of manifestations.[30]

> *Li* is possessed of (the?) six virtues of the sage: venerating [heaven's] body/ies (*ti*) through the comprehension of *qi*; venerating [heaven's] patterns (*wen*) through the synthesis of numbers; venerating [heaven's] signs (*xiang*) through the investigation (*kao*) of categories; venerating [heaven's] seasons/time through the initiation of affairs (agriculture and sacrifice); venerating [heaven's] source through the reading (*zhan*) of things past; and venerating [heaven's] course through the knowledge of things to come.
>
> [The sages'] great patrimony has conveyed it [unto us], and blessing and blight are born therefrom – and that is the reason why the true gentleman (*junzi*) arises therefrom, why [he] takes consultation of it in the pursuance of his affairs and why [he] would rather accept his fate than violate it. Now, for the sake of keeping to heaven and conforming to earth, surveying (*kui*) the seasons/time and spreading the teachings, promulgating [*li*] at the Bright Hall and acting as the pole [star] of the people – [for this] there is nothing greater than the monthly ordinances (*yue ling*). When [this] the great affair of the god king is complete, [his] special capacity towards the world-under-heaven is replete. Beyond this, nothing of the teeming taboos and negligent prohibitions (hemerology) does the true gentleman know (*HHS zhi* 3, 3057).

Sagecraft is a joint endeavour, and it is *li* men like Cai and Liu – Cai and Liu remind us – to whom the burden of *sagesse* falls. It is *they* who apprehend the substance, principle, past and future of the cosmos, and it is *they* who shall usher mankind into peace and prosperity by this knowledge. All that is required of the emperor is the silence and the fixity of a distant star. The legend of the sage kings,

[25] See Hucker (1985), entries 1981, 4453 and 6018.

[26] See *SJ* 26.1255–60; *HHS zhi* 3, 3055–8; *Song shu*, 12.227–8; *WS* 107A.2659; *JS* 17.497–8; *Sui shu*, 17.415–16.

[27] See the memorials prefacing the Luminous Inception *li* (237 CE), the Epochal Excellence *li* (445 CE) and the Orthodox Glory *li* (523 CE) in *Song shu*, 12.232–3, 260–2; *WS* 107A.2662–3; *JS* 18.535–6.

[28] See Chapter 4 below.

[29] See the '*Li* Numbers' chapter of Xu Gan's (170–218 CE) *Zhong lun*, cited later in this section.

[30] On the authorship of the Han monographs, see Mansvelt Beck (1990, 56–63).

as *li* men tell it, tends to validate a very particular world view. When it comes to sagecraft, Liu Hong and Cai Yong would have the weight of history come down against the follies of hemerology – *on their side*, that is to say, of the apparently contentious divide that we see in historiographic and bibliographic classification.[31] Indeed, the more that *li* men identify with the sages, the more the sages, in turn, begin to look like *li* men. Consider Xu Gan's (170–218 CE) version of the legend:

> As to the sage kings' creation (*zao*) of *li* numbers in the past, [they] inspected (*cha*) the operation of regulators and pitch pipes (*ji lü*), observed (*guan*) the movement of the 'rotating mechanism' (*yun ji*), plumbed the consecutive culminations of the stars, and apprehended the varying lengths of light and shadow. Thereupon, they constructed instruments to level them, erected gnomons to measure (*ce*) them, set water clocks to investigate (*kao*) them and spread out counting rods to catch them. Thereafter, the origin and [era] heads were aligned [in the past] above, and the medial [*qi*] and new moons were straightened [in the present] below; cold and hot proceeded in proper order, and the four seasons did not err.[32] Now, *li* numbers are the means by which former kings announced the periods of taking life and ordered the rhythms of the initiation of affairs – that which ensures that the people of the myriad states are not remiss in their patrimonies (*Zhong lun*, 2.14a–b).[33]

No longer the inaccessible paragons of spontaneous gnosis we see in the 'Appended Statements', these are sages made in the *li* man's image – inventors, observers, mathematicians and advocates of modern technologies.[34] The sages are admittedly nothing more than fetishes, fashioned for worship by human hands, but what's curious is the *way* of seeing and knowing that they have been fashioned here to exemplify – a way, after Kitcher's (1993, 3) 'SCIENTIFIC METHOD' in the introduction, that I like to call 'the sagely method'.

1.1.3 Terminology

Tianwen, albeit the more miscellaneous of the two categories, is a word that easily translates: 'skywriting', or 'heavenly patterns'. *Li*, on the other hand, can refer to any number of things: a/the 'calendar' or 'astronomical table', a/the 'procedure text' from which these are generated, certain 'tables' and 'sequences' found therein, a/the 'astronomical system' from which the procedure text is distilled, and the study of any of the above.[35] In the classics,

31 On the values reflected in the division and hierarchical organisation of the genres of 'Numbers and Procedures' in the *Book of Han*, see Kalinowski (2004).

32 In Han terms, an 'origin' (*yuan*) is the period it takes for all *li* cycles to coincide anew; a 'head' (*shou*) denotes the beginning of a lesser period, like an 'era' (*ji*), in which everything but the sexagenary day-count coincides; 'medial *qi*' (*zhong qi*) are twelve equal divisions of the solar year, counted from winter solstice, which later intercalation practices link to specific months or 'new moons' of the calendar year. For the twelve medial *qi*, see Table 1.1 and Appendix, Figure 2, below.

33 On the *Zhong lun*, see Makeham (2002).

34 On more orthodox, philosophical approaches to sage gnosis, see Brown & Bergeton (2008).

35 Martzloff (2009, esp. 367–72) and Sivin (2009, 38–40).

furthermore, *li* is a symbol of royal succession, and it is also used as a verb. What are we to make of this polyvalence?

Written variously as 曆, 歷, 厤 and 厤, the word *li* (**lek*) appears to derive from a root meaning of 'sequence'.[36] The second-century CE lexicography *Shuowen jiezi* glosses the phonophoric 秝 as 'sparse and regular' 稀疏適, and it glosses 厤 as 'to put in order' 治.[37] A pictograph of grain stalks standing side by side, 秝 suggests that the underlying metaphor is that of planting grain in 'orderly' (evenly spaced) rows. The addition of the semantic classifier 止 (foot) gives us the most common member of this word family, 歷, which means 'to pass (sequentially) through' units of time or space (as if through rows of grain in a well-kept field). Derived from 日 (sun/star), 曆 extends this sense of 'sequencing' to heavenly objects. One speaks of 'pacing' (*bu*) planets in the context of procedure texts, but rather than lose ourselves in the implications of celestial fields and rows of time, let us recall that what is tilled in astronomy is, more often than not, rows of numbers on a desk. The classical lexicography *Erya*, to this end, glosses the graph 厤 as 'calculation' (*suan*) and 'numbers' (*shu*).[38]

'*Li* numbers', like 'heavenly patterns', are regularities incipient in nature. 'Heavenly patterns' reside in *xiang* 象 ('signs' or 'simulacra'), celestial objects whose forms mirror the myriad things of earth. *Xiang* both *resemble* and *resonate with* their earthly counterparts: a 'broom star' (comet) *looks* like a broom and *heralds* 'sweeping' change, and what happens in Celestial Kitchen (Tianchu, Dra) speaks to the fate of the imperial food service.[39] '*Li* numbers', on the other hand, are manifest in *ji* 紀 ('net threads', 'thread-ends' or 'mark points'). *Ji* are latticeworks of order perceptible only from specific angles as we move around them in time and space. Pankenier (2015) traces the word through a family of weaving-related metaphors, exploring the spatial implications of its original sense as 'mesh'. Extended to *li*'s 'orderly field of grain', I invite the reader to think of *ji* thus, in four dimensions: standing still, a *ji* is the point from which a row of grain falls into visual alignment; 'passing through', it is the undulation between symmetry and sprawl as individual rows come in and out of alignment before your eyes. A *ji*, in the technical vocabulary of the Quarter-remainder *li* (85/86 CE) refers to both the coincidence of

36 My inspiration for this translation comes from He Yan (d. 249), who glosses the term '*li* numbers' as it occurs in the context of Yao's abdication (Section 1.1.2) as *lieci* 列次 ('sequence') in *Lunyu zhushu*, 10.9a (comm.). Unless otherwise noted, phonological reconstructions are as per the Later Han Chinese of Schuessler (2007).

37 *Shuowen jiezi zhu*, 7A.55b, 9B.20b.

38 *Erya zhushu*, 1.6a. On the etymology of *li*, see Karlgren (1957, 227) and Schuessler (2007, 353). On the somewhat curious usage of '*li*' in epigraphic sources from the Western Zhou, see Chao Fulin (2008) and Li Xueqin (2003, 69). I thank Edward Shaughnessy for bringing these articles to my attention. On the terms *shu* 'number' and *suan* 'calculation', see Chemla & Guo (2004, 984–6, 988–9).

39 On *xiang*, see Schafer (1977, esp. 54–6).

midnight, new moon, winter solstice and day *jiazi*.$_{01}$, *and* the periodicity by which that coincidence recurs. This 'era', as we tend to translate it in the history of astronomy, is represented in the *lü* 18,800 and 1,520.

What are *lü*? *Lü* 率, as distinct from quantities, are numbers that have no meaning except as they relate to one another by proportion. In dealing with circles, Zu Chongzhi (429–500 CE) offers that 'the tight *lü* [are] circle diameter 113, circle circumference 355' (*Sui shu*, 16.388). Neither number means anything in itself – '355 *what*?' – but together they give you a rate with which to find a circumference from a diameter or vice versa ($\frac{355}{113}$= 3.141593). Given a diameter, one does not 'multiply by π'; rather, one multiplies by the 'circumference *lü*' and divides by the 'diameter *lü*'. In other words, the *lü* is not the 'ratio' but one of two numbers that you place therein. Such is the nature of '*li* numbers' as they appear in procedure texts, e.g. the Quarter-remainder *li* of 85/86 CE (cited in *HHS zhi* 3, 3058–9):

Era divisor (*ji fa*):	1,520
Era lunations (*ji yue*):	18,800
Obscuration divisor (*bu fa*):	76
Obscuration lunations (*bu yue*):	940
Rule divisor (*zhang fa*):	19
Rule month (*zhang yue*):	235
Circuits of heaven (*zhou tian*):	1,461
Sun/day divisor (*ri fa*):	4
Obscuration days (*bu ri*):	27,759

In time, an 'era' is the resonance period of 1,520 years : 18,000 lunations : 555,180 days : 9,253 sexagenary cycles necessary for winter solstice and new moon to reoccur on midnight on the same sexagenary day. This is based on the 'rule' of 235 months : 19 years, whereby winter solstice and new moon coincide. With a year of 365¼ days (1,461:4), one 'rule' makes 6,939¾ days, so it takes an 'obscuration' of 4 rules : 76 years : 940 months : 27,759 days for the two to coincide at midnight. *In space*, the sun travels one *du* per day to return after one year to the same position in the sky, which makes the circumference of heaven 365¼ *du* (1,461:4). The moon catches the sun 235 times in 19 years, making 235 +19 circuits through the stars, and thus the moon travels $\frac{254}{19}$ circuits per year, $\frac{254}{19}$ *du* per day. There are $29\frac{499}{940}$ (27,759:940) days in a *yue* ('month'), which is also the number of *du* (minus a circuit) that the *yue* ('moon') has travelled in that time. Structurally, Sivin (1969) asks us to think of these numbers like the train of gears driving a clock; functionally, the thing with *lü* is that you can mix and match to turn any one thing into another, be it big or small, sun or moon, time, position or velocity. Time and space are 'threads' of the same 'net', and '*li* numbers' belong to neither – they are *lü*, the magical gateway between kinds.[40]

'*Li* numbers', like 'heavenly patterns', constitute a universal order running deeper than matters of 'astronomy' or 'astrology'. The sages, as we recall, decoded the technologies, social patterns and bureaucratic institutions of civilisation from 'heavenly patterns'.[41] Where the *Book of Documents* comes to the sage kings' abdication by '*li* numbers', the (pseudo) Kong Anguo (d. *c.*100 BCE) commentary glosses the latter as 'the *dao* of heaven', which Kong Yingda (574–648 CE) elaborates to mean 'the number(s) (*shu*) of imperial ascension as sequentially revolved through (*li yun*) by heaven' (*Shangshu zhushu*, 4.56a). What the two Kongs evidently have in mind here is not the transfer of regalia but a sequence of royal succession preordained by time. We see this early on with attempts to legitimise regime change via the three concordances and five virtues, and we see it develop in mid-Han classicism to a sophisticated operation of calculating end times and dynastic collapse.[42] The mandate of heaven, in short, is only good until one's '*li* numbers' are up.

'*Li* numbers', like 'heavenly patterns', are also human artefacts – models fashioned by man in imitation and pursuit of an ineffable order. Open a book of *tianwen*, and one is greeted by written facts and figures – by descriptions, measurements and formulae for identifying and interpreting individual *xiang*. So too when faced with '*li* numbers' in the classics do early imperial scholars understand the ancients to be invoking the products of human artifice. Yao's legendary order 'to *li* and *xiang* the sun, moon and stars' (Section 1.1.2), Sima Qian (*c.*145–*c.*86 BCE) translates as 'to number (*shu*) and model (*fa*) the sun, moon and stars' (*SJ* 1.16). Where the 'Great Plan' chapter of the *Book of Documents* lists '*li* numbers' alongside the year, moon, sun and stars as one of the 'five regulators' of sagely rule, furthermore, Kong Yingda offers the following gloss:

> '*Li* numbers': the numbers of the timing of the *li*-accounted *qi* and new moons that come from the calculation (*suan*) of the sun and moon's travel of their paths – that which goes into the *li* of one year (*Shangshu zhushu*, 12.171b (comm.)).

Here, in Kong's interpretation, the sage king is evidently relying on *knowledge* of nature's *li* as counted, measured, computed, modelled and transmitted by

[40] On *lü* in the context of *suan* 'mathematics', see Chemla & Guo (2004, 119–219, 956–9). On the broader philosophical implications of *lü*, see Chemla (2010). For a longer explanation of the cycles and relationships offered here, see Eberhard & Mueller (1936, 204–19) and Cullen (1996, 20–7, 197–200). On the *du*, see Section 2.1.2 below. My presentation of *lü* as operative in '*li* numbers' is based on the more substantial work of Karine Chemla's forthcoming chapter 'Conjunctions between the sun and the moon, and pursuit problems: mathematical reasoning in Chinese writings on astral sciences', in *Mathematical Practices in Relation to Astral Sciences*, ed. M. Husson, K. Chemla, A. Keller and J. Steele.

[41] See the *Book of Changes* 'Appended Statements' as cited in Section 1.1.2 above (and continued in Chapter 6).

[42] On the calendrics of cataclysm, see Nielsen (2009–10) and Bokenkamp (1994).

human agents – the sort of '*li* numbers' we saw quoted from the Quarter-remainder *li* procedure text.

Unpacked, the origins and entailments of this terminology should clarify why our subjects grouped together some of the objects and activities that they did. This does not bring us to a definition, nor a catch-all translation, nor should it. First, a grapheme's meaning depends on *use*. In the *Book of Documents* alone we find *li* as a noun and a transitive verb, both alone and in compounds, and graphemes and compounds alike are glossed differently according to context. Second, a grapheme's *use* changes over time. *Li* no longer functioned as a verb in early imperial parlance, which is probably what compels Sima Qian to *translate* the *Book of Documents*' object-taking *li* as *shu* ('to number/count'). If it is a definition and a translation we want, therefore, we must go about it case by case like Kong Yingda and Sima Qian. It is in this spirit that I gathered every instance of the grapheme *li* in our period by exhaustive database and manual search and classified them one by one to produce a stratigraphy of context-determined usages from all available linguistic information. The results of this analysis are presented in dictionary-entry form in Table 1.2.

So how are we to talk about the history of astronomy in early imperial China? Modern categories pose an obvious problem. More than anything, static and often undefined anachronisms like 'science' obscure the categories that our historical subjects actually used, and they obscure how those categories changed.[43] Premodern categories are of more obvious utility for exploring premodern thought, but they pose their own problems, particularly in translation. '*Li*', for example, suggests itself as an apt category through which to approach the history of astronomy in China, but not when sinologists persist in its unilateral rendering as 'calendar'/'calendrics' based on the word's modern definition and its purported function of 'granting the seasons'.[44] To reduce everything in Table 1.2 to 'calendars' results in nonsense that obscures and devalues what our historical subjects are doing. Consider the following summary that we might make of the history of lunar anomaly (Sections 1.2.2 and 3.3.3 below): 'Liu Hong's calendar (procedure text) introduced a calendar (table/algorithm) for lunar motion that was immediately recognised in calendrics (mathematical astronomy) for its accordance with the celestial calendar (astronomical phenomena), the result of which was to divorce the calendar (civil calendar) therefrom'. If *li* is more than 'calendars', then I suggest that we

[43] For a detailed treatment on the problem of 'science' in the history of astronomy in the ancient world, I refer the reader to Rochberg (2004).

[44] The terminological distinctions developed in this section, it is worth noting, were inspired by the typologies of '*li*' presented in Sivin (2009, 38–40) and Martzloff (2009, 367–72). Prior to these studies, historians of astronomy like Nathan Sivin and Christopher Cullen had already variously rendered the term as 'astronomy', '(astronomical) system', etc., but 'calendar' otherwise remains the norm.

Table 1.2 *The vocabulary of* li *(origins to the eighth century* CE*)*

曆/歷/歴/厤 ***li*****:** ❶* a celestial 'sequence' (*Documents*, 11.13b); ❷* a/the 'calendar' or 'astronomical table' embodying said sequence (*Zuo Tradition*, Xiang 27; Ai 12); ❸ a/the 'procedure text' from which such tables are generated (*HS j.* 21); ❹ 'tables' and 'sequences' operative within such procedures (Supernal Icon *li*); ❺ a/the 'calendro-astronomical system' as independent of text (*HS j.* 21); ❻* 'to calculate' or 'to sequence' any of the above (*Documents*, 1.8b; uncommon); ❼ the study of any of the above (*HS j.* 21).

曆數 ***li shu*** **('*li* numbers'):** ❶* numbers inherent in heaven and reproduced by man (*Documents*, 11.13b); ❷* one's lot in the cosmically preordained sequence of royal succession (*Documents*, 3.12a); ❸ numbers forming the body of a *li* text/system; ❹ a *li* text/system (*HS* 6.212); ❺ the study of *li* (*HS* 58.2634); ❻ '*li* and mathematics' (*Sui shu*, 34.1026).

天曆 ***tian li*** **('heavenly *li*'):** ❶ the state astronomical system (*SJ* 130.3285; uncommon); ❷ abb. 'the *li* numbers of heaven' 天之曆數, i.e. one's lot in the cosmically preordained sequence of royal succession (*Taiping jing hejiao*, 137.707).

天數 ***tian shu*** **('heavenly numbers'):** ❶ numbers inherent in heaven and their study (*SJ* 27.1343; uncommon).

曆術 ***li shu*** **('*li* procedures'):** ❶ a *li* text/system (*SJ j.* 26); ❷ a 'sequence procedure' for computing latitude or equation of centre (Luminous Inception *li*, in *Song shu, j.* 12); ❸ the study of *li* (*WS* 48.1068).

曆法 ***li fa*** **('*li* method'):** ❶ a *li* text/system (*Song shu*, 12.230).

曆算 ***li suan*** **('*li* calculation'):** ❶ the study of *li* (*HS* 12.258); ❷ to perform *li*-calculation (*Sui shu*, 18.479); ❸ '*li* and mathematics' (*JTS* 47.2039).

星曆 ***xing li*** **('star/planet *li*'):** ❶ stellar/planetary sequences inherent in nature (*Guanzi*, 41.703); ❷ undefined responsibility of the Prefect Grand Clerk (*HS* 99.4170); ❸ undefined mysterious knowledge (*HS* 62.2732; common); ❹ the study of *li* (*JTS* 66.2463).

年曆 ***nian li*** **('civil year *li*'):** ❶ a/the civil calendar (*Zhong lun*, B.13a); ❷ annals (*JTS* 149.4030).

曆日 ***liri*** **('*li* day'):** ❶ a civil day-calendar (134 BCE; uncommon); ❷ 'sequence day', i.e. the number of days entered into the lunar speed or latitude sequence (Supernal Icon *li*).

日曆 ***rili*** **('day *li*'):** ❶ a civil day-calendar (*Lunheng*, 70.994).

曆書 ***li shu*** **('*li* book'):** ❶ a monograph on *li* (*SJ j.* 26; uncommon); ❷ an/the almanac (Song and later).

具注曆 ***ju zhu li*** **('annotated *li*'):** ❶ an almanac (excavated examples from Dunhuang, see *Dunhuang tianwen lifa wenxian jijiao*).

* Asterisks mark usages precedented in the pre-Qin classics. In parentheses, I offer the earliest unambiguous instance of a given usage, followed by a note on its frequency thereafter (up to the eighth century CE).

either leave it ambiguous, in its untranslated state, or appeal to a modern term, in a disambiguating role, to gloss it according to context. If *li* is more than calendars, moreover, I suggest that we prepare ourselves for the possibility that it is about more than 'granting seasons' too.

1.2 History of Events

Now that we have a clearer sense of the constituent elements of the astral sciences of the early imperial period, I must remind the reader that the present work is not intended as a *survey* of these elements, but a focused exploration of how they fed into evolving narratives about 'observing the signs' and 'granting the seasons' in historical time. *Li* is where most of the action is in this regard, so while we will return to *tianwen* in Chapters 2 and 5 below, let us focus for the remainder of this chapter on *li, li* reform and the gentlemen of *li* as circumscribed by our primary sources. To get our bearings, we begin here with a history of events *c.*221 BCE–220 CE as viewed through the lens of government policy and, then, private practice.

1.2.1 A History of Policy

The Zhuanxu Li *of 221* BCE The first documented *li* reform is that conducted by the third-century BCE Kingdom of Qin. In the lead-up to his rule as First Emperor of Qin (r. 221–210 BCE), King Ying Zheng (r. 246–222 BCE) is said to have invoked the five virtues to signal the immanence of his dominion over the Zhou's former territories. In symbolic terms, *Zhou* : *fire*, and *water* > *fire*, so the Qin was obliged to declare itself 'water' and make the gestures appropriate thereto: the theme of court costume and insignia was changed to black and sixes (: *water*), the Yellow River was renamed 'Virtue Waters' (Deshui), and the beginning of the civil year was moved to month X (see Table 1.1). This apparently went hand-in-hand with the implementation of a/the Zhuanxu *li*, about which we know very little except what we can reconstruct from excavated calendars and later sources.[45]

The Grand Inception Li *of 104* BCE The Qin Zhuanxu *li* underwent modification in the first few decades of Han rule, but it was not until the reign of Emperor Wu (r. 141–87 BCE) that calls for new regalia found political traction.[46] Proponents variously cited three reasons for reform:

1. The three concordances (see Table 1.1) dictate that the Han declare itself 'earth' (> *water*) and readopt the Xia calendar, because *Xia* → *Shang* → *Zhou* → ~~*Qin*~~ → *Han* (: *Xia*).
2. Lunar phenomena predicted by the court *li* were 'behind heaven'.
3. The discovery in 113 BCE of an ancient *ding* tripod beneath the altar of the Sovereign of Earth and the (*li*-calculated) concurrence in 105 BCE of winter solstice and new moon on midnight, day *jiazi*$_{.01}$ (25 December) would

[45] *SJ* 6.237 places the Qin's water-virtue reforms in 221 BCE, clues from excavated sources indicate that the change in the civil calendar went back to at least Ying Zheng's first year as king in 246 BCE; see Huang Yi-long (2001) and Zhang & Zhang (2006).

[46] Zhang Peiyu (2007) brings up-to-date excavated sources to bear on the question of the Qin civil calendar, the early Western Han adoption and modification of it, and the tenuousness of either's connection to the Zhuanxu *li* as known from received sources. Chen Meidong (1995, 516–18) furthermore makes the case that such discrepancies might derive from the method of 'borrowing a half-day' (*jie ban ri*) from syzygy to adjust for lag discussed in 104 BCE in *HS* 21A.976.

reproduce for Emperor Wu the conditions that permitted the Yellow Emperor (: *earth*) to achieve transcendence and, thus, merited symbolic action.

It was a petition in 104 BCE from Grand Palace Grandee Gongsun Qing (the architect of reason 3), Hu Sui and Grand Clerk Sima Qian that finally prompted the emperor to action.[47] After soliciting a second opinion from Grandee Secretary Ni Kuan, Emperor Wu issued an edict inaugurating the Grand Inception era (104–101 BCE) in public testimony to his aspirations for immortality. He then ordered the original petitioners to deliberate with Gentleman-in-Attendance Zun and Great Director of Stars She Xing about the creation of a 'Grand Inception *li*'. After a series of observations to determine the basic elements of this *li*, they ultimately declined on account that they 'couldn't do the maths' (*HS* 21A.975), recommending that the emperor recruit *li*-workers (*zhi li*) more up to the challenge.

The emperor assembled a team of more than twenty led by *Li*-worker Deng Ping, Major Ke of Changle, Watchman Yijun of Jiuquan, methodman (*fangshi*) Tang Du (Sima Tan (*c*.165–110 BCE) and Qian's teacher), Luoxia Hong of Ba Commandery, and '*li*-workers from among the folk' (*HS* 21A.975). Luoxia and Deng arrived independently at the same solution: the derivation of fundamental lunar parameters from pitch-pipe and *Book of Changes* numerology. These were verified by observational and computational testing; handed to Grand Clerk Sima Qian, who eliminated seventeen of its competitors by reason of being 'particularly loose (*shu*) and far off (*yuan*)' (*HS* 21A.976); and submitted to a further round of testing. After review, Eunuch Chunyu Lingqu confirmed the 'tightness' (*mi*) of Deng Ping's Grand Inception *li* and submitted it to the throne. In month V, the emperor ordered the *li*'s institution, promoted Deng to Assistant to the Grand Clerk, and returned the beginning of the civil year to month I (Grand Inception 1 thus began in X and ended in XII for a total of fifteen months).[48]

The Grand Inception *li*, like its predecessor, underwent modification. In 78 BCE, Prefect Grand Clerk Zhang Shouwang petitioned the throne to address 'maladjustment of *yin-yang*' (*HS* 21A.978) by implementing pitch-pipe and *li* adjustments – adjustments, he claims, made by the Yellow Emperor and in use since the founding of the Han. Perplexed, Emperor Zhao (r. 86–74 BCE) ordered Zhang's subordinate Xianyu Wangren, the head of *li* services, to interrogate him on the matter, but Zhang refused to submit himself to questioning. Xianyu petitioned to conduct lunar and seasonal observations at the Shanglin Pure Terrace with the Grand Minister of Agriculture and a staff of over twenty with the aim of testing his director's solution for the state *li* against

[47] For more on Sima Qian vis-à-vis the history of astronomy, see Kawahara (1996, 129–47).
[48] On the reform of 104 BCE, see Cullen (1993).

ten others. The throne approved, assigning the Counsellor-in-Chief, an imperial censor, a general-in-chief and a clerk to the General of the Right to assist with 'watching' (*hou*) and 'ranking' (*ke*).

Running from 78 to 74 BCE, Xianyu's observation programme concluded with three noteworthy results. First,

> The Grand Inception *li* was number one; the Grand Inception *li* made by Xu Wanju of Jimo and Xu Yu of Chang'an was also number one (*HS* 21A.978).

Second, the Grand Clerk's 'Yellow Emperor *li*' was both historically and empirically refuted, leading to accusations of 'defying the *dao* of heaven' and 'great irreverence', prosecution for which was remitted by imperial edict. Third, the rather complete fixed-star measurements that they made were apparently incorporated into *Mr Shi's Star Canon*.[49]

Next, under the reign of Emperor Ping (r. 1 BCE–5 CE), Liu Xin (*c*.50 BCE–23 CE) elaborated the basic framework of the Grand Inception *li* with numerology and planetary models into the extant Triple Concordance *li*. His aim, we are told, was 'to explain the *Spring and Autumn Annals*' (*HS* 21A.979), and there is no direct evidence that his *li* saw state service or marked the sort of 'reform' (*gai*) seen in 104 BCE. What we do know is this: Liu completed the Triple Concordance *li* after his transfer to the newly created post of Xi-He, in the Clerk's Office, and amid sweeping reforms of state ritual and regalia surrounding the regent Wang Mang's (*c*.45 BCE–23 CE) consolidation of power.[50]

Nothing to See Here, 9 CE Whatever the connection to Liu's work on pitch-pipes and archaeo-astronomy, Wang Mang inaugurated his 'New' (Xin) Dynasty (9–23 CE) with a new start to the civil year: month I was 'established' at *chou*.$_{B02}$ (16 January ± 15 days) in accordance with 'the Yin dynasty calendar'. This was different from the Qin and Han, who had maintained the same 'Xia dynasty calendar' (month I : *yin*.$_{B03}$, i.e. 14 February ± 15 days) but moved

[49] The relationship between Xianyu Wangren's survey and *Mr Shi's Star Canon* is discussed further in Chapter 2 below.

[50] The exact sequence of events here is difficult to reconstruct. Named after the Xi and He brothers, whom Sage King Yao charged with *li*, Emperor Ping established the post of Xi-He in spring of 1 BCE to 'promulgate indoctrination, prohibit excessive sacrifices and banish [lascivious music]' (*HS* 12.351). The 'Shizhe Hezhong suo ducha zhaoshu sishi yueling wushi tiao' edict places Liu Xin in that office by 4 CE (see Chapter 3 below). The order of Liu's biography, furthermore, intimates that he conducted his pitch-pipe research and authored the 'Triple Concordance *li* and genealogy' by order of Empress Dowager Wang Zhengjun (71 BCE–13 CE) *after* he had been appointed to that post (*HS* 36.1972). For a biography of Liu Xin and his father, Xiang, see Xu Xingwu (2005). On Liu Xin's *lü-li* work, see Teboul (1983), Kawahara (1991), Vogel (1994), Kawahara (1996, 148–95) and Cullen (2017, 32–137). Prof. Cullen generously provide me with a copy of his book while my own was in copy-editing stage; I apologise to him and to the reader for not having the time to better integrate his exemplary contribution and its translations into the present work.

New Year's Day from I-1 to X-1 and back to I-1. In European terms, this is similar to how *December* became the twelfth month; what Wang Mang did, by contrast, was name December 'January'. Why? Because paper beats rock, and Yin beats Xia (: *Han*) (Wang also called 'earth' over the Han's 'fire', which, although it had called 'earth' in only 104 BCE, had since changed its mind). In terms of practicalities, the civil calendar skipped in 9 CE from Inaugural Beginning 3-XI to Establishment of the State 1-I, only to be reset in 24 CE with the insertion of an extra month XII.

There are two remarkable features about this 'first month reform' (*gai zheng*). First, received literature is utterly silent on the topic. As both the earliest and longest-running instance of such reform in Chinese history, it should have set some sort of precedent meriting discussion by later generations; it was instead purged from cultural memory. Indeed, our only textual sources for this reform are edict excerpts engraved on bronze metrological standards that have archaeologically survived from his reign.[51] Second, this episode highlights the generally incoherent and ineffectual nature of politico-scholastic debate on classical institutions and imperial ideology. One pontificates about changing the first month, but we never hear about the three emperors who did (as we will in Chapters 3 and 4 below). One dramatises the priority of regalia, but we're never explained why it's only 'pretenders' who make it theirs. One speaks about the immanence of concordance and virtue cycles, but we never talk about the flipping, skipping, picking and choosing by which *every* dynast arrives at the symbols appropriate to his needs.[52]

The Han was back, Liu Xiu (b. 5 BCE, r. 25–57 CE) having re-established the dynasty's rule from the new capital at Luoyang. After more than a century, however, its *li* was running noticeably behind the observed moon and (less noticeably) ahead of the observed sun. In 32 CE, the Chamberlain for the Imperial stud and the Superior Grand Master of the Palace petitioned to no effect. In 62 CE, expectant appointee Yang Cen memorialised a prediction of a lunar eclipse a full day prior to full moon by the 'official *li*' (*guan li*) – and on 5-VII-15 (7 September 62 CE) heaven proved him right. Emperor Ming (r. 58–75 CE) ordered the Clerk's Office to collect further data from lunar phases, which, over the course of five months, proved Yang consistently on the mark. On 20 December, the emperor ordered Yang to predict the *position* of lunar phases that expectant appointees Zhang Sheng, Jing

[51] See *Qin Han jinwen huibian*, 198, 201, 208.

[52] Consider what we have seen thus far: (1) by 221 BCE, the Qin had justified calling 'water' by switching from the 'production' to the 'conquest' sequence of the five agents; (2) in 104 CE, the Han called 'earth' to conquer Qin's 'water' while also calling the concordance 'Xia' (≠ *earth*) to claim their legitimate inheritance of the *Zhou* mandate, thus (contradictorily) invalidating the Qin's; (3) the Han then changed to 'fire' (: *Zhou*) while maintaining the concordance 'Xia'; (4) in 9 CE, Wang Mang justified his takeover of the Han (once 'earth', now 'fire') *not* by calling the tainted agent 'water' but by appealing to the *production sequence* to claim 'earth' while at once calling the concordance 'Yin' to conquer the Han's 'Xia'. For a taste of the political and scholastic gymnastics involved in such decisions, see Sukhu (2005–6).

Fang, Bao Ye et al. may test these against a 'quarter-remainder method' (*sifen fa*). After a year's trial, in 64 CE, it was found that 'Sheng et al. bull's-eyed (*zhong*) six more items than Cen' (*HHS zhi* 2, 3025). In 66 CE, as nothing happened, expectant appointee Dong Meng submitted from the Clerk's Office yet another plaint about current policy. The emperor kicked it to the Three Excellencies to deliberate with 'those who understood *li*' (*zhi-li zhe*) at the Ministry of Rites, but the debate ran into the summer of 67 CE without resolution. Finally, in an edict of 3 December 69 CE, Emperor Ming announced that Zhang Sheng et al.'s 'quarter-remainder method' would replace Yang Cen's in generating substitute lunar data for the failing 'official *li*' of 104 BCE.[53]

The Quarter-Remainder Li *of 85* CE In 85 CE, faced with systematic errors of five *du* and between three-quarters of a day and a full day for lunar and solar phenomena, Emperor Zhang (r. 76–88 CE) was compelled by the intransigence of both the *li* and its director to turn to the latter's technician subordinates – *li*-workers Bian Xin, Li Fan, et al. The emperor received, reviewed and authorised their new Han Quarter-remainder *li* by II-4 of that very year (18 March 85 CE). His lengthy edict of that date cites the following reasons for policy reform:

1. The classics, warp and weft, show the sage kings to have effected cosmic harmony by marching to heaven's beat, so it must be the *li*'s disruption of the sacrificial schedule that is to blame for the state of affairs in Emperor Zhang's day.
2. The sages knew heaven by repeated experiential investigation thereof, thus 'every 300 years the Dipper *li* must be reformed', and this was evidently such a time.[54]
3. The Quarter-remainder *li*, based on a solar year of 365¼ days, has scriptural precedent in the weft texts.[55]

The matter was settled, except that the emperor began to worry about the scriptural precedent for the *li*'s arrangement of big and small civil months. Having 're-evaluated [his] sage thoughts' (*HHS zhi* 2, 3027), the emperor ordered General of the Gentlemen-of-the-Household Jia Kui (30–101 CE) to inquire about the matter from Bian Xin, Li Fan, and eight others, the list of which included two further *li* workers, subordinates of the Defender-in-Chief, the Minister over the Masses and the suite of the heir-apparent, as well as the eighth-order nobleman Su Tong from Julu. Justified in his suspicions, Emperor Zhang ordered the *li* 'corrected' (*zheng*) prior to its debut in 86 CE, accusing its

[53] See Cullen (2007a, 245–7) for this episode.

[54] Specifically, Emperor Zhang cites the *Spring and Autumn Annals* weft *Bao qian tu* on this point; see *HHS zhi* 2, 3026.

[55] On weft-text *li*, see Takeda (1989). On the place of weft texts in Eastern Han *li* debate, see Cullen (2007a).

creators of 'tunnel vision' (*xuejian*) and 'validating experience without text[ual precedent]' (*wu wen zheng yan*) (*HHS zhi* 2, 3027).[56]

The Quarter-remainder *li*, like its predecessors, underwent subsequent modification during its service. In 102 CE, for example, expectant appointee Huo Rong successfully petitioned from the Clerk's Office that water clocks' day–night indicator rods be swapped as a function of solar declination rather than time. Likewise, the solar table appended to the received version of the Quarter-remainder procedure text was compiled in 174 CE from contemporary gnomon data.[57]

This brings us up to the end of the Eastern Han (25–220 CE). We shall continue beyond this date in subsequent chapters, but I believe that the 465 years outlined here are sufficient to illustrate features typical of the 'reform history' approach. First of all, reform history tends to conflate distinct categories of state policy: 'first month reform' (*gai zheng*) and '*li* reform' (*gai li*). These do sometimes coincide, as in 104 BCE, but they involve different questions, and different competences, let alone *different words*. Second, our three 'reforms' – the edicts of 85 CE, ~~9 CE~~, 104 BCE, and *c.*246 BCE – do mark milestones in *li* practice, but they are milestones akin to the awarding of a scientific prize. Indeed, the disjunction between the headings and the contents of this section goes to show how policy reform, as the political recognition of work done, precludes neither the continuation of that work nor that of the competition. Third, reform history tends to lodge with the emperor more historical agency than those actually authoring and executing the policies that he approves, thus placing the cart before the horse.[58]

1.2.2 A History of Practice

Geng Shouchang: Follow the 'Yellow Road' The earliest *li* at our disposal rely on a mean sun and moon travelling at constant equatorial rates. In 52 BCE, Grand Minister of Agriculture and Palace Assistant Secretary Geng Shouchang memorialised that he had built a 'diagram sight' (*tu yi*) by which he had measured the rate of lunar and solar progress along the 'red road' (the equator) to differ between the solstices and equinoxes. He concluded that the reason for this was the red road's obliquity to their true path of constant motion – the 'yellow road' (the ecliptic). Using the yellow road, he concluded

[56] On the reform of 85 BCE, see Ōhashi (1982).

[57] On Huo Rong's petition and the solar table of 174 CE in the current Quarter-remainder *li* (*HHS zhi* 3, 3077–81), see Cullen (2007b; 2017, 224–32).

[58] The period outlined in Section 1.2.1 is treated at greater length in Yabuuti (1969, 21–45), Kawahara (1996, 115–287), Chen Meidong (2003, 103–217) and Zhang Peiyu et al. (2008, 375–597). For distinctions like those that I draw here between 'first month reform' versus '*li* reform' and '*li* reform' versus 'procedure modification' (*gai shu*), see Niu Weixing (2004).

that 'the sun falls neither ahead nor behind, nor do quarter and full moons err by one day; it is tighter and closer than using the red road and [thus] suitable for [official] implementation' (*HHS zhi* 2, 3029). Now lost, the *Book of Han* bibliographic monograph records a *Geng [Shou]chang's Silk Diagram of Lunar Motion* and a *Geng [Shou]chang's Lunar Motion-du* in 232 and two fascicles respectively, attesting to his substantial written output on the subject (*HS* 30.1766).

Jia Kui: from 'Yellow Road' to 'Nine Roads' Once the idea of the yellow road took hold, people began to measure it, to measure *with* it, and, finally, to realise that the moon moves faster and slower there too. It was to address this newly measurable inequality that Liu Xiang (79–8 BCE), his son Liu Xin, and anonymous weft-text authors began nurturing the idea of the 'nine roads' (*jiu dao*). Early accounts are difficult to reconcile, as authors variously use the term to describe the moon's anomalistic period, coloured roads travelled seasonally thereby, and a 171-year adjustment of the Grand Inception *li*'s mean lunar lag. The first coherent discourse on the topic is that of Jia Kui in 92 CE. The moon's daily progress varies, Jia explains, because 'the path(s) travelled by the moon has far and near, egress and ingress', the 'speedy place' (*ji chu*, i.e. perigee) precessing at a rate of three *du* per month such that 'the nine roads return/repeat once every nine solar years' (cited in *HHS zhi* 2, 3030). The nine roads, in other words, is an answer to the 8.85-year precession of the lunar line of apsides.[59]

Whatever the excitement about these new roads, the throne had yet to get involved. In 92 CE, Jia Kui petitioned Emperor He (r. 89–105 CE) to equip the observatory with a yellow-road instrument so that the Clerk's Office might pursue the sort of testing conducted independently by 'Fu An et al.' (*HHS zhi* 2, 3028–9) and Minister Geng. Nothing came of this in Jia's lifetime. In 103 CE, at last, the emperor commissioned a great instrument in bronze and ordered the Clerk's Office to 'watch [the lunar phases] via the nine-roads method, [and they] verified that there was no misstep [between them]' (*HHS zhi* 2, 3027).[60] This went nowhere: the order to use the instrument was 'rarely heeded', and the nine roads were 'abandoned and no longer studied' (*HHS zhi* 2, 3030).

Publicly, the nine roads would have been a tough sell. Emperor He took the throne on 30 January 89 CE; he was aged nine, and his father's *li* was only three.

[59] I would like to thank Christopher Cullen and Yin Shoufu for kindly sharing their notes with me on this topic, especially as their insights led me to the studies of Chen Jiujin (1982), Wang Shengli (1982), Ōhashi (1997) and Qu Anjing (2008, 331–6). Now see also Cullen (2017, 120–1).

[60] *HHS zhi* 2, 3030, dates the testing only to 'the Eternal Epoch reign (89–105 CE)', but one imagines that it followed the completion of the instrument built for that purpose. On the 'Grand Clerk yellow-road bronze sight' of 103 CE, see Chapter 2 below.

On 9 September 89 CE (1-VII^2-16), a lunar eclipse occurred a month ahead of schedule, and on 5 March 90 CE (2-I-16), it happened again. Eighth-order nobleman Zong Gan of Meng had predicted the latter four days in advance, forcing the young emperor's hand by public spectacle. On the recommendation of the Prefect Grand Clerk, the emperor made Zong an expectant appointee to the Clerk's Office, where his 'method' (*fa*) replaced the eclipse predictions of the official *li* until 147 CE.[61] In 92 CE, having just patched the new *li*, the court might reasonably hesitate before investing bronze and man-hours into exposing problems whose solutions were linked to the old Grand Inception *li*.[62] The nine roads met resistance on another front as well. At the collective deliberation of 123 CE, gentlemen of the Masters of Writing Zhang Heng (78–139 CE) and Zhou Xing – both 'able at *li*' (*neng li*) – weighed in that 'having investigated (*kao*) the past and analysed (*jiao*) the present by way of reference to instrument [observational] notes (*yi zhu*), [we] consider the nine-road method the most tight (*mi*)' (*HHS zhi* 2, 3034).[63] Still, subsequent opinion ran forty to two against them on the ground that 'using the nine roads for new moons, the [civil] month would see three bigs and two smalls in a row' (ibid., 3034), upsetting the civil calendar's neat alternation of thirty- and twenty-nine-day months.[64]

Privately, people continued to advance the Grand Inception *li* and nine roads well beyond the Quarter-remainder's rise from 'the grassy marshes' (*cao ze*) into government service, 'these two schools forever cleaving to their [respective] procedures' (*HHS zhi* 2, 3033). At court, we see subsequent petitions for its reinstatement by one Liang Feng of Henan in 123 CE, by Gentleman-in-Attendance of the Masters of Writing Bian Shao in 143 CE, and finally by Former *Li*-Worker Gentleman Zong Zheng of the Kingdom of Liang and Member of the Suite of the Heir-Apparent Feng Xun, separately, in the 170s CE.[65] In a nutshell, revivalists appealed to claims of verified 'tightness'

[61] Ōhashi (1982, 7–11) argues that what Zong Gan's 'method' did was to slide the 135-lunation 'coincidence' (*hui*) cycle's template of twenty-three (potential) eclipses forward, into better agreement with observation, which comports with the sort of cycle-tweaking hypothesised in Sivin (1969, 33–52). Zong's method did the job until 147 CE, when it too slid systematically behind by one lunation, provoking subsequent debate between 174 and 181 CE.

[62] As described at the end of Liu Xin's Triple Concordance *li* in *HS* 21B.1007, the nine roads ($= 3^2$) experience 'lesser termination' (*xiao zhong*) in 171 years ($= 3^2 \times 19$) and 'greater termination' (*da zhong*) in 1,539 years ($= 3^4 \times 19$), the latter of which corresponds with a 'concordance' (*tong*) – the coincidence of new moon, winter solstice and midnight – as constructed from a 'rule' (*zhang*) of 235 months : 19 years and a mean lunation of $29\frac{43}{81}$ days. These cycles are irreconcilable with the Han Quarter-remainder *li*'s 'obscuration' (*bu*) of 76 years ($= 2^2 \times 19$) and 'era' (*ji*) of 1,520 years ($= 2^4 \times 5 \times 19$) as constructed from the 'rule' and a year of 365¼ days.

[63] For more on the important figure of Zhang Heng, see Xu Jie (1999) and Lien (2011).

[64] On the problem of 'three bigs and two smalls', see Section 3.3 below.

[65] In order, see *HHS zhi* 2, 3034, 3035, 3030. Note also the mention in *HHS zhi* 2, 3033, of inaction in implementing the 171-year nine-road stopgap for the Grand Inception *li*'s lunar lag proposed for 68 CE.

(*mi*) and the *li*'s connection with past prosperity.[66] The first two proposals were struck down in collective deliberation, and the latter via court-approved testing.

Liu Hong: From 'Nine Roads' to 'Slow–Fast' Born a scion of the King of Lu, near Mount Tai, Liu Hong was a military colonel with a knack for *li*. The Prefect Grand Clerk recruited him as talent sometime between 158 and 167 CE, bringing Liu into the world of high-level practice and policy-making at the Clerk's Office and among the gentlemen of the palace. This may have involved him in the gnomonics programme behind the solar table of 174 CE, but Liu was already back on the Shandong peninsula by the date of its completion, memorialising, as chief clerk of Changshan, a 'Seven Luminaries Procedure' (*Qiyao shu*) addressing failures in Zong Gan's eclipse patch. In 177–8 CE, after leave to mourn his father (twenty-seven months) and transfer to deliverer of accounts, Liu was recalled to the capital as gentleman-of-the-palace at the recommendation of Cai Yong. Cai needed his help finishing the *lü-li* monograph taken on as part of the Later Han history project at the Eastern Observatory (a sort of institute for advanced studies founded for the purposes of women's education), because, as the monograph explains, 'Yong could compose text . . . and Hong could do the calculations' (*HHS zhi* 3, 3082).[67] By the winter of 178 CE, the two went their separate ways – Cai exiled to the northern frontier, and Liu, watchman of the wall, then internuncio, retained to deliberate on eclipse prediction – only to cross paths in Kuaiji, where Liu was installed as chief commandant in the 180s.

It was while commandant of the Kuaiji Eastern Regiment that Liu Hong finished research begun in Luoyang into his Supernal Icon *li* by 'referencing and comparing' (*can jiao*) forerunners, 'testing' (*ke*) lunar phenomena, and 'absorbing [himself] in inner contemplation for more than twenty years' (*JS* 17.499). His masterpiece complete, he was recalled to the capital in the late spring of 189 CE, presumably to discuss reform. The timing could not have been worse, however, as Emperor Ling's (r. 168–89 CE) death and Dong Zhuo's (d. 192 CE) *coup d'état* forced him to turn back halfway. Liu Hong continued to work on his Supernal Icon *li* until at least 206 CE through subsequent governorships on the Shandong peninsula. Back on the peninsula, Liu also transmitted his work in progress by personal exchange to fellow enthusiasts – to Zheng Xuan (127–200 CE), near Shanyang, in 196 CE and to Xu Yue of Donglai, at Mount Tai – who, in turn, spread the versions (and commentaries) that they had received, by personal exchange, to each of the Three Kingdoms (220–80 CE). It was in 223 CE, after his death in Han office, and back near Kuaiji where it had

[66] Also at play in this deliberation, though less relevant to the topic at hand, was the matter of the canonicity and coherence of 'high-origin' (*shangyuan*) selection, for which I refer the reader to Cullen (2007a).

[67] On the intellectual life of the Eastern Observatory, see Goodman (2005).

all begun, that Liu's *li* finally won a place in state policy, albeit as the emblem of Sun Wu independence.[68]

Liu Hong's Supernal Icon *li* is remembered as marking what later experts He Chengtian (370–447 CE), Zhen Luan (fl. 535–70 CE) and Li Chunfeng (602–70 CE) describe as a 'turn' (*zhuan*) in 'tightness' (*mi*),[69] establishing '[Liu] Hong's procedures', in Li Chunfeng's words, as 'the exemplar (*shibiao*) of calculation for the subsequent age' (*JS* 17.503). *Politically*, the Supernal Icon *li* was a debacle, but no one really remembers the politics.[70] It was 'tightness' that was at stake, and Li Chunfeng explains how this was achieved:

> What [Liu Hong] did was establish numbers (*shu*) based on the [*Book of*] *Changes* [such that] they called out to one another in hidden motion and sought each other out from secret parts – and [at this he] named it the 'Supernal Icon *li*'. Also, [he] created (*chuangzhi*) the solar/daily motion slow–fast (i.e. equation of centre) while concurrently investigating (*kao*) lunar motion, [concluding that] *yin* and *yang* (i.e. negative and positive latitude) cross inside an outside the yellow road, and that the sun travels on the yellow road, experiencing advance and retreat in terms of red-road lodge *du* (i.e. a reduction to the equator) – and only with this was there a turn (*zhuan*) towards the fine and tight (*jing mi*) relative to prior methods (*fa*) (JS 17.498).

So Liu Hong's innovation came down to 'numbers' (*shu*) and 'procedures' (*shu*). In terms of 'numbers', Liu appeals to *Changes* number mysticism in the very title – Qian ䷀ ('Supernal'), the first hexagram, being the *xiang* ('image/sign') of heaven – and in deriving the moon's anomalistic period from 'the total number of heaven and earth [divination rods] (i.e. 55)' (*JS* 17.510).[71] In a similar vein, whereas his predecessors built their *lü* upon the 235-month : 19-year 'rule' (*zhang*) by simple factors of twos and threes, Liu Hong builds his from rather extravagant strings of prime numbers – a 'supernal' of 1,778 (= 2 × 19 × 31), a 'circuits of heaven' (*zhou tian*) of 215,130 (= 2 × 3 × 5 × 71 × 101), etc. – which significantly obscures how it is, mathematically, that they fit together.[72] In terms of 'procedures', he placed the moon on the 'white road' (*bai dao*) and the sun on the yellow road, working out

[68] On Liu Hong's life and the transmission of the Supernal Icon *li*, see Chen Meidong (1986) and Morgan (2015).

[69] *HHS zhi* 3, 3081–2 (comm.), *Shushu jiyi*, 2a (comm.), *JS* 17.498.

[70] For example, the Supernal Icon *li*'s date of institution is a matter of some confusion. *SGZ* 47.1129 records that 'in [Yellow Militarism] 2-I (223 CE) . . . the Quarter-remainder was reformed (*gai*) and the Supernal Icon *li* was used (*yong*)', but this passage has escaped attention in all but Yabuuti (1963, 453), Chen Zungui (2006, Table 50), Qu Anjing (2008, 629) and Martzloff (2009, 350) (cf. Chen Meidong (1986) and Sivin (2009, Table 2.1)). Historians tend to equate *li* with *li* reform and, thus, the date of authorship/submission with the date of institution (in a single line of reforms). Zhu Wenxin (1933, Table 2), for example, dates the Supernal Icon *li*'s use to 206–37 CE, beginning at the text's closure and ending at the authorship and institution of the Cao Wei's Luminous Inception *li*. We see similar confusion about Supernal Icon 'reform' (*gai*) in pre-modern sources, as laid out in Chen Meidong (1986, 130–1).

[71] On the origins of this number in *Book of Changes* divination-rod sorting, see Rutt (1996, 151–201).

[72] On *lü* and '*li* numbers', see, once again, Section 1.1.3 above. On the Luminous Inception *li*'s (237 CE) 'choix des nombres "bizarres"', relating to Liu Hong's *lü*, see Martzloff (2009, 117 n13).

how to calculate the latitude of the one and correct the right ascension of the other. More importantly, he worked out a method for 'fixing' (*ding*) the true moon via a table of daily speed corrections over a $27\frac{3303}{5969}$ day 'slow–fast sequence' (*chi-ji li*), or anomalistic month. It would appear that this was a radical improvement over nine-road methods, as the latter does not survive, and as subsequent *li* follow Liu's procedures.[73]

Were we to restrict ourselves to policy reform we would have to begin the history of lunar anomaly with the adoption of the (derivative) Luminous Inception *li* in 237 CE (Chapter 4 below) and pass off the likes of Liu Hong as a failure thwarted by bureaucracy and royal inaction. It is hard to imagine, in this light, why *anyone* would have persisted all this time, let alone a Grand Minister, a general or a governor, who presumably had better things to do. Liu Hong, like Geng Shouchang and Jia Kui before him, did not get what he had petitioned for in a timely manner – not in his lifetime – but is that really *all* that he was asking for, and *all* that he was getting out of *li*? When we focus on the way that a problem like lunar anomaly was pursued as independent of 'reform', one sees the outlines of a discourse and a community in which actors' private activities begin to make sense. This is admittedly hard to do, considering that most of what we know comes down through dynastic histories, but it is worth picking through the chronicles, memorials and biographies to piece together what was going on beyond (and before) state policy. One suspects that Minister Geng had *private motives* for building a 'diagram sight' for his own home – motives as simple as answering questions that others like him were asking about the moon. One suspects that Governor Liu had similar reasons for sitting down with private numbers men in the provinces to share with them his life's work. The more one focuses on the individual, the more one begins to wonder if the history of astronomy in China was not built of moments such as these.

1.3 Cast of Characters

1.3.1 Institutions

What makes a king a *sage king*, the classics tell us, is that he 'observes the signs and grants the seasons' so as to ensure peace on earth and harmony above. This was his most sacred task, and it was not for him to do. The sage king acts by inaction, the classics say, so he is not to *do* anything of the sort. In the astral sciences, as in war, it is to the sage king to have the 'great plan' (*hong fan*), and to his officer corps to work out the simple day-to-day. Anyway, all that was

[73] On Liu Hong's lunar theory, see Cullen (2002); on Chinese lunar theory more generally, see Qu Anjing (2008, 308–89).

asked of them was utopia, and *that*, the classics affirm, is the natural product of equal parts astronomy and hard work.

Such was the idyllic past that our historical subjects imagined and longed to re-create. There were some details to sort out, but the mission was clear, and so too was the apparatus: the Grand Clerk, subordinate to the Ministry of Rites, who presided over an office of technicians in omen reading, timekeeping, and calculation. Such was the Clerk's Office of the ancient past, as documented in the *Offices of Zhou*, and so too was it in Han times. The *Book of Later Han*, for example, offers us the following description of the post:

> [The Prefect Grand Clerk] handles the seasons/time of heaven and sequence of the stars (*xing li*). Near the end of each year, [he] memorialises the new year's *li* (civil calendar). For all state matters of sacrifices, funerals and weddings, he handles the memorialising of auspicious dates and seasonal prohibitions. For every time that the state experiences an auspicious [omen-]response (*ying*) or calamitous anomaly, [he] handles the recording of it (*HHS zhi* 25, 3572).

The dual task of observation and time-control is reflected in the organisation of this office as documented there and in 'Han offices' literature (see Table 1.3).[74] The Prefect Grand Clerk had three assistants, each presiding over a separate physical site: (1) the Grand Clerk's Office at the Ministry of Rites, staffed by *li*-workers, diviners, hemerologists and ritualists; (2) the Bright Hall, where the emperor 'acts as the pole star of the people' by issuing the calendar and performing monthly rituals; and (3) the Numinous Terrace observatory, staffed by 'watchmen' (*hou*) of astronomical, meteorological and harmonic phenomena.[75] The names, structure and affiliation changed over time, but the Eastern Han apparatus is an adequate reference for the purposes of this book.[76]

Within the Clerk's Office, the actual work of 'observing and granting' fell to several dozen 'expectant appointees' – a theoretically provisional title for those, like Liu Hong and Zong Gan, who had been conscripted for their special talents.[77] Experts were *conscripted*, one notes, because the Clerk's Office did not produce its own talent until the sixth century CE, when erudites (*boshi*) and students (*sheng*) were added to its personnel.[78] As to the management, the

[74] *HHS zhi* 25, 3572; *Han guan liu zhong*.

[75] On the Bright Hall, see Section 3.2.2 below. On the Numinous Terrace and observatories of the Han, see Chen & Zhang (2008, 38–41). For a site report of the Eastern Han Numinous Terrace complex unearthed in 1974–5, see *Kaogu* 1977.1: 54–7.

[76] For a complete history of the Clerk's Office, see Deane (1989) and Chen & Zhang (2008).

[77] On the role of the expectant appointee in the Clerk's Office, see Lai Swee Fo (2003) and Cullen (2007a, 246). On the peculiarly non-provisional nature of this title in the context of the Clerk's Office, see Sugimoto (1973, 86–7) and Yang Hongnian (1985, 124–5).

[78] See *Sui shu*, 28.775, and *Tang liu dian*, 10.13a–16a (comm.); cf. Chen & Zhang (2008, 297–304). Interestingly, Chen Meidong (2007, 17–32) notes how the Clerk's Office continued nationwide searches and conscription of talent in tandem with the increasing restrictions and bans of the Tang.

Table 1.3 *Organisation of the Eastern Han Clerk's Office*[*]

Office	Pay grade (*shi* of grain)
皇帝 Emperor	∞
三公 Three Excellencies	10,000
九卿 Nine Ministers	2,000
太常 Grand Master of Ceremonies	2,000
太史令 Prefect Grand Clerk	600
太史丞 Assistant to the Grand Clerk	200
待詔 Expectant appointees (× 37)	–
治曆 *Li*-workers (× 6)	
龜卜 Tortoise diviners (× 3)	
廬宅 Hut and residence [diviners] (× 3)	
日時 Day and time [diviners] (× 4)	
易筮 *Changes* milfoil [diviners] (× 3)	
典禳 Directors of apotropaic rites (× 2)	
籍氏 Mr Ji [tradition experts] (× 3)	
許氏 Mr Xu [tradition experts] (× 3)	
典昌氏 Mr Dian and Chang [tradition experts] (× 3)	
嘉法 Method experts (× 2)	
請雨 Rain supplicants (× 2)	
解事 Elucidators (× 2)	
醫 Physicians (× 2)	–
明堂丞 Assistant for the Hall of Light (ritual complex)	200
靈台丞 Assistant for the Numinous Terrace (observatory)	200
待詔 Expectant appointees (× 42)	–
候星 Watchmen of the stars (× 14)	
候日 Watchmen of the sun (× 2)	
候風 Watchmen of the wind (× 3)	
候氣 Watchmen of the *qi* (× 12)	
候晷景 Watchmen of the gnomon shadow (× 3)	
候鍾律 Watchmen of the bells and pitch-pipes (× 7)	
舍人 Member of the suite (× 1)	

* The personnel and pay grades listed here are as per the *Book of Later Han* monograph on Han offices and fragments of 'Han offices' literature cited in Li Xian's (654–84 CE) commentary thereto. The Eastern Han apparatus offers a helpful framework within which to place the official titles of our period, as it remained largely unchanged through the Three Kingdoms (220–80 CE) and Jin dynasty (265–420 CE) and echoes titles mentioned in Western Han sources. Note, however, that some of these titles change, and that 'Han office' literature omits the offices of Xi-He 羲和, his four 'masters' 子 and the 'director of stars' 典星, which we encounter (albeit rarely) in our sources.

Prefect Grand Clerk and his assistants were *sometimes* appointed by virtue of technical expertise – some promoted from expectant appointees, and others transferred from some unrelated office (see Table 1.4). Others, like Zhang

Table 1.4 *Career paths in the Han Clerk's Office*[*]

Grade	Title
司馬談	**Sima Tan (d. 110 BCE)**
200	Ass't to Grand Clerk
600	Prefect Grand Clerk
鄧平	**Deng Ping (fl. 104 BCE)**
< 200	*Li*-maker
200	Asst' to Grand Clerk
劉歆	**Liu Xin (*c.*50 BCE–23 CE)**
	Expct. appt., Eunuch Office
	ⓢGentleman of the palace gate
	collator, imp. library
eq 2,000	ⓜChf. cmdt. of imp. equipages
2,000	ⓜCol. of the capital rampart
eq 2,000	Palace attendant
eq 1,000	Grand palace grandee
eq 2,000	ⓜChf. cmdt. cavalry
eq 2,000	Imp. Household grandee
eq 600	Erudite
2,000	ⓡGovernor of Henei
2,000	ⓡGovernor of Zhuo Cmndry
< 2,000	ⓜChf. cmdt. dependant states
2,000	ⓢBureau head of the left
2,000	ⓜCol. of the capital rampart
≤ 600	Xi-He
2,000	Governor of the capital
♛	Marquis of Hongxiu
eq 2,000	Imperial Household grandee
10,000	Master of state
eq 2,000	ⓜChf. cmdt. cavalry
> 2,000	ⓜGen. resolute cavalry
尹咸	**Yin Xian (fl. 26 BCE–5 CE)**
600	Prefect Grand Clerk
10,000	Chancellor
2,000	Grand Minister of Agriculture
鮑鄴	**Bao Ye (fl. 64–77 CE)**
< 200	Expct. appt. to Grand Clerk
200	Music ass't to grand M.C.
宗紺	**Zong Gan (fl. 89 CE)**
♛	Grandee of the 8th order
< 200	Expct. appt. to Grand Clerk
張衡	**Zhang Heng (78–139 CE)**
	Recm. filially pious and incorrupt
eq 300	Gentleman-of-the-palace
600	Prefect Grand Clerk
300	Gentleman of M. of Writing
600	Prefect Grand Clerk
eq 2,000	Palace attendant
單颺	**Shan Yang (fl. 170–3 CE)**
	Recm. filially pious and incorrupt
600	Prefect Grand Clerk
eq 2,000	Palace attendant
2000	ⓡGovernor of Hanzhong
–	relieved of office
600	Master of writing
劉洪	**Liu Hong (fl. 158–206 CE)**
	ⓜColonel
< 200	Expct. appt. to Grand Clerk (?)
eq 300	Gentleman-of-the-palace
600	ⓡChief clerk of Changshan
600	ⓡOffl. handing up accounts
eq 300	Gentleman-of-the-palace
	Historian, East. Observatory
eq 600	Internuncio
600	ⓜWatchman of the city wall
eq 2,000	ⓜChf. cmdt. Kuaiji E. Regiment
2,000	ⓡChancellor, Marq. of Qucheng
2,000	ⓡGovernor of Shanyang

[*] I give titles in chronological order, distinguishing military (ⓜ), regional (ⓡ), supernumerary (ⓢ) and noble (♛) titles from the civil administrative posts in the capital. Sources are as follows: for Sima Tan, *SJ* 130.3288 (comm.); for Deng Ping, *HS* 21A.975–8; for Liu Xin, *HS j*. 36, 99; for Yin Xian, *HS* 19B.856, 30.1701, 36.1967; for Bao Ye, *HHS zhi* 3, 3025, *zhi 2*, 3015 (comm.), *Sui shu*, 15.352; for Zong Gan, *HHS zhi* 2, 3040; for Zhang Heng, *HHS j*. 59; *zhi 2*, 3034; for Shan Yang, *HHS* 60B.1990, 82B.2733; for Liu Hong, Chen Meidong (1986).

Shouwang, were criminally incompetent in this regard. Nepotism is a potential explanation, but Sima Qian's prefectorship is the only known case of inheritance in this period.[79] What we *do* see, particularly at the founding of a dynasty, is the appointment of 'king makers' like Liu Xin – figures central to a founding emperor's legitimacy-building project via *yin-yang*, occult arts, *tianwen* omenology and/or classical state ritual.[80] Anyway, the top posts were likely more administrative than technical.

Mostly what the Clerk's Office did was perform its job description, but the throne handed its staff the occasional special task by way of edict (*zhao*). The majority of these edicts charged the Clerk's Office with testing memorialised procedure texts, i.e. conducting technical analysis, comparison of predicted and observed results, and/or multi-year live trials. More rarely, as we saw in the edicts of 104 BCE and 85 CE, a staff member was ordered to *produce* a new *li* with the assistance of inside and/or outside expertise. Some of these edicts come from out of the blue, but most were issued upon petition, thus effectively signing off on the chargee or some other official's proposed plan of action. To reiterate, it was not the Grand Clerk's job to innovate, but innovations *were* sent to him to judge, some of them originating from his staff.[81]

Anything beyond the Grand Clerk's job description went through the throne, which is to say that it passed through the imperial secretariat and its 'Masters of Writing' (*shangshu*). The emperor received petitions asking for all sorts of different things, which the Masters of Writing were responsible for sorting and presenting to him in order of urgency. Once an edict had been decided, it was the Masters of Writing who wrote it, ensuring that the emperor spoke on topics as diverse as water management, foreign tribes and mathematics with the credibility and erudition expected of a sage. In other words, the emperor's job was to say 'OK' (*ke*); that of his secretariat was to work out the details.[82] This, needless to say, demanded a great breadth of knowledge from the latter, which is probably why we find former Grand Clerks promoted to their ranks (Table 1.4) and writers, like Zhou Xing and Bian Shao in Section 1.2.2 above, who were personally active in expert debate.

Most petitions concerning *li* policy were decided between the secretariat, the Clerk's Office and the emperor, but contentious requests were put up for debate,

[79] The hereditary post is a phenomenon that emerged at the Clerk's Office sometime between the Jin and Tang, by which point it became prevalent. On the situation in the Tang, see Lai Swee Fo (2003) and Jiang & Niu (2001, 105–33).

[80] On Liu Xin's role in Wang Mang's rise to power, see Xu Xingwu (2005, 432–78). The similar case of Gaotang Long and Xu Zhi is treated in Chapter 4 below. Jiang Xiaoyuan (1991, 50–2) and Sivin (2009, 1–33, 133–50) give special emphasis to this criterion in later periods.

[81] On the role of outside knowledge vis-à-vis the Clerk's Office, see Chen Meidong (2007, esp. 17–32), Cullen (2007a) and Sivin (2009, 58–9).

[82] On Masters of Writing and the process of memorialisation and imperial decision making, see Bielenstein (1980, 48–9), Giele (2006) and Olberding (2012).

be it a closed 'deliberation' (*yi*) amongst the Three Excellencies or a public one open to 'the hundred officials'. Cullen (2007a, 252–4) reveals that the voices in such debates are rather mixed, the institution providing 'an arena for orally delivered persuasion before a mass audience' where, in the words of the *Book of Later Han*, 'policy is settled in accordance with the many' (*HHS* 45.1519).[83] Serendipitously, when it is scaremongering and 'untested' metaphysical hogwash that has slipped past the secretariat's desk, it is the voice of the numbers man that invariably rises to the fore, taking down his opponent on credentials, hard data, internal logic, classical learning and, sometimes, metaphysics. This was serious business; in an empire of bean counters, where clerical account errors were charged as 'theft' (*dao*), getting the throne to act on bad *li* numbers was a public embarrassment and an offence punishable by hard labour or death.[84]

The institution of the petition, edict and deliberation encouraged involvement by a considerable spread of actors. In the history of *li* from 206 BCE to 420 CE, the *Book of Han*, *Latter Han*, and *Jin* monographs identify a total of 117 people by name. Six are emperors. Of the remaining 111 names, eighty-three are further identified by office, nine by place of origin, five by noble title and three by family affiliation. Of the eighty-three office holders, thirty-six belong to the Clerk's Office, eight belong to the secretariat, and thirty-nine hold unrelated palace (nineteen), civil (eleven), military (six) and academic (three) positions. All told, acting Clerk's Office personnel constitute only 31 per cent of those named as contributing to state *li* policy in this period.

1.3.2 Individuals

The institutional framework gives us an idea of the playing field upon which the 'observing of signs and granting of seasons' was conducted. It is crucial that we distinguish between positions and players, of course, because the players cycle positions. There is a world of difference between 'the' astronomer royale when it is a Zhang Shouwang or a Zhang Heng playing the part; the difference between a Grand Clerk or Master of Writing Zhang Heng, however, is not so great as to merit labelling him a 'former astronomer'. Either way, the process was such that *li* policy was open to all types to author and decide, which made the question of rank and official

[83] On *yi* deliberation, see also Nagata (1972).

[84] For laws governing the accounts and account errors of local clerks, see for example Hulsewé (1985, 163). In the period covered in Section 1.2 above, we see Grand Clerk Zhang Shouwang, in *c.*74 BCE, and Gentleman of the Palace Feng Guang and Deliverer of Accounts Chen Huang, in 176 CE, handed respective sentences of death and 'firewood gathering' (*guixin*) for *successfully* petitioning the throne with misleading claims about *li*, though all three did receive clemency by imperial order; see *HS* 21A.978 and *HHS zhi* 2, 3040.

function somewhat irrelevant to the advancement of knowledge. More than anything, it was *expertise* with which actors were concerned, and expertise was sought where it was found.

It is probably not a coincidence that *li* men label one another by level and manner of engagement: '*li* lineage expert' (*li jia*), 'procedure lineage expert' (*shu jia*), 'one who knows *li*' (*zhi li zhe*), 'one good at *li*' (*shan li zhe*), 'one who constructs *li*' (*zao li zhe*), 'one who works on *li*' (*zhi li zhe*), 'transmitter of heaven's numbers' (*chuan tianshu zhe*) and 'gentleman of *li* calculation' (*lisuan zhi shi*). The modern historian would be wise to recognise the benefits of such appellations, for they foreground what mattered in this context without reducing someone with a professional and intellectual life as complex as Liu Xin's to 'astronomer'. 'Expert', moreover, is probably a better term than 'specialist', as will be apparent in Chapter 5 below, because we are dealing with polymaths whose names occur across the specialities into which the modern academy divides the world.[85]

Above all else, the men and women of the astral sciences were members of gentry (*shi*) culture. More specifically, they belonged to a well-born office-holding class with access to education in core sacred and philosophical texts, which inculcated in them, in the words of Lloyd and Sivin (2002, 19), a 'nostalgia for an imagined hierarchy . . . that has already slipped away' and a belief in 'an unending governmental quest to control society from the top down'. Whether or not two gentlemen shared the same expertise(s), they shared an idiom of myth, values, etiquette and cultural tradition that marked them as members of the same exclusive community. It was on his ability to comport himself as such, they explain, that the number man's credibility foremost depended:

> Those whose livelihood comes from being able to predict the future or to determine what has gone wrong in the human body must satisfy their clients that they have access to special knowledge not open to everyone else. Chinese diviners and physicians did not generally stake their authority on the metaphysical foundations of what they knew, nor on the formal rigor with which they presented it . . . Their qualifications tended to be social. Because expertise was not inherently problematic, it was initiation that separated insiders and outsiders, and gentlemanly behavior that marked the superior insider.[86]

So how did one go about learning the sort of 'special knowledge' that distinguished a 'gentleman of *li* calculation' from any other? Calculation (*suan*) was a subject that was compulsory for the gentlemen and county clerk alike; by affiliation, their curriculum may therefore have exposed the average numerate man to certain fundamentals of *li* (Cullen 2009). Among *experts*, specifically, one discerns four educational patterns. First, you have true 'lineage experts'

[85] On the polymathy of the cultural elite of this period, see Goodman (2005; 2010).
[86] Lloyd & Sivin (2002, 205). Particularly helpful to think with in this regard is Shapin's (1994) sociological analysis of gentlemanly culture in Enlightenment science.

like Sima Qian, Liu Xin and Zong Gan's grandsons Cheng and Zheng (fl. 174–80 CE), who were born into family traditions. Second, you have those like Zhang Heng and Zheng Xuan, who learned *li* from the Imperial Academy, as well as those like Zheng Xing (fl. 14–33 CE) who learned it from (Liu Xin's) 'lectures' (*jiang*) at a private provincial school.[87] Third, you have elites tutored by social inferiors like Sima Qian, and methodman Tang Du, Empress Dowager Deng Sui (81–121 CE), and literata Ban Zhao (44/49–118/121 CE).[88] Fourth, you also have gentlemen like Zheng Xuan who 'received' (*shou*) it in private exchange with peers possessed of different skills.[89] None of these avenues were mutually exclusive, as is evident from the examples given here.[90]

The bigger question is perhaps where someone like Chief Commandant Liu Hong found the time to 'test quarter and full moons' (*JS* 17.499) while holding down a day job. The obvious answer is 'at night'. This is where Liu Xiang found the time: 'nightly observing stars and lodges [he] sometimes didn't sleep until the break of dawn' (*HS* 36.1963).[91] Fu An and Geng Shouchang had commissioned instruments for their estates, and one imagines that a gentleman enthusiast with an estate, let alone a school of live-in students, may well have had help.[92] We know that Zhou Qun (fl. *c*.210 CE), an assistant to the governor of Yizhou, for example, 'constructed a small tower at the centre of his courtyard, and as his family was rich, with many slaves (*nu*), [he] regularly ordered slaves to take shifts atop the tower to watch for heavenly [phenomena]' (*SGZ* 42.1020). A sabbatical could also be productive, be it the legal twenty-seven-month mourning period for one's father, or more dramatic cases of exile. There was Zhang Zixin (d. 577 CE), for example, who 'went into hiding on a sea isle where he devoted himself for more than thirty years to the observation, by means of [armillary] sphere instrument, of data on the differences and changes in the sun, moon and five [planets]' (*Sui shu*, 20.561).

[87] For instances of *li* learning at state and private schools, see *HHS* 35.1207 (Zheng Xuan), *HHS* 36.1217 (Liu Xin and Zheng Xing), *HHS* 59.1897 (Zhang Heng), *HHS* 82A.2719–21 (Liao Fu and Fan Ying), and possibly *HHS* 30A.1074 (Yang Tong, fl. 76 CE). Note that Zheng Xing, in turn, taught what he learned from 'lecture' to his son Zheng Zhong (d. 83 CE); see *HHS* 36.1224.

[88] See *SJ* 26.1260, 27.1349, 112.2965, and *HHS* 10A.424.

[89] See Section 1.2.2 above. Note also the case of Chancellor Zhai Fangjin (d. *c*.7 BCE), who taught the *Zuo Tradition* to Liu Xin (who would approach the classic through the astral sciences) and 'stellar/planetary *li*' to the acting mayor of Chang'an, Tian Zhongshu (*HS* 36.1967, 84.3421).

[90] The examples given here are treated in further detail in Morgan (2015, 566–9). For more on the history of *li* education, see Lee (2000, 512–41) and Chen & Zhang (2008, 249–352).

[91] I thank Jesse Chapman for bringing this quote to my attention.

[92] This is how Tycho Brahe (1546–1601) ran Uraniborg, as detailed in Christianson (2000), and it seems conceivable to me that the culture of retainership at local courts like King Liu An of Huainan's (*c*.179–122 BCE) and the enormous live-in schools that we see in the Han might well have fostered in the astral sciences a similar sort of ad hoc private institution for technical research. On patronage and retainer culture, see Lloyd & Sivin (2002, 28–34). On the organisation of private schools, see Yu Shulin (1966) and Lee (2000, 54–7).

The sort of 'testing' and 'inner contemplation' that led to a Supernal Icon *li* need not have involved 'observation' in the sense that we may expect; *li*, if we remember from Mencius' words in the Introduction, was a 'sitting' science practised by day. This is how most astronomies work. The Sanskritist, for example, would tell you that her sources attribute their knowledge to oral tradition that holds divine revelation above the human senses.[93] The Hellenist would tell you that the genius of the *Almagest* is the elegance with which it constructs its geometry from logical proof and but a handful of data drawn largely from Babylonian records.[94] The sociologist would tell you that so too in laboratory science is observation necessarily mediated by 'inscription' – 'transformations through which an entity becomes materialised into a sign, an archive, a document, a piece of paper, a trace' (Latour 1999, 306–7).

All 'one who constructs *li*' (*zao li zhe*) really needs is *data*, and there seems to have been plenty of that available in written form. Consider Grand Clerk Sima Qian's following approach to 'observation':

> In the old Gan [De] and Shi [Shen] methods for *li*-computing the five [planets] only [Mars] had retrogradation. They took the [asterisms] it guards in retrograde, the retrogradation of other [planets], and the veilings and eclipses of sun and moon all as [objects of] omen interpretation (*zhan*) . . . I have observed (*guan*) the clerk's records (*shiji*) and examined (*kao*) past events, and in a hundred years the five [planets] have never once emerged without going into retrograde . . . this is their great measure (*SJ* 27.1349–50).

Chronicles like the *Spring and Autumn Annals* contain dated eclipse records running back to the early first millennium BCE, which proved central to *li* research from the time of Sima Qian on, and it was the job of the Clerk's Office to keep compiling them into the historical period.[95] This is clearly what Sima Qian is referring to by 'clerk's records' (*shiji*) and 'past events' (*xingshi*) in this context, while later sources make similar reference to 'Clerk's Office watch notes' (*Shiguan hou zhu*), 'watch notes' (*hou zhu*), 'sight notes' (*yi zhu*), 'notes' (*zhu*), 'note records' (*zhu ji*) and 'record notes' (*ji zhu*).[96] Outside and lower-ranking officials regularly cite Clerk's Office 'notes' in their interactions, which suggests that access to observatory databanks was, among gentlemen, relatively easy to come by.[97] With 'procedures' *and* 'numbers' in hand, you had all you needed to change the world.

[93] Plofker (2009, 113).

[94] On these points, see variously Graßhoff (1990, 198–216), Jones (2005) and Jones & Duke (2005). I thank Alexander Jones for directing my attention to these studies.

[95] On Liu Xin's work on *Spring and Autumn Annals* astronomical records, see Cullen (2001). For more examples of actors deploying these records in *li* research and debate, see *JS* 18.563–7; *Song shu*, 13.308, 314; *Sui shu*, 17.418, 424–6, 430.

[96] *HHS zhi* 2, 3027, 3029, 3030, 3034, 3039, 3041, 3042; *JS* 17.498, 18.564; *Song shu*, 12.290, 13.309, 311, 312, 315; *Sui shu*, 18.460.

[97] In this period we see the following outsiders cite these 'notes': General of the Gentlemen of the Household Jia Kui (*HHS zhi* 2, 3027, 3030); Gentleman Consultant Cai Yong (*HHS zhi* 2, 3039);

1.3.3 Values

The motivation of the state in all this is simple enough to understand: it was unthinkable to do otherwise. Even in sage-time, as we read in Section 1.1.2 above, *failure* to 'grant the seasons' would unravel the carefully maintained order by which the son of heaven holds the natural and social world together, inviting famine, plague and radical political correctives. One element of this order was metaphysical/symbolic – 'Rocktober' – but the other was predictive accuracy in a very modern sense. The old refrain goes that the fate of agriculture depended on the calendar, as if what the farmer needed was a timetable accurate to a thousandth of a day to determine when to plant beans. This is preposterous, as Jiang Xiaoyuan (1991, 115–18; 1992b, 167–72) shows, but that does not change the fact that this is very much what the premodern state *believed*. It did not envision this process the way that the twentieth-century historian of science might – the farmer, seeds in hand, watching the second hand of a clock – but rather in terms of ritual timing. Consider Emperor Zhang's edict of 85 CE:

> The Quarter-remainder reckons the establishment of spring.$_{\mathrm{Q01}}$ to fall one day prior to the [official] establishment of spring.$_{\mathrm{Q01}}$, so if one were to open prisons and end great punishments by [the latter], one would violate the *qi* (cosmic energies) and distance oneself from any prospect of peace or harmony (*HHS zhi* 2, 3026).[98]

As a side note, Emperor Zhang's ritual empiricism reveals an interesting tautology: the establishment of spring.$_{\mathrm{Q01}}$ falls three *qi* ($\approx$ 45.7 days) after winter solstice.$_{\mathrm{Q22}}$; the only way to empirically determine when this occurs was by gnomon shadow or meridian star (*zhong xing*), which were not accurate to the day; one must therefore depend on *calculation* to determine its date; this raises the question of which *li* calculations to use; and this brings us back to shadows and meridian transits.[99] One also notes that the establishment of spring.$_{\mathrm{Q01}}$ is a culture-specific construct of the civil calendar with no correspondence to any physical phenomena as observed or imagined in other language-cultures.

The state was locked to certain institutions; the question, then, is not why the state continued to do what it had always done but why *the individual* thought to join in. Advanced *li* was anything but compulsory – except, perhaps, for those born into family traditions – so, for most, it would have been a choice. It is doubtful that financial ambitions motivated that choice. The Clerk's Office had all of four pay-graded positions; these were graded rather poorly as official salaries

Palace Attendant Han Yue, Erudite Cai Jiao, Watchman of the Wall Liu Hong, and Gentleman of the Palace of the Right Chen Tiao (*HHS zhi* 2, 3041).

[98] On this point, and this particular edict, see Cullen (2007a, 241; 2009, 597–8).

[99] On the complexity of determining solstitial and equinoctial points via these methods, see Cullen (2007b) and Sivin (2009, 259–62).

go, and they only really served to trampoline those transferred from outside the bureau (see Tables 1.3 and 1.4). It may well have been some sort of religious/ideological conviction in sagecraft that motivated that choice: the chance to embody Cai Yong and Liu Hong's 'six virtues of the sage' (Section 1.1.2 above) and contribute to the latter's utopian vision. For that matter, it may well have been for the status which other gentleman accorded this pursuit, as we shall see in Chapters 4 and 5 below. Speculation aside, the simplest explanation is that which our sources actually name: some people are simply 'fond of *li*' (*hao li*).[100]

Whatever their expertise or motivation, we see gentlemen appeal to a common set of values in assessing the merits of a given *li*. Good *li* is expected to meet a combination of four criteria: (1) empirically verified accuracy, (2) precedence in scripture or 'master's method' (*shifa*), (3) embodiment of the occult metaphysical/numerological order, and (4) symbolic legitimisation of the imperial bloodline. The perfect *li* is one that meets all four: *li*, after all, is the study of an order inherent in nature, part of which is manifest, part of which is hidden, and all of which conspires in promoting the right bloodline. That said, actors lay different emphases on these criteria depending on whether the subject at hand is 'first month reform' or '*li* reform'. When it is the symbolism of the civil calendar year in question, matters of legitimation come first, as argued through scriptural precedence and metaphysics; 'accuracy', as such, does not come up, because 'first month reform' is not a quantitative matter to be resolved by data. When it comes to eclipse prediction, on the other hand, accuracy comes first.

What is 'accuracy', and how is it 'verified'? Hashimoto (1979) identifies a rhetoric of accuracy in our sources clustered around the metaphors of striking (*zhong* 中, 'to hit the mark', versus *shi* 失, 'to miss the mark'), accordance (*he* 合, 'to match', versus *cha* 差, 'to differ'), distance (*jin* 近, 'close', versus *yuan* 遠, 'far/off'), grouping (*mi* 密, 'tight', versus *shu* 疏, 'loose') and quality (*jing* 精, 'fine', versus *cu* 粗, 'coarse'). In terms of *precision*, we also see actors identify things as 'fine' (*xi* 細), 'detailed' (*xiang* 詳), 'circumspect' (*shen* 審) and 'cursory' (*lüe* 略). It is meaningless to simply *claim* that something is 'tight', of course, unless one has proof to support that claim. As the monograph of Cai Yong and Liu Hong puts it:

> For procedures (*shu*), one doesn't revise (*gai*) what isn't in error (*cha*), and one doesn't use what isn't verified (*yan*). The way of heaven is perfect and subtle, its *du* and numbers being hard to fix; there is a variety of procedures and methods, the net threads of *li* (*li ji*) being not [just] one. Without verification (*yan*), there is no way to know that it is correct (*shi* 是); without erring (*cha*), there is no way to know that it is amiss (*shi*). If it is amiss (*shi*), you then revise (*gai*) it; if it is correct (*shi*), you then use it – this is called 'holding truly to the middle way' (*HHS zhi* 2, 3041, citing *Analects* XX.1).[101]

[100] See, for example, *Song shu*, 12.260–1, translated in Section 4.4 below, and *NQS* 16.207.
[101] Tr. modified from Sivin (1986, 158).

Table 1.5 *The vocabulary of examination*

	Examine	Calculate	Compare	Match	Model	Verify	Rank
ji 稽	X	X		X	X		
kao 考	X						
jiao 校	X	X	X				
xiao 效効	X		X		X	X	
jian 檢	X				X		
yan 驗	X					X	
cha 察	X						
shen 審	X						
can 參	X						
ke 課	X		X			X	X

Verification relied on data from 'observation' (*guan* 觀), 'watching' (*hou* 候), 'measurement' (*ce* 測) and the sort of observational 'note-records' mentioned in Section 1.3.2 above. Regardless of whether the putative object that it takes is experiential or textual, the vocabulary of examination presented in Table 1.5 is inevitably the same, suggesting an equivalence of the two in practice. It would be tempting to translate all of these words as 'examine', but several bear further context-specific connotations of 'comparison', 'collation', 'modelling', 'verification' and 'ranking'. Actors rarely qualify what they mean by these terms, but we will ourselves 'examine' the process they imply in Chapter 4 below.

'Tightness' is the one value to which *everyone* appeals in matters of *li*. The source of narrative tension, therefore, is not how to weigh 'tightness' against some other criterion but how to adjudicate competing and 'unverified' claims thereto. In the absence of a public spectacle like Zong Gan's eclipse predictions, a new idea might appear more credible when associated with a known quantity. Some weave quotations from the classics and weft texts into their claims, but when it's not 'tight', we hear things like 'though alternative origin [dates] may not be elaborated in the charts and prophecies, each [has fostered] the procedures of its own expert lineage (*jia*) and thus must have had validity (*xiao*) in its own time' (*HHS zhi* 2, 3026).[102] Others stake their credibility on a 'master's method' (*shifa*) inherited from someone of better renown, but when it's not 'tight', we hear things like 'though [he] has a master's method, it is the same as if he had none at all' (*HHS zhi* 2, 3043).[103] *No one*, by contrast, takes such exception to what is proven 'tight' ('though it is tight . . . ').

[102] Tr. modified from Cullen (2007a, 241). [103] Tr. modified from Cullen (2007a, 247, 259).

A new idea might likewise resonate with its audience if it presents itself as capturing something of the occult – the way that Liu Hong supposedly 'established numbers based on the *Changes*' (Section 1.2.2 above). Still, very few *li* men actually appeal to this strategy, and those who do are not necessarily recognised for their efforts.[104] Consider Jia Kui's official assessment of one of the Grand Clerk's former staff:

During the Eternal Tranquillity reign (57–75 CE), there was an edict ordering Zhang Long, former expectant appointee to the Grand Clerk, to (predictively) note the added hour of [lunar phases] and lunar eclipse according to the Quarter-remainder method. Long said that he was able to use the nine, six, seven and eight lines from the *Changes* to know the extent of lunar motion.[105] [We] now know Long's notes to have missed the mark (*shi*) in most cases. [I,] Your servant, made Long retrodict [added hours] noted by former hands, and they did not correspond, sometimes [even] falling on different days; he was even further off (*yuan*) in failing to hit the mark (*zhong*) in heaven, [being off by a matter of] up to more than ten *du* (*HHS zhi* 2, 3030).

The difference here between a Liu Hong and a Zhang Long is simple: it is not that Zhang Long *tried* to predict lunar motion through *Book of Changes* numerology, it is that he *failed*, miserably and consistently.

1.4 Conclusion

Twentieth-century scholars spoke of Chinese astronomy as a science defined by secrecy and hereditary office. Its 'official character' determining its 'practical nature', some thought, this science was hostage to 'the blighting hand of bureaucracy, offering technicians security in a hierarchy which had nothing to gain from theory' (Sivin 1969, 4), even if the sword at its neck was double-edged: 'from early times Chinese astronomy had benefited from State support, but the semi-secrecy which it involved was to some extent a disadvantage' (Needham 1959, 193). It would be easy to dismiss this as wrong in light of the pluralistic ecosystem that reveals itself here through thick description. We have distinguished *two* 'sciences', in actors' terms, and further distinguished those from several others. We have separated the practice from the policy of these things, and we have recognised how each tells a different history. We have discerned the individual from the state, and we have a sense of the shared values and disparate motives weaving everything together. Some 69 per cent of those contributing to *li* policy from 206 BCE to 420 CE are bureau outsiders, and of more than a hundred names surveyed only one or two are related – this not

[104] On the minimal role of numerology in the history of astronomy in China, see Chen Meidong (2007, 551–65) and Martzloff (2009, 38–44). Liu Xin's Triple Concordance *li* of *c*.5 CE represents the greatest effort in this regard for the period in question, on which see Teboul (1983), Kawahara (1991; 1996, 148–95) and Vogel (1994).

[105] On the *Changes* hexagram line numbers 'nine, six, seven and eight', see Rutt (1996, 151–201).

a world defined by *secrecy*, as I have already intimated, but one defined by *sharing*. If that makes Sivin and Needham wrong, however, that is the fault of how we have thought to pose the question.[106]

They are not wrong – or not quite – because *we* have been considering the Han, and secrecy is something that happened later, piecemeal and over time. Whereas the Qin banned everything *except* technical literature in 213 BCE, and whereas the Han lent institutional status to weft and prophecy literature, later courts began to eye the astral sciences with increasing paranoia. The first to go was *tianwen*, the private practice of which was repeatedly banned from 267 CE on. As Lü Zongli (2003) points out, these bans were concentrated in the reigns of founding emperors trying to control the discourse surrounding their ascension, and it is only when the bans *stop* after the Sui (589–618 CE) that it really begins to disappear. The Tang (619–907 CE) marked a similar period of transition for *li*, during which the Clerk's Office began training its own technicians and passing administrative posts through family lineages now increasingly banned from outside contact. Things were such by the year 1600 that Matteo Ricci (1552–1610) reports, 'In China it is forbidden under pain of death to study mathematics without the king's authorisation', adding 'but this law is no longer observed'.[107] *This* is the situation that we are tempted to read back onto the Han, but it is as out of place there as would be a discussion of the Jesuit mission.

What we must disabuse ourselves of, then, is not the claim that Chinese astronomy *was* secret but the very notion of a transhistorical entity called 'Chinese astronomy' upon which one could read that claim back (or my analyses forward) in the first place. We should not assume 'China' but work instead to substantiate the textual, mythical, conceptual, historical, institutional, social and epistemological framework within which a given community is performing a given activity, and we should do so century by century as we move. The point of this chapter is not, then, to answer questions about 'what the Chinese thought' about heaven; it is to set the stage and introduce the cast that we may watch what plays out as conflicting elements collide. What concrete solutions does the technocrat work out to milk utopia from astronomy? How far does this project get? What happens when men of 'tightness' devote themselves to a single project for several centuries on end? What sort of new dimensions (and directions) does that project take on? How is this new thing reconciled with its past? And what relationship does *any of this* bear with the world of scholastics and political sloganeers?

[106] To be fair, note that one finds a significantly more nuanced approach to this question in Sivin (2009, esp. 58–9), and that Needham (1959, 186–94) walks this characterisation back with counterexamples of private study in the Song (960–1279).

[107] Cited in Needham (1959, 194). For further discussion of state bans in the history of astronomy in China, see Jiang Xiaoyuan (1991, 52–7) and Chen Meidong (2007, 23–32).

2 Observing the Signs

The astral sciences are conducted here, in the world below, and they begin with the act of seeing. More specifically, it is 'observing' (*guan*), 'measuring' (*ce*), 'checking' (*jiao*) and 'verifying' (*yan*) to which our sources appeal to keep 'in reverent accordance with prodigious heaven' (*qin ruo haotian*), and these are practices of seeing that are learned and mediated by human artifice. They are mediated by what we 'watch' (*hou*) for – by the questions we think to ask – but they are mediated too by the tools through which we see. In 1106, Yao Shunfu claimed to have first 'checked and measured' (*jiao ce*) the lodges via armillary sphere to reveal the 'true numbers' (*zhen shu*) of their fractional widths, but Guo Shoujing (1231–1316) later dismissed his data as 'yielding to personal inclination' (*Yuan shi*, 164.3850). Was what Yao saw the 'true numbers', as he claims, or just those that he wanted to find? Was it the make of the instrument that stood in the way of 'truth', or the way that it was employed? We will examine the process for arbitrating the 'tightness' (*mi*) of predictive models in Chapter 4, but for now we must content ourselves with an altogether more basic problem.

There are no numbers in the sky. Fuxi may have read them there in scintillating *wen* of light, but all we mortals have to go on are the numbers that we write upon it ourselves. Observation, that is to say, is as much an act of inscription as it is an act of seeing. Consider, for example, the following 'observation' adduced in support of *li* reform in 444 CE:

> It was inspected (*jian*) that at full moon, 11-VI-16, the moon was eclipsed. The added hour (*jia shi*) was *mao*. The eclipse began on day 15 in the second chant of the fourth watch, at the beginning of *chou*, and the eclipse was complete in the fourth chant, at the end of Hall 15 *du*. The Luminous Inception [*li*] had the sun at Baseboard 3 *du* that day. Examining (*kao*) it by means of opposition to the place of lunar eclipse, the sun should have been at Wings 15 *du* and a half that day (*Song shu*, 12.262).

This is not what one 'sees' in India, Mesopotamia or modern France, because co-ordinates like '*mao*' are meaningless beyond a culture-specific frame of reference. These numbers are 'seen', moreover, because they are *looked for*, the question in 444 CE being whether the winter solstice is truly at Dipper.$_{L08}$

21¼ *du*. They are only 'seen' through the lens of *theory* – the position of the sun here being extrapolated from the moon – and theory, in turn, is beholden to *history*, for prior to Jiang Ji's insight of 384 CE, this could not be 'seen'.[1] What is 'seen', lastly, is *read* from a physical instrument – the *du* of a sphere, the *ke* of a water clock – and it is *written down* on what becomes the real object of 'testing' (*kao*) and 'checking/collation' (*jiao*). To 'observe' is to write numbers into the stars, and the stars into numbers, and it is a practice of seeing that is mediated as much by calculation, ink and mechanical aids as it is by the questions that inspire us to look.

Our main source for the history of astronomical instrumentation in this period is the *tianwen* monographs of the *Book of Song*, *Book of Jin* and *Book of Sui*, which place the topic under 'Heaven's Form' (Tianti). One might say that these are different versions of the same text, seeing that Li Chunfeng's (602–70 CE) monograph for the Sui (581–618 CE) recycles that for the Jin (265–420 CE), which appropriates Shen Yue's (441–513 CE) for the Liu Song (420–79 CE). *This*, in turn, Shen draws from He Chengtian's (370–447 CE) now lost monograph in Xu Yuan's (395–475 CE) history, which was inspired by Cai Yong (133–92 CE). Leaving the matter of historiographic filiation to Chapter 5, we shall focus here on Li Chunfeng's treatment of the topic in the *Book of Sui*. Being at the top of a complicated food chain, Li's is the most developed and opinionated.[2]

'Heaven's form' brings us to a grey area between *tianwen* and *li*. As discussed in Chapter 1, the core of *tianwen* literature had always been catalogues and annals of 'omen-readings' (*zhan*). *These* were what the sages meant by 'observing the signs' when Sima Qian (*c*.145–*c*.86 BCE) wrote *his* monograph, and so too for Ban Zhao (44/49–118/121 CE) and Ma Xu (fl. 111–41 CE) in the *Book of Han*; it is what they meant for Cai Yong and Qiao Zhou (199–270 CE), and so too for Fan Ye (398–446 CE), who included *theirs* in the *Book of Later Han*.[3] He Chengtian's addition of the heading 'Heaven's Form' to his fifth-century monograph thus represents a significant break with tradition. He not only adds a third leg to the tripod – or third wheel to the bicycle, as the case may be – he also adds one of considerably different size, shape and colour. Wholly unrelated to omenology, 'heaven's form' is the study of instrument-making, instrument-based data collection and data-based theorisation of optics and celestial mechanics. 'Heavenly Patterns' is perhaps where this new heading would seem to belong – on the 'observing-the-signs' end of the spectrum –

[1] See Cullen (1977, 309–14).

[2] On Li Chunfeng's authorship of the *Book of Sui* and *Book of Jin* 'Tianwen zhi', see *JTS* 79.2718, *Shitong*, 12.19b–20a, and Howard L. Goodman's chapter in Chaussende, Morgan & Chemla (forthcoming).

[3] On the authorship of the *Book of Han* monograph, see *HHS* 84.2784–5; for the *Book of Later Han* monograph, see *JS* 11.278. See also Mansvelt Beck (1990, 111–30).

but it is interesting to note how contemporary bibliographies waffle on how to classify the relevant literature. Instruments are for taking measurements, after all, and measurements feed into *li*.[4]

It is this new, quantitative *tianwen* that interests us, for it stands to reveal how *li* experts saw – and how they *chose to see* – the things they did. Our inquiry will proceed in two stages. In Section 2.1, we will survey the repertoire of instruments at the expert's disposal, the point of which is to establishing what was available to whom, when, what they *could* do with it and, more importantly, what they *actually did* do with it (and when). We are interested in invention, but only insofar it coloured the way that our subjects thought about the meaning of their practices: we do not of course *believe* the story of a prehistorical armillary sphere hewn from jade by ancient god kings, but we do believe that the historical instrument is incomprehensible without it. In Section 2.2, we will then reflect upon the conceptual revolution with which modern and premodern scholars alike attribute the armillary sphere. Reviewing the depths to which instrument and cosmos are intertwined, we ask whether it was not the world view that justified the instrument as much as it was the other way around. The armillary sphere became the obsession of the millennium, but the gentleman experts responsible for its exaltation we see in the same breath mourning its futility in the face of scarcity, loss and misappropriation. There were no spheres, no spheres good enough, and nothing to challenge the sphere, except of course *in practice*, which leads us in Section 2.3 to ask what it actually means to 'observe the signs' and how that meaning changed.

2.1 Instrument Repertoire

We have been writing histories of instrumentation in China since the fifth century CE; save the occasional archaeological find, most of what is left for the historian to pick through are the writings of his medieval analogue.[5] Li Chunfeng's 'Heavenly Patterns Monograph' in the *Book of Sui* is a history that brims with bias, as we shall see in Chapter 5 below, but viewing this period through one man's eyes gives us the double advantage of focus and insight back into the man. In order to foreground this source, I have divided this section into the headings used by Li Chunfeng himself: 'Sphere Heaven Sight' (*huntian yi*), 'Sphere Heaven Effigy' (*huntian xiang*), 'Umbrella Diagram' (*gai tu*), 'Gnomon Shadow' (*gui ying*) and 'Leak Notch' (*lou ke*). I have reversed the

[4] For the *Gnomon of Zhou*, bibliographers vacillate between 'Heavenly Patterns' and '*Li* and Mathematics', for example, while placing gnomon and water clock titles under '*Li* and Mathematics'; see *Sui shu*, 34.1018, 1025; *JTS* 47.2036, 2038; *XTS* 59.1543–4, 1546–8.

[5] For modern studies on the history of astronomical instrumentation in China, see Maspero (1939), Needham (1959, 284–390), Needham, Wang & Derek (1986), Cullen (1996, 101–29), Pan Nai (2005), Wu & Quan (2008) and Sivin (2009, 171–225, 561–72).

headings, moving from time to space and from two dimensions to three, but I have done so to reflect the developmental order between them – the developmental order upon which both Li and modern scholars agree. Where I deviate from Li Chunfeng's framework is in separating instrument from cosmos, but these shall be reunited in Section 2.2 below. Note that I abbreviate compound decimal length measures like '1 *zhang* 4 *chi* 6 *cun* 1 *fen*' to '$1^{z}4^{ch}6^{c}1^{f}$' and that I give metric equivalents as per the inflationary rates reconstructed in Qiu Guangming (1992), where 1^{ch} = 23.1–33.3 cm.

2.1.1 Water Clock

In its simplest manifestation, the water clock is a 'leaky jar' (*lou hu*) that marks the passage of time via the flow of water. The instrument appears in the *Offices of Zhou* as operated by the 'jar bearer' on the Grand Clerk's (legendary) Western Zhou (1045–771 BCE) staff. Li Chunfeng offers that it goes back even earlier – that 'the Yellow Emperor invented (*chuang*) the observation (*guan*) of leaking water and fabricated a/the vessel upon the principle with which to divide the day and night' (*Sui shu*, 19.526) – and, on this point, he is simply following the Liang era (502–57 CE) *Leak Notch Classic*.[6] In terms of 'solid' evidence, archaeologists have unearthed bronze exemplars from Western Han (206 BCE–9 CE) sites, such as that in Figure 2.1, which help place the *terminus post quem* of the simple mono-vascular design at the beginning of the early imperial period.

As to the professional model, Zhang Heng (78–139 CE), in *System for Turning the Sphere Heaven Sight by Leak Water* (*Loushui zhuan huntian yi zhi*), offers us the following description of his personal creation:

> The vessels are made of bronze; they are placed one above the other at different levels and filled with pure water. Each opens at the bottom in a hole, where jade dragon baby [spouts] spit leak water into two jars – right for night and left for day (cited in *Chuxue ji*, 25.2a).
>
> On the lids (of each inflow receiver) are furthermore cast in gilt bronze an immortal, dwelling on the left jar, and an attendant for [his] assistance, dwelling on the right jar. Both [statuettes] hold an arrow with the left hand and indicate the notch [with] the right hand so as to distinguish [what is] early and late by heaven's time (cited in *Wenxuan zhu*, 56.25a (comm.)).[7]

This description gives us a fairly clear idea of how this worked, as illustrated in Figure 2.1. In the receiver floats the 'arrow' (*jian*) – an indicator-rod graduated along its length by 'notches' (*ke*) and anchored at one end to a bob. The lid is

[6] The Liang *Louke jing*, cited in *Chuxue ji*, 25.1a, gives 'The rise of the leak notch began in the days of Xuanyuan', another name for the Yellow Emperor.
[7] Cf. Needham (1959, 320).

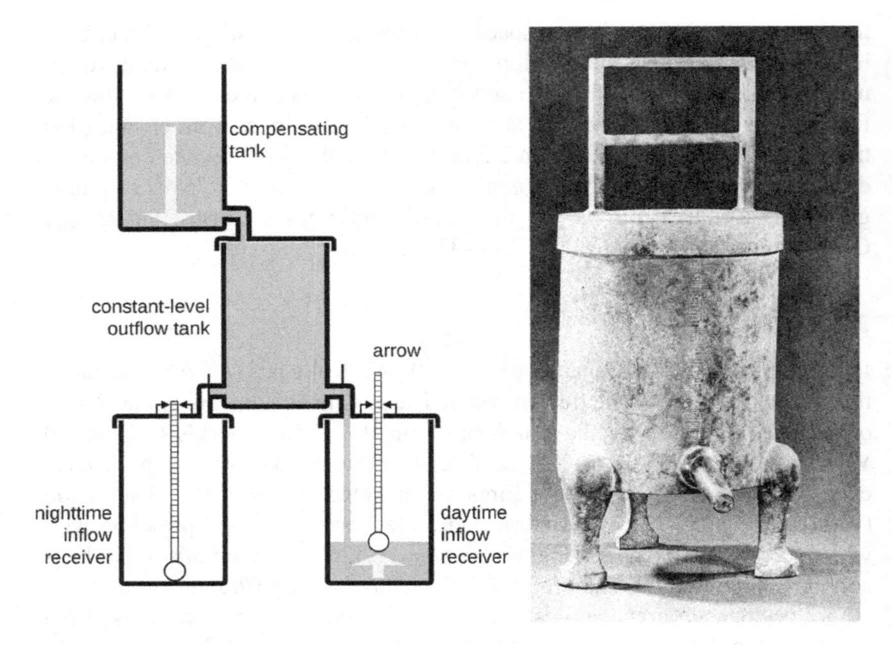

Figure 2.1 Early imperial water clocks. Zhang Heng's poly-vascular design (author's interpretation). Right: Yikezhao-meng outflow vessel (*Kaogu* 1978.2, plate XI).

fitted with a guide that keeps the 'arrow' upright and marks the number of 'notches' by which it has risen over time. Zhang's device features separate inflow receivers, allowing one to switch at the transition of day to night, and night to day, in order to maintain the uninterrupted operation of the device (rather than literally lose time emptying the receiver and switching arrows). Lastly, Zhang's is a poly-vascular installation, meaning that a reservoir (and potentially additional compensating tanks) has been added to the primary outflow vessel.

The advantage of the poly-vascular design is that it keeps a level pressure head on the primary outflow vessel such that the latter 'spits' water at an even rate. As to historical benchmarks, in Sun Zhuo's water clock inscription of 360 CE we see Zhang Heng's design expanded to a chain of 'stacked vessels in three tiers' (*Taiping yulan*, 2.13b). This made a difference, and the laboratory experiments of Hua Tongxu (1991) help to quantify it. Hua uses water vessels built to Yan Su's (991–1040) historical specifications, bypassing the 'arrow', to test the accuracy of mono-vascular and poly-vascular designs as a function of outflow volume over time. Three of his findings are particularly relevant here:

first, Western Han exemplars such as that from Yikezhao-meng (Figure 2.1) are too small and too quick to be anything other than minute timers (≈ 12 minutes); second, over a twenty-four-hour period the standard error of the mono-vascular chain is 6m37s, that for a two-tier poly-vascular chain is 18 seconds, and that for a three-tier chain is 14.4 seconds; third, the line of diminishing returns sets in at three tiers.

Hua Tongxu's experiments were conducted in a controlled laboratory environment to eliminate the effects of corrosion, sediment and temperature on the flow of water. His justification for this is in part that we know early timekeepers to have had appropriate countermeasures. As to corrosion, the preference for jade valves is not an accident. As to sediment, the height of the valve in Figure 2.1 is an effective solution. As to viscosity, Huan Tan (*c.*43 BCE–28 CE) gives us an idea of why the *Offices of Zhou* has the jar bearer 'heat a cauldron of water with fire in the winter' (*Zhouli zhushu*, 30.461b):

> As a gentleman-of-the-palace, I was in charge of the leak notch. The *du* would differ as soon as it was dry or humid, cold or warm, but one was able to right (*zheng*) it if checked (*can*) by day against the gnomon shadow and checked (*can*) by night against the stellar lodges (cited in *Chuxue ji*, 25.2b; *Taiping yulan*, 2.13a).

The lack of systematic seasonal errors in eclipse timing found by Steele (2000, 161–216) *c.*500–1625 CE would seem to confirm that the problem of viscosity was under control.

In terms of application, the 'leaky jar' is nothing without the 'arrows'. In civil timekeeping, the day–night arrow pair covered the length of a *ri* nychthemeron (the twenty-four-hour period from midnight to midnight), dividing the *ri* into daylight (*zhou*) and nighttime (*ye*) hours. Li Chunfeng offers the following description of the 'ancient system' inherited by the Han:

> A total of 100 notches were split between day and night, with 40 day notches to 60 night notches at winter solstice, 60 day notches to 40 night notches at summer solstice, and 50 day notches to 50 night notches at the spring and autumn equinoxes. It is bright two notches and a half before the sun has risen, and it is dark two notches and a half after it has completely set, so five notches are subtracted from the night and added to the day leak [pot] – we refer to this as 'dusk' and 'dawn'. Both drip notches increase and decrease in accordance with the [twenty-four] *qi*. Between winter and summer solstice there is a total difference between the length of day and night of 20 notches, for each difference of one notch there being a new arrow [pair]. Starting in both directions from winter solstice, there is a total of 41 [changes of] arrow [pairs] (*Sui shu*, 19.526).

The policy concerning the number and schedule of 'arrow' changes saw modification over time; other than short-lived experiments with 96, 108 and 120, however, the 100-notch day remained the standard for civil and astronomical timekeeping.[8]

[8] On civil 'arrow' policy, see Chen Meidong (2006).

It is from 'notch' time, the *li* literature explains, that one calculates the twelve or twenty-four 'added hours' of the civil nychthemeron and the five seasonal 'watches' of the night (such as announced by city drum and bell towers).

In practice, one sees a mixture of 'notches', 'hours' and 'watches' in civil and astronomical usage, not to mention the persistence of seasonal daylight hours in popular language (e.g. 'cock's cry' and 'level dawn').[9] Still, one cannot overemphasise the ubiquity of the 'leak notch' in daily life. 'Notch' time, as it were, appears throughout the standard histories – not just in edicts and norms for office use but in biographical anecdotes and state ritual.[10] It is there in the archaeological record, which provides us with both the hardware and documentation of its use in such mundane administrative procedures as postal delivery (Ma Yi 2006). One also sees it in religious literature, where precision timekeeping played a role in practices such as alchemy (Ren Song 2008). Indeed, we are dealing with a milieu in which gentlemen and county clerks alike were used to managing affairs down to the 'notch' (14m24s), and one can only imagine how this milieu informed expectations towards government *li*. In 102 CE, for example, the Han court issued an order that civil 'arrow' changes be pegged to solar declination. The problem with the 'ancient system', the edict explains, is that 'the official drip missed the mark (*shi*) of heaven in upwards of three notches', while the alternative 'reduces the disparity, [and its] tightness and closeness have been verified' (cited in *HHS zhi* 2, 3033). The sort of solar tables that one used in this regard give fortnightly 'arrow' pairs to a precision of a tenth of a notch (1m26.4s).[11]

2.1.2 *Gnomon*

Rivalling the 'leaky jar' in terms of simplicity, the gnomon (*gui* or *bi*) consists of a pole (*biao*) anchored perpendicularly to a shadow template (*gui*). As to origins, the gnomon, like the water clock, appears in the *Offices of Zhou*, and the *Gnomon of Zhou* credits the Duke of Zhou (r. 1042–1036 BCE) with having learned it from the Shang (?–1046 BCE) to understand how 'in antiquity [Fu]xi established the *li du* (successive/astronomical degrees) of the circuits of heaven' (*Zhoubi suanjing*, 1.1b). This, in addition to mention of 'solstices' (*zhi*, lit. 'extreme') in other classics, was sufficient to convince experts like Li Chunfeng that 'long ago, the Duke of Zhou measured gnomon shadows at Yangcheng so as to investigate (*can-kao*) the net threads (*ji*) of *li*' (*Sui shu*, 19.523). Archaeologically, one stick tends to look like all the others, but the

[9] The 'added hours' are illustrated in the Appendix. On 'added hours' and night 'watches', see Qu Anjing (1994). On the (potentially) seasonal sixteen-, eighteen-, twenty-eight- and thirty-two-hour schemes found in early excavated sources, see Li Tianhong (2012).

[10] See, for example, Hua Tongxu (1991, 40–3).

[11] See Zhang Peiyu et al. (2008, 37–42).

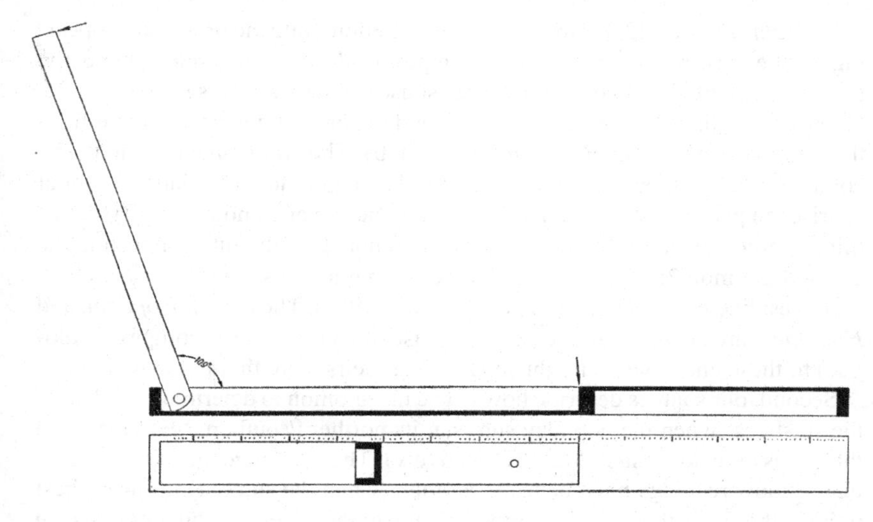

Figure 2.2 Han survey gnomon from Shibeicun tomb 1 (*Kaogu* 1966.1: 18, fig. 6).

bronze survey instrument recovered from Shibeicun tomb 1 in Yizheng, Jiangsu, does provides us tangible evidence of Han gnomonics (Figure 2.2).

The professional model does not appear to have experienced much by way of technological innovation over the early imperial period. By and large, experts stuck to classical specifications: 'The Zhou gnomon was eight *chi* long (184.8 cm)' (*Zhoubi suanjing*, 1.26a, 37b), and so too, for example, was the bronze gnomon constructed for the Chang'an observatory in 101 BCE.[12] Li Chunfeng mentions that Zu Geng's (fl. 504–25 CE) gnomon at the Liang astronomical office (also eight *chi*) featured a template base fitted with water channels for 'getting it level and upright' (*Sui shu*, 19.523), though the use of (independent) water levels for the task goes back once again to the *Offices of Zhou*.[13]

Its simplicity lent the gnomon to a variety of attested uses. First and foremost, our sources describe how to use the gnomon to find the cardinal directions. In the context of construction, for example, the *Offices of Zhou* mentions the use of a compass (*gui*) with gnomon shadows taken at sunrise and sunset

[12] The *Sanfu huangtu*, 5.1b, places at the Chang'an observatory a bronze gnomon eight *chi* tall, 1.2 *chi* wide, with a thirteen-*chi* shadow template and the inscription 'Made in Grand Inception 4 (101 BCE)'.

[13] *Zhouli zhushu*, 10.153b–155a, 41.642a–b. Note that the *Offices of Zhou* makes no direct mention of the eight-*chi* standard, only the '*chi* and five *cun*' solstitial shadow one would expect such a gnomon to cast in the Zhou heartland at Luoyang (34° 45′ N) at summer solstice.

(*Zhouli zhushu*, 41.642a). This would seem to adumbrate the procedure appearing in the *Gnomon of Zhou* (*sans* compass): (1) draw a circle around the gnomon, (2) mark where sunrise and sunset shadows intersect that circle, (3) bisect the line between those points, and (4) draw a north–south line from there to the gnomon (*Zhoubi suanjing*, 2.5a–b). The *Huainanzi* (139 BCE), by contrast, starts with an east–west line: (1) plant a gnomon; (2) plant another at sunrise ten paces west on the line connecting the sun and gnomon 1; (3) plant a third gnomon at sunset ten paces east of gnomon 2 on the line connecting the sun and gnomon 2; (4) bisect the line between gnomons 1 and 3; (5) draw an east–west line from there to gnomon 2 (Cullen 1976). The *Gnomon of Zhou* and *Huainanzi* are silent about the necessary use of a compass (or compass analogue) to these ends, which might suggest that theirs were thought experiments.

Second, our sources describe how to use the gnomon to determine the date of the solstices (when the noonday sun is at its northern/southern 'extreme', and the day is at its longest/shortest, respectively). The challenge, in any culture, is that shadow readings must be taken at noon (when the sun reaches its highest point in the sky at the meridian, directly south of the observer, and thus casts the shortest shadow), whereas the solstice may fall at *any hour*. The length of the tropical year (365.242 days), furthermore, ensures that the solstice falls at *different hours* in subsequent years. In the words of the *Book of Later Han*,

> In the derivation of *li* numbers, one sets up the sight (?) and gnomon to compare (*jiao*) solar shadows. When the [noon] shadow is longest then the sun is most distant: this is the starting point of the *du* of heaven [i.e. the winter solstice]. The sun leaves its starting point, and makes a circuit in a year, but the [noon] shadow does not repeat. After four circuits, i.e. 1,461 days, the shadow repeats its starting [value], and this is the conclusion of the [whole pattern of] the sun's movement. If you divide the days by the circuits, you get 365 and ¼ *du*, which is the number of days in a solar year (*sui*). The sun moves one *du* in a day, so this is also [counted in] the number of *du* in [one circuit of] heaven (for a diurnal rotation of 366¼ *du*).[14]

Of foremost concern was the winter solstice, 'when the [noon] shadow is longest', and to find the date and hour of solstice one had to rely on interpolation. Early sources are silent about how this worked, but Zhou Cong (eleventh century CE) tells us that 'the *li* procedures of the Jin and Han mostly took middle [values] between gnomon measurements before and after [the solstice] – this, for its part, exceeded half a day (*ri*) in error' (*Song shi*, 76.1763). That doesn't tell us much. What we *do* know is how Zu Chongzhi (429–500 CE) later went about the problem: (1) measure noon shadow a on one side of the solstice and measurements b and c on the other side, where $b < a < c$; (2) interpolate to find the date and time between b and c when the shadow length equals a; (3) subtract to find the time difference between the two points, before and after the

[14] *HHS zhi* 3, 3057; tr. modified from Cullen (1996, 96).

solstice, where the shadow equals *a*; (4) halve the interval, and count to the midway point.[15] The analysis of historical *li* reveals that the introduction of Zu Chongzhi's procedure coincided with a drop in solstice determination errors from two to three days to forty to fifty notches (Chen Meidong 1995, 50–79).

The rest of the twenty-four *qi* (see Appendix) were artefacts of interpolation. One cannot observe the instant of excited insects.$_{\text{Q03}}$ in nature – not the way that counterparts in Luoyang and Alexandria could come to the same date for winter solstice.$_{\text{Q22}}$ – so one must look where *li* has placed it, and early *li* placed the *qi* at intervals of a twenty-fourth of a year from winter solstice.$_{\text{Q22}}$. Then and only then, with a calculated time and date, could one 'measure' excited insects.$_{\text{Q03}}$. We know that such activity went into the Luoyang solar table of 174 CE, He Chengtian's Jiankang table of 445 CE and Yixing's (683–727 CE) Luoyang table of 727 CE, but few ever went to the effort of *measuring* shadows. The *Gnomon of Zhou*, for example, is quite explicit about how it derives intermediary shadow lengths via interpolation – '[over] all the eight nodes and twenty-four *qi*, [the gnomon shadow] is reduced/increased by 99 *fen* and $\frac{1}{6}$ *fen* per *qi*' (*Zhoubi suanjing*, 2.22b) – while other sources experiment with 96 *fen* and $99\frac{2}{6}$ *fen* per *qi* (Wu Jiabi 2007).[16] Mostly, though, actors were content to rely upon their predecessors' tables rather than reinvent the wheel.[17]

Third, our sources attest to the use of a gnomon–water clock combo to time meridian transits. The *Book of Han*, for example, recounts the initial step of the 104 BCE *li* reform thus:

[They] fixed east and west, erected gnomon and sight, and dropped the leak notch in pursuit of the distances between the twenty-eight lodges of the four quadrants (of the sky) and, in the end, to fix the new and dark moons (first and last days of the month), the equinoxes and solstices, and the steps and distances of quarter and full moons (*HS* 21A.975).

[15] See Sivin (2009, 259–62).

[16] Note that this passage of the *Gnomon of Zhou* is an insertion by the commentator Zhao Ying 嬰/Shuang 爽, as noted in Cullen (1996, 196). Cullen (1996, 158–62) argues for dating this commentary to 263/80 CE based on three considerations: one, the *terminus ante quem* supplied by Zhen Luan's (fl. 535–70 CE) subcommentary; two, textual parallels with Liu Hui's 263 CE commentary to the *Nine Chapters of Mathematical Procedures* that Cullen argues must have been taken *from* Liu Hui *by* Zhao Ying/Shuang (rather than vice versa or from a third source); and three, that Zhao Ying/Shuang's citation of the Supernal Icon *li* places him in Sun-Wu (222–80 CE), where it saw government implementation (see Chapter 4 below). This last point is not, in my opinion, particularly strong, given the number of authors like Li Chunfeng who discuss and cite from the Supernal Icon *li* in later times; I would furthermore note that Cui Hong's (478–525 CE) *Shiliuguo chunqiu*, 71.1b, cites a memorial of one Zhao Ying 嬰/Shi 奭 of Lanchi (modern Gansu) in 314 CE, which suggests that our commentator was actually active in the fourth-century north-west (cf. *Jin shu*, 68.2227).

[17] Zhang Peiyu et al. (2008, 34–6); cf. Chen Meidong (1995, 126–64).

The *Book of Later Han* likewise mentions 'dropping the leak and counting the notch to examine the centred stars' (*HHS zhi* 3, 3056). Presumably they were working in *du* of right ascension:

***Du* 度 ('measure/crossing'):** a linear measure, convertible with terrestrial distances, used in the context of the astral sciences (and that context only) as a measure of the circumference of a great circle whereby one *du* equals the distance travelled by the mean sun in one day, and the 'circuit of heaven' thus equals the length in days of the tropical year (*sui*).

Cullen (1996, 33–66) argues that the *du* and the measurement of the twenty-eight lodges (see Appendix) likely originated from clocking meridian transits. Our sources are silent about what this entailed, but one imagines that it would have worked like this: (1) note notch time *a* when the 'reference star' (*ju xing*) of Horn.$_{L01}$ is 'centred' behind the gnomon; (2) note notch time *b* when that of Neck.$_{L02}$ is 'centred'; (3) subtract *a* from *b* for the difference *c* in notch time; (4) convert time difference *c* into *du* distance *d* by the rule of three, i.e. multiply *c* by the number of *du* the sky turns in one day (365¼ + 1) and divide by the number of notches in the same period (ibid., 100). From there, one can measure an object's right ascension in 'lodge-entry *du*' (*ruxiu du*) via the same procedure, as illustrated in Figure 2.3.

Chinese sources have little to say about the gnomon's commonplace applications; where they shine is in observational procedures that are circular, physically impossible, or both. The *Gnomon of Zhou*, for example, has one measuring star shadows and counting *du* of right ascension from a circle 365¼ *chi* (85.8 metres) in circumference drawn around the gnomon *in the plane of the horizon*. More famously, it describes how to use the 'shadow rule' – one *cun* (2.35 cm) at the observer's hands corresponds to 1,000 *li* (415.80 km) in heaven – to deduce figures for the height of heaven, the distance of the subpolar point, the distance of the subsolar point at the solstices and equinoxes, and so on – all of which directly depend on the height of the gnomon used. The *Huainanzi* describes the same exercise (using a ten-*chi* gnomon!), and it details yet another that, while geometrically sound, has one discerning one-*cun* (2.3 cm) increments from a distance of half a *li* (208 metres). Stressing the hypothetical nature of these procedures, Cullen (1976; 1996, 111–28) attributes them to parochialism, whimsy and anti-sphere polemics.

Medieval scholars were no less critical of these texts. Li Chunfeng's monograph cites a petition of 604 CE by Liu Zhuo (544–610 CE) that casts the matter in the starkest of terms:

Shadow, leak and polar distance can [all] be extrapolated (*tui*) from the sphere, their being not *different things* but the hundred bones of a common body. The truth (*zhen*) of *this* is already verified (*yan*), and the falsehood (*wei*) of *that* is self-evident, [so] how is it that the bright sun (of truth) has yet to shine and that the torchlight (of benightedness) [burns] unextinguished – that the reasoning (*li*) is there and yet left to ruin? Is not this grievous? (*Sui shu*, 19.521).

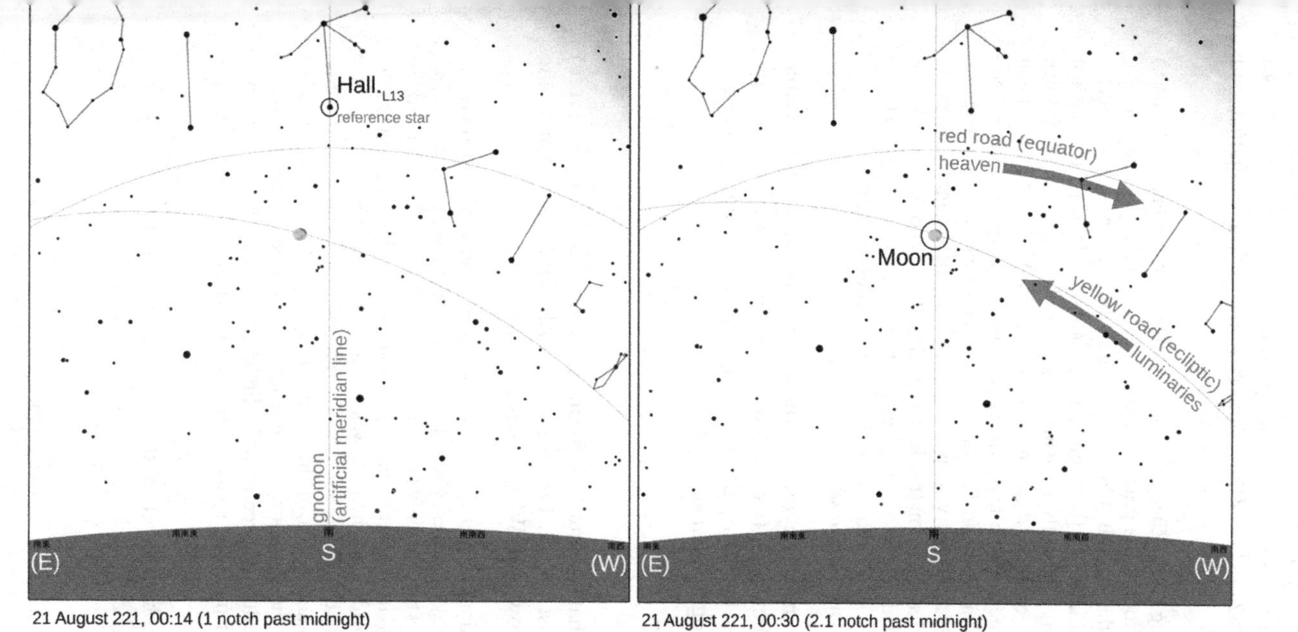

Figure 2.3 Clocking meridian transits with a gnomon. In this example we are measuring the position of the moon in 'lodge-entry *du*' of right ascension at Luoyang on 21 August 221 CE. First, stand north of the gnomon, facing south, such that the gnomon forms an artificial meridian (the great circle running due south through the celestial pole and the zenith). Second, note the time at which the reference star of the appropriate lodge is first 'centred' (*zhong*) at culmination behind the gnomon (Hall.$_{L13}$ at 1 notch past midnight). Third, note the time at which the moon is 'centred' (2.1 notches past midnight). Fourth, find the difference (1.1 notches). Fifth, convert notches into *du* by multiplying by the number of *du* in a diurnal circuit and dividing by the number of notches in a day ($1.1 \times \frac{366\frac{1}{4}}{100} \approx 4$ *du*). Rounding, the moon's right ascension in lodge-entry *du* is thus 'Hall.$_{L13}$ 4 *du*' at the time of observation. Images modified from *StellaNavigator10* with the help of Hirose Shō.

After recounting how he 'established a procedure to reform and correct the old sphere' with, among other things, 'the [gnomon] shadows of the two solstices' (*Sui shu*, 19.521), Liu Zhuo launches into a tirade against bad gnomonics:

> Previous scholars like Zhang Heng, Zheng Xuan (127–200 CE), Wang Fan (228–66 CE) and Lu Ji (188–219 CE) all took it [for granted] that the shadow differs by one *cun* for a thousand *li*. [They all] say that the sub-solar point in the south (the Tropic of Cancer) is 15,000 *li* (6,885 km) [from Luoyang], [but though] the gnomon shadow [at Luoyang that they used to derive this figure] is exactly the same, the [figures for] the height for heaven [at which they arrive] are somehow different! Investigate (*kao*) it via computational methods (*suan fa*), and [one will see that] this is certainly unacceptable. That a *cun* difference [corresponds to] a thousand *li* is likewise devoid of canonical explanation (precedent) – [this] is clearly an arbitrary supposition (*yiduan*) which cannot be taken for granted. Now, in Jiao[zhou] and Aizhou (modern Vietnam), there is no shadow north of the gnomon [at summer solstice], [by which we can] estimate that in less than 10,000 *li* (4,590 km) one goes south past the sub-solar point. This is [proof that] a thousand *li* per *cun* is not the true (*shi*) difference (*Sui shu*, 19.521–2).

In the *Book of Sui*, Li Chunfeng devotes the subsection on gnomonics to similar critiques of the 'insufficiency' (*wei zu*) of *Gnomon of Zhou* cosmology. Subsequent monographs, for their part, make the history of gnomonics one of *qi* determination and geographical surveys.[18]

2.1.3 Diagram

It is not entirely clear what Li Chunfeng means by 'umbrella diagram' (*gai tu*) because he devotes most of this heading to Liu Zhuo's screed against the *Gnomon of Zhou*'s shadow rule. What he does offer is this:

> In the past, when the sage kings corrected the *li* and elucidated time, they created a round umbrella to diagram the array of lodges. The pole was at its centre, and one rotated (*hui*) it in order to observe (*guan*) the heavenly signs. It was divided into 365 *du* and ¼ *du* so as to fix the number of days (in a year) . . . Here I note that after the Opening Sovereignty era (581–600 CE), the world under heaven was unified, and the Numinous Terrace (observatory) used the [Tuoba] Wei (386–535 CE) iron sphere heaven sight (*huntian yi*) to measure the excess and deficit of the seven luminaries, and [they] used a/the umbrella diagram to array the star seats (constellations) and divide the *du* fractions of the distance of both the yellow and red roads (ecliptic and equator) from the twenty-eight lodges – no one ever made the switch to the sphere effigy (*hun xiang*) [for this task]! (*Sui shu*, 19.520).

Earlier in the monograph, it is clear that what Li has in mind by such a 'diagram' (*tu*) is a star chart:

[18] On geographical gnomon surveys, see Beer et al. (1961).

> Yu Jicai (d. 603 CE) et al. were ordered to reference (*can*) and compare (*jiao*) the old diagrams, official and private, of the [previous dynasties] and of Zu Geng and Sun Senghua (d. 538/539 CE), to cut their sizes, correct their looseness/tightness, and to create an umbrella diagram on the basis of the Three Experts' star positions.[19] Laterally, it spread the starting divisions; discriminatingly, it indicated the constant *du*, and it was possessed of both red and yellow roads in two compass [circles], inside and out. The suspended signs, manifest and brilliant, the circling divisions, spacious and sequenced, the obscurity and plainness of the stars, the luminous curve of the Heavenly Han (the Milky Way) – bent like the vaulted grey (the sky), it would become the orthodox model (*Sui shu*, 19.504–5).

If we read 'umbrella' as a reference to the planar heaven of the *Gnomon of Zhou* (Section 2.2 below), Li's description would seem to invoke the sort of two-dimensional chart we see in the later the 'Tianwen tu' stele in Figure 2.4, but this raises the question of how to 'rotate' (*hui*) such a thing to 'observe (*guan*) the heavenly signs'. Qin Jianming (2008) suggests that we instead read 'umbrella' as it is. Citing substantial precedence for the decoration of chariot cover (*gai*) roofs with constellations, Qin suggests that the 'umbrella diagram' may refer to a star chart drawn on the interior of a real umbrella – its surface, 'bent like the vaulted grey', imitating the effect of a planisphere *sans* stereographic projection. Whether flat or curved, of course, a star chart is not something that one 'observes' *through*, like a telescope, but a product and object of observation.

As to *looking through*, let us recall from Section 1.2.2 that the *Book of Later Han* mentions the use of a 'diagram sight' (*tu yi*) for data collection. In 52 BCE, we see Geng Shouchang use such a device to '*du*-measure solar and lunar motion, and investigate (*kao*) and verify (*yan*) the manner of heaven's revolution'. The result was to quantify the moon's red-road (equatorial) speed at the solstices and equinoxes and to prove the yellow road (ecliptic) 'tighter and closer than using the red road' (*HHS zhi* 2, 3029). In 76/88 CE, we see additional reference to the use of a 'diagram sight, gnomon and leak [notch]' in the 'examination and correction of the *li du*' (ibid., 3037). Lastly, in 175 CE, we see Cai Yong invoke 'the use of the sphere heaven [and] diagram sight to examine (*jian*) *tianwen*' (ibid., 3039) in order to counter an argument from textual authority. This is the last historical mention of such an instrument.

Cullen (1996, 61–3) speculates that this 'diagram sight' may have been a board mounted in the plane of the equator – a board with a graduated circle around which one inserted sighting pegs in alignment with the celestial object 'sighted' and the gnomon at the circle's centre. Archaeology has provided us with two contemporary artefacts matching this design: the lacquer lodge disc from Shuanggudui tomb 1 in Fuyang, Anhui, and a variety of Han 'sundials' (Figure 2.4).[20] Sealed

[19] The 'Three Experts' (*san jia*) of 'star canon' catalogue literature refer to the likely apocryphal pre-imperial figures of Shi Shen, Gan De and Wuxian, on which see Sun & Kistemaker (1997).

[20] For the provenance of the 'sundials', see Li Jiancheng (1989) and White & Millman (1938); for the Shuanggudui disk, see *Wenwu* 1978.8: 12–31.

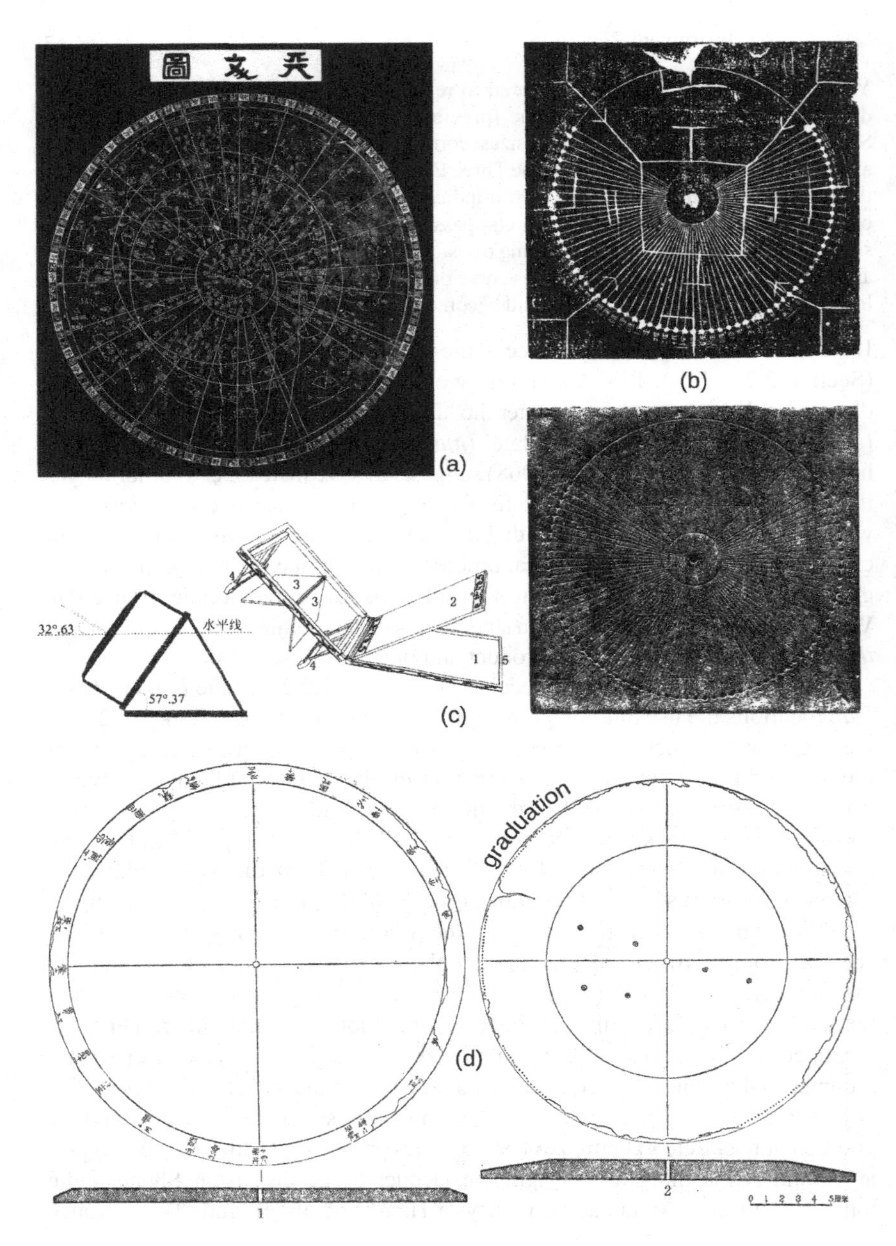

Figure 2.4 ‘Umbrella diagram’ candidates. Top left (a): Suzhou ‘Tianwen tu’ stele star chart, 1247 (Rufus 1945, plate 1A). Top right (b): ‘Sundial’ from Duan Fang collection and (below) Royal Ontario Museum (White & Millman 1938, 418, fig. 1, and plate XII). Bottom left: (d) Shuanggudui tomb 1 (165 BCE) lacquer disc and (c) equatorial mount (Shi, Fang & Han, 2012, 7, fig. 10, 9, fig. 14, 10, fig. 17).

in the tomb of Marquis Xiaohou Zao of Ruyin in 165 BCE, the lodge disc comprises two lacquer plates joined at their centres by a pin allowing for the rotation of one disc vis-à-vis the other. The upper plate (23 cm, or one *chi*, in diameter) features the seven stars of the Northern Dipper (Beidou; UMa), a pair of perpendicular lines running through the pivot and, around the edge, graduation holes appropriate to the number of du in a 'circuit', while the base plate (25.5 cm in diameter) is marked with the names and *du*-widths of the twenty-eight lodges. Shi, Fang and Han (2012) have identified a folding lacquer contraption alongside the disc as its equatorial mount, addressing Cullen's (1980–1, 34–6) original reservations in identifying the device as an observational instrument. Likewise mounted in the plane of the equator, Cullen (1996, 63) notes, the oddly graduated 'sundials' meet the specifications of an observational 'diagram sight'.

There is a world of difference between the artefacts depicted in Figure 2.4, and it is unclear which, if any, Li Chunfeng has in mind by this label. Umbrellas and lodge discs lend themselves to 'rotation'; umbrellas and star charts allow the 'arraying' of stars, roads and milky ways; but only the disc and 'sundial' allow for observation, right? The problem goes away when we begin to question what 'observation' – as an actors' category – actually means.

2.1.4 Sphere

In the *Book of Sui*, Li Chunfeng divides sphere instruments under two headings: 'Sphere Heaven Sight' (Huntian yi) and 'Sphere Heaven Effigy' (Huntian xiang). The difference, he explains, comes down to the 'traverse' (*heng*) sighting tube:

> The sphere heaven sight is constructed with engine and traverse. Not only in its at once moving and static state does it replicate the true situation (*qing*) of the two *yi* (*yin* and *yang*), the complete rotation of the transverse tube allows investigation of the fractions of the three lights (the sun, moon and stars). It is that by which one estimates and corrects the lodge *du* and levels and paces excess and void – a method handed down from antiquity (*Sui shu*, 19.517).
>
> The sphere heaven effigy is constructed with engine and no traverse . . . It is inferior to the sphere sight, which has in addition a traverse tube – the thing that [allows] the measure and estimation of sun and moon and the division and pacing of stars and *du* (ibid., 19.519).

It is telling that Li Chunfeng feels the need to define his terms so clearly. Etymologically, the word *hun* 渾/混 relates to the idea of a 'confused' and 'undifferentiated' state of matter at the beginning of time. 'Undifferentiated confusion' is the metaphor behind the celestial sphere, and it is masterfully appropriate as concerns the history of the instrument as well. The term *yi* 儀 ('sight') derives from the graduated sight/rangefinder pegs of early missile weapons, which, extended to the armillary sphere, came to stand for sighting pegs, graduated rings and the instrument itself. The term *xiang* 象 ('effigy'), if we recall from Section 1.1.3, refers to a 'simulacrum' linking something in the world of man to a truth beyond his ken. In modern terminology, *huntian yi* (or *hun yi*) refers to an observational or demonstrational armillary sphere, while *huntian xiang* (or *hun xiang*) refers to a celestial globe; early imperial sources, however, often label the demonstrational sphere a *xiang*, and either non-observational apparatus an *yi*, hence Li Chunfeng's insistence on a rectification of names.[21]

Adding to the confusion is the term *xuanji yuheng* 璿璣玉衡, 'rotating [jade] mechanism and jade traverse' (?). The *Book of Documents* records that, upon his ascension, Sage King Shun 'attended to the *xuanji yuheng* so as to order the seven governors/government matters' (*Shangshu zhushu*, 3.35b). For centuries, this line has been read as evidence of a prehistoric jade armillary sphere. The *Spring and Autumn* weft *Wen yao gou*, for example, tells us that 'when Yao of Tang took the throne, the Xi and He [brothers] established the sphere sight' (cited in *JS* 11.284). Early sources more reliably attribute 'sphere heaven' to Luoxia Hong (fl. 104 BCE), and Li Chunfeng is quick to address the anachronism. He identifies the *locus classicus* of this myth as Ma Rong (79–166 CE), who 'first conceived of saying that the [*xuanji*] *yuheng* was the sphere heaven sight' (*Sui shu*, 19.515–16), directing the reader to the *Records of the Grand Clerk* and *Book of Han*'s identification of Xuan, Ji and Yuheng as the stars of the Northern Dipper (Beidou, UMa).[22] The idea, however, never went away, and *xuanji yuheng* begins to refer to the 'sphere sight' from at least the third century CE on.[23]

Sphere Computers As an appendix to the 'sight', Li Chunfeng's minimalist account of the 'effigy' focuses on the fifth and sixth centuries CE. The *word* is no earlier than Wang Fan's *The Sphere Heaven Effigy Explained* (*Huntian xiang shuo*), though the instrument itself clearly goes back as early as Zhang Heng. Under 'Sphere Heaven Sight', Li Chunfeng attributes Zhang with the following installation:

In Emperor Huan, Prolongation of Brightness 7 (164 CE), Prefect Grand Clerk Zhang Heng redesigned [the 103 CE 'bronze sight'] in bronze with four *fen* (9.4 mm) to the *du*, for a circuits of heaven of $1^{z}4^{ch}6^{c}1^{f}$ (343.34 cm). It was placed in a sealed-tight (*mi*) chamber and rotated by means of leak (water clock) water. The person charged with watching called it out from behind closed doors to announce to the observers of heaven (*guan tian zhe*) of the Numinous Terrace (observatory) the add[ed hour] (?) of the *xuanji*, that such-and-such star was first visible, that such-and-such star was already

[21] On the terminology of sphere instruments, see Li Zhichao (2014, 171–7) and Wang Yumin (2015).
[22] Note that Li Chunfeng is not consistent on this point, as we see in *XTS* 31.806 and Section 5.2.1.
[23] On *xuanji yuheng*, see Cullen & Farrer (1983).

centred (culminated), and that such-and-such star was currently setting – all of which were like matching [the two halves of] a tally (*Sui shu*, 19.516–17).

Li Chunfeng dates this device to twenty-five years after its maker's death in 139 CE, but Li's attribution parallels an undated attribution from Ge Hong (283–343 CE) some centuries earlier that would seem to confirm its historicity.[24] Li, to be clear, places this device under 'Sphere Heaven Sight' as a *counterexample*; elsewhere he cites Wang Fan identifying 'Zhang Heng's redesign' as an 'effigy' (*JS* 11.288) and rails against those who would confuse it for a 'sight' (Section 2.2.2 below). The fact that the device was *indoors*, separate from the activity of 'watching', means that it is strictly *non-observational*, in the sense stressed by Li Chunfeng, but this is something more than a *demonstrational aid* – it is what Arai (1989, 325) calls a 'computer'.

Ironically, it is to Zhang Heng's *The Sphere Heaven Sight* (*Huntian yi*) that we must turn to understand his work with the 'effigy'. As preserved in Liu Zhao's (fl. 502/557 CE) commentary to the *Book of Later Han* and Gautama Siddhārtha's (fl. 729 CE) *Kaiyuan zhanjing*, the point of *The Sphere Heaven Sight* is to derive an algorithm for converting between ecliptic and equatorial lodge-entry *du*. The treatise begins with the roads, poles and measures of 'sphere heaven', but it becomes clear that what it is describing is a material 'sight':

The first row running horizontally up on top are the numbers of yellow-road advance–retreat. What one *should* do is *du*-measure these over days and months via a/the bronze sight – *then* could [they] be known – [but as] this would take a year at the sight to complete, and [as] there would furthermore be overcast and rainy [days] interspersed therein, it would be difficult to bring to successful completion (*HHS zhi* 3, 3076 (comm.)).

The 'bronze sight' (*tong yi*) likely refers to the 'Grand Clerk yellow-road bronze sight' commissioned for the Luoyang observatory in 103 CE, to which Zhang Heng would have had access as Prefect Grand Clerk *c.*106/123 and 125/132 CE. The 'advance–retreat numbers' (*jin tui shu*) are corrections added to ecliptical lodge *du* to render the equivalent in equatorial units (i.e. the reduction to the equator). Rather than '*du*' this by observing the sky, *The Sphere Heaven Sight* instructs the reader to 'make a small sphere complete with red road and yellow road, then allocate (graduate) each with 365 *du* and ¼ *du* and make sure to align their relative values starting from the position of winter solstice.$_{Q22}$' (ibid.). Next, the reader is to fix the ends of a bamboo strip to each pole such that it runs snugly against the side of the sphere; he is then to

[24] For the date of Zhang Heng's death, see his biography in *HHS* 59.1939. Ge Hong's near-identical description of the device is introduced and translated in Section 5.3.1. Ge Hong, one notes, does not give Zhang Heng's water-driven sphere a date. For more on the device, see Needham, Wang & Derek (1986, 110–13).

'make 182 *du* and ⅝ [*du*]' (ibid.) from pole to pole along its centreline. This graduated rotating meridian line, finally, allows the user to physically measure the polar distance (*qu ji*) and the yellow-/red-road differential at each *du* along the ecliptic.

Measuring this differential, the text explains that one gets zero at the solstices.$_{Q22/10}$ and equinoxes.$_{Q04/16}$ and +3 *du* at the establishments.$_{Q01/07/13/19}$. Between upper and lower limits, one 'advances' from 0 to 3, and 'retreats' from 3 to 0, via linear interpolation, to arrive at the correction appropriate to a particular lodge-entry *du* – its 'advance–retreat number'. This is easy enough, but it involves rather complex fractions when dealing with twenty-fourths (*qi*) of a solar year (and/as circuits of heaven) of $365\frac{385}{1539}$ or $365\frac{455}{1843}$ days. Instead, *The Sphere Heaven Sight* has the user round: one counts *qi* in a 15–15–16-day pattern, and one counts 'advance–retreat' in ¼ *du* increments achieved in three- and four-day blocks. In order to avoid a 368-*du* circuit, Liu Hong (fl. 167–206) explains, one resets this at the solstices.$_{Q22/10}$ and equinoxes.$_{Q04/16}$, and the three-day blocks are inserted where the remainder of the 'advance–retreat numbers' reaches ½.[25] In brief, *The Sphere Heaven Sight* offers the reader a shortcut for an algorithm; the algorithm is a shortcut for physical measurement, and the measurement of the 'effigy' is in turn a shortcut for observation with a 'sight'. One finds an additional shortcut in solar tables from 174 CE on, which append 'advance–retreat numbers' to every *qi*.

Returning to the *Book of Sui*, Li Chunfeng devotes the heading 'Sphere Heaven Effigy' to three devices. The first is a wooden effigy from the late Liang palace library, which Li describes as 'round like a pellet, several spans large, with axes at the north and south heads, and its entire body laid out with the twenty-eight lodges, the Three Expert [catalogue] stars, the yellow and red road, as well as the Heavenly Han, etc.' (*Sui shu*, 19.519). This is clearly a celestial globe – the sort of device that Li would have the Sui–Tang observatory use in place of the 'umbrella diagram'. The second is a bronze 'sphere sight' constructed at imperial order by Prefect Grand Clerk Qian Lezhi in 436 CE at '5 *fen* (12.35 mm) per *du* for a diameter of 6^{ch}–$^{c}8^{f}$ and less (150.24 cm) and a circuit of $1^{z}8^{ch}2^{c}6^{f}$ and less (451.08 cm)' (ibid.). The third is a 'small sphere heaven' built by Qian in 440 CE at '2 *fen* (4.94 mm) per *du* for a diameter of $2^{ch}2^{c}$ (54.34 cm) and a circuit of $6^{ch}6^{c}$ (163.02 cm)', which was fitted with stars in tricolour pearls, a yellow road and pivots for rotation; 'and', Li adds, 'the earth was at its centre' (ibid., 19.520).

The Qian Lezhi spheres pose something of a conundrum. First, how could the 440 CE 'sphere heaven' have both stars on the outside and 'earth' on the inside? In other words, is this an armillary sphere or a celestial globe? As to the

[25] On Zhang Heng's *Sphere Heaven Sight*, see Arai (1989), Cullen (2000) and Lien (2012).

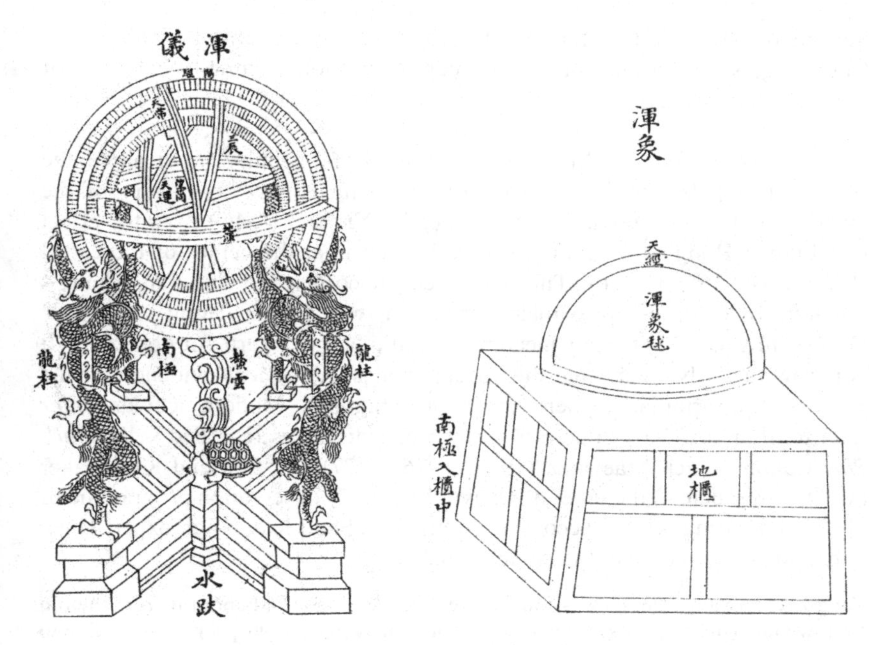

Figure 2.5 Sphere instruments. Left: 'sphere sight'. Right: 'sphere effigy'. Source: *Xin yixiang fayao*, A.9a, B.2a.

436 CE 'sphere sight', 'earth was inside heaven, unmoving, and he established *rings* (*gui*) for the yellow and red road and *rings* for the north and south poles, upon which were arrayed the twenty-eight lodges and Northern Dipper and pole star' (ibid., 19.519). This points to a hollow, armillary construction with a limited array of relevant 'stars' engraved upon its rings, which may well explain the 'sphere heaven', but Li Chunfeng is clearly perplexed about the 436 CE 'sight':

> As to this construction, it could be taken for a sphere sight, but the sight is absent a traverse tube inside; it could be taken for a sphere effigy, but the earth is not on the outside [see Figure 2.5]. This references both models (*fa*) but constitutes a separate form (*ti*). Gathering from the device's use, it would seem to belong to the current (i.e. tradition) of sphere effigies (ibid., 19.519).

This, matched with his description of the wooden sphere, would seem to indicate that to Li Chunfeng the word 'effigy' implies a solid globe. Li insists that 'the sight and effigy are two [distinct] devices with nothing whatsoever to do with one another' (ibid., 19.519), but the line between them is harder to draw than he lets on. The problem is the *non-observational armillary sphere* of the type built by Zhang Heng and Qian Lezhi and identified variously in later literature as 'sight' *and* 'effigy'. This, to Li Chunfeng, is 'a separate form' – an exception without a name. What, then, of the orthodox 'sight'?

Sphere Sights After refuting the second-century myth about the prehistorical jade armillary sphere, Li Chunfeng begins his history of the 'sphere sight' from Luoxia Hong. Li cites Yu Xi (fl. 307–46 CE) to the effect that Luoxia Hong 'turned a/the sphere heaven in/at the centre of the earth' (*Sui shu*, 19.516) *c.*104 BCE. This would seem to correspond with Yang Xiong's (53 BCE–18 CE) more proximate attribution that 'Luoxia Hong *ying*-ed it' (*Yangzi fayan*, 10.1b), *ying* meaning 'build', 'align', 'turn' or 'operate'. It is not particularly clear what this means, nor is subsequent evidence for the Western Han armillary sphere any less ambiguous. Modern scholars *infer* the use of an armillary sphere from the polar distances recorded in *Mr Shi's Star Canon*, which Maeyama (1975; 1976; 1977) and Sun and Kistemaker (1997, 9–69) date to the time of Xianyu Wangren's observation programme of 78–75 BCE. Yang Xiong, according to his friend and leak notch expert Huan Tan, apparently met with the device's maker:

> Yang Ziyun (Yang Xiong) was fond of *tianwen*, and he asked about it from the old Yellow Gate (palace) artisan who made sphere heaven(s), [who] said, 'I was able to make the thing when I was young, [but I] was merely following the stipulated *chi* and *cun* (dimensions) [that I was handed] without really understanding what they meant. Little by little, I gradually got better [at it], but now [I am] seventy, old and about to die by the time that [I] have only begun to understand myself. Now my son receives training [on how to] make it, and [he] too shall repeat the years like me, being old and about to die by the time [he] understands himself' (cited in *Taiping yulan*, 2.11a–b).

Beyond this, there is the 'Grand Clerk yellow-road bronze sight' of 103 CE, introduced in Section 1.2.2 above, of which all we have is the following description:

> With Horn.$_{\text{L01}}$ as 13 *du*, Neck.$_{\text{L02}}$ [as] 10 . . . for a total of 365 *du* and ¼ *du*. The winter solstice was at Dipper.$_{\text{L08}}$ 19 *du* and ¼ *du*. The Clerk's Office perimetered (?) solar and lunar motion and checked quarter and full moons, and though it was tight and close, it was not used for noting the sun/days. As to the sights (*yi*), the yellow road and *du* (equator ring) rotated, [but it] was difficult to watch (*hou*) [with], which is why the matter (the order to use it) was rarely heeded (*HHS zhi* 2, 3029–30).

In the *Book of Sui*, Li Chunfeng focuses instead on two later instruments: a bronze 'sight' constructed by Kong Ting in 323 CE for the Xiongnu Liu Zhao (304–29 CE) observatory at Chang'an and an iron 'sight' made by Prefect Grand Clerk Chao Chong (fl. 395–402 CE) and Artisan Xie Lan in 398 CE for the Xianbei Tuoba Wei court at Pingcheng.

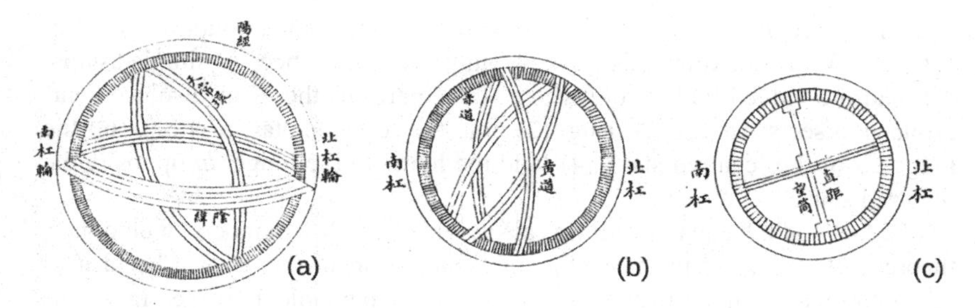

Figure 2.6 Observational armillary sphere component groups. Left to right: (a) six-joint sight (*liuhe yi*), (b) three-chronogram sight (*sanchen yi*), (c) four-displacement sight (*siyou yi*). Source: *Xin yixiang fayao*, A.11a,13a, 14a.

Li Chunfeng offers a lengthy description of the Xiongnu model, studied in Maspero (1939, 303–52), Li Hai (1994) and Morgan (2016a). In brief, the Xiongnu sphere featured two of three 'sight groups' (*yi*) typical to later Chinese models (see Figure 2.6). The first was the 'six-joint sight' (*liuhe yi*) – a fixed outer cage 'joining' a horizon, meridian and equator ring at six points (and to the platform). Aligned at the horizon and celestial pole, the outer cage provided a fixed co-ordinate grid within which to turn interior rings. At the meridian, the Xiongnu sphere featured 'double ring-circles joined [parallel] to one another with roughly three *cun* (9.09 cm) between them', allowing an unobscured line of sight to the meridian. Its (single) horizon and meridian rings, one notes, were 'divided around the circumference into *du* numbers and inscribed with the positions of the corners and chronograms' (*Sui shu*, 19.517–18). The second component group was the 'four-[directional] displacement sight' (*siyou yi*), a meridian ring turning east–west around the polar axis and fitted with a sighting tube that pivoted north–south through its centre. On the Xiongnu sphere, the rotating meridian ring was likewise formed of 'double circles joined [parallel] to one another', between which pivoted 'a traverse eight *chi* (242.4 cm) in length, through the centre of which was a [sighting] hole one *cun* (3.03 cm) in diameter' (ibid.).

What the Xiongnu sphere *lacked*, in later terms, was a 'three-chronogram sight' (*sanchen yi*): an intermediary equator and ecliptic ring that rotated on the polar axis into alignment with heaven's 'roads' to allow the reading of lodge-entry *du*. The fixed equatorial ring of the 'six-joint' group would have provided instead the sort of 'hour angle' more familiar from Mediterranean astronomy. We see evidence of this measure in eclipse observations from 221–3 CE (Chapter 4), from which we learn that it was called the 'added hour' and counted in twenty-four 'corners and chronograms' from the midnight meridian line (see Appendix). This is *not*, however, an equatorial measure that one finds in any *li* procedure text. If the measurements made thereby were to be of any use to *li*, therefore, one would need to convert between 'added hours' and lodge-entry *du*. We do not know this worked, but it might have been as simple as this: (1) read the 'added hour' from where (the centre of) the 'four-displacement' group intersects the equator ring, (2) note the 'centred star', (3) look up the lodge *du* of the 'centred star', (4) 'add the hour' to the lodge *du* opposite the 'centred star'.

The fact that the equator ring was also divided into '*du* numbers' would have theoretically obviated the need for unit conversion in this case. The idea of adding *du* read from an instrument to *du* read from a table, however, raises the question whether these were the same *du*. The number of *du* in a 'circuit' varied according to one's definition of the solar year – $365\frac{385}{1539}$ (Grand Inception *li*), 365¼ (Quarter-remainder *li*), $365\frac{145}{589}$ (Supernal Icon *li*), $365\frac{455}{1843}$ (Luminous Inception *li*), etc. No instrument maker could be expected to graduate a circle to such precision, let alone *re-graduate* it with every *li* reform, so we should expect some sort of compromise in practice. We do not know how many *du* the Xiongnu 'sphere sight' had, but we might infer the number 365¼ from what we have seen in the second- to fifth-century spheres surveyed thus far (see Table 2.1). After the fourth century, however, we find different numbers. According to the *Old Book of Tang*, Li Chunfeng's sight of 633 CE was graduated with '365 *du* in warp (RA) and weft (declination)' (*JTS* 79.2718), while Monk Yixing's sight of 725 CE featured 'extremes (solstices) drawn in two places . . . the *du*-numbers of the circuits of heaven . . . the 100 notches . . . and the 360 rods, levelled with the reigning hexagrams' (*JTS* 35.1297–8). Given the importance of the *du* in *li* procedure texts, it might seem odd that 'sight' makers divided rings into units of 24, 100, 360, 365 and/or 365¼, but one could always convert between units. Anyway, even at 365¼, the material *du* was at best an approximation of its mathematical counterpart (Morgan 2016a).

As to the Xianbei sphere of 398 CE, Li reports that, excepting the base, 'everything else was largely the same as [Xiongnu Emperor] Liu Yao [of Zhao's] sight' (*Sui shu*, 19.518). Li would know. First, as Prefect Grand Clerk, he reports that '[I, His Majesty's] historian/clerk servant went to the observation terrace to visit the sphere sight and saw (*jian*) that it was the original constructed by [Tuoba] Wei Prefect Grand Clerk Chao Chong and that it was made of iron' (*Sui shu*, 19.505). It is less clear how Li Chunfeng knows about the Xiongnu sphere. The Sui 'obtained the sphere sight instruments of the [Liu] Song clan' from the Chen (557–89 CE) capital at Luoyang in 589 CE (*Sui shu*, 19.504), suggesting that it may well have made it into Tang hands, and Li Chunfeng, for his part, seems to speak from personal experience: 'Inspection (*jian*) of the engraving [reveals that] it was constructed by Clerk's Office Assistant Kong Ting of Nanyang in year 6 of the [Xiongnu] imposter Liu Yao's Glory Inception reign (323 CE)' (*Sui shu*, 19.518).

2.2 The Sphere World

2.2.1 *The Philosophical Sphere*

Of all the instruments at their disposal, early imperial experts exhibit something of a monomania with the 'sphere heaven'. It was 'the bright sun (of truth)', in the words of Liu Zhuo, from which 'shadow, leak and polar distance can [all] be extrapolated' (Section 2.1.2), and 'with regard to observation (*kui*) and measurement (*ce*)', Li Chunfeng says of the 'four-displacement' model, 'it did truly everything that one could desire' (*Sui shu*, 17.518). Sinologists, for their part, tend to speak of it in equally glowing terms, Needham (1959, 339) calling it 'the indispensable instrument of all astronomers for the determination of celestial positions before the invention of the telescope'.[26] This might sound odd to a historian of Mediterranean or Indian astronomy, where the sphere was mostly relegated to demonstration, but Chinese sources are rather more sanguine about the object's potential.[27]

Cullen (1980–1, 36–9; 1996, 53–9) places the sphere at the centre of a 'revolution in practice' that left the concept, instrumentation and world of *du* measurement forever altered. Having originated as a measure of time in clocking meridian transits, and having achieved concrete spatial form on diagrams and reckoning devices (Section 2.1.3), it was not until someone played with aligning a 'diagram sight' to the sky, he imagines, that heaven was first seen in all three dimensions: a circular road divided into macrocosmic *du* running half above and half below the earth and, from there, *a sphere*, the north–south of which might lend itself to the same measure. Whatever the exact sequence of events, by the first century BCE we see (1) the sphere cosmos, (2) the sphere instrument and (3) the measurement of declination and longitude (thereby) in *du*. This comes together all at once, and the relationship is such that one is never sure whether a given reference to 'sphere heaven' is to its macro- or microcosmic instantiation.[28]

This was a 'revolution' in the very Kuhnian sense that 'when paradigms change, the world itself changes with them' (Kuhn 1996, 111). The sphere englobed all within. It allowed actors to discover and measure things unimaginable to the second-century BCE mind: the *distances* of stars, the true path of the luminaries, the precise obliquity of that path and its effects on measured (equatorial) speed. It offered thinkers a simple and comprehensive explanation of the mysteries of the phenomenal world: the apparent rising and setting of celestial objects; the temperature, shadows and daylight hours of the seasons;

[26] Cf. Sun Xiaochun (2015, 2127).

[27] On the role of the armillary sphere in other traditions, see, for example, Sayili (1960), Ōhashi (1994, 259–73) and Sarma (2009, 14–16).

[28] On the 'sphere revolution', see also Liu Jinyi (1984) and Li Zhichao (2014, 104–28).

Table 2.1 *Ring dimensions and graduation of known sphere instruments, 1–1000* CE*

							Diameter		Circumference				Unit size	
No		Name	Obs	Owner	Maker	City	*chi*	(cm)	*chi*	(cm)	π	Grad.	*fen*	(mm)
1	–	Circle sight, red-road sight	Y	state	–	Chang'an	–	–	–	–	–	–	–	–
2	103	Yellow-road bronze sight	Y	state	–	Luoyang	8.0000	(188)	†25.1327	(590.6)	–	365¼ *du*	†6.88	(16.2)
3	< 139	Small sphere	N	priv.	Zhang Heng	Luoyang	–	–	–	–	–	365¼ *du*	–	–
4	?164	Sphere heaven sight	N	state	Zhang Heng	Luoyang	†4.6505	(109.3)	14.6100	(343.3)	–	†365¼ *du*	4.00	(9.4)
5	< 219	Sphere effigy	N	–	Lu Ji	Lujiang	–	–	–	–	–	–	–	–
6	< 266	Sphere effigy	N	–	Wang Fan	Jianye	†3.4879	(83.7)	10.9575	(263.0)	–	†365¼ *du*	3.00	(7.2)
7	< 280	Sphere heaven	N	priv.	Ge Heng	–	–	–	–	–	–	–	–	–
8	323	Sphere sight	Y	st. / roy.	Kong Ting	Chang'an	8.0000	(195.2)	24.0000	(585.6)	†3	24 hrs	100	(244)
												? *du*	–	–
9	398	Sphere sight (iron)	Y	state	Chao and Xie	Pingcheng	"	"	"	"	"	"	"	"
10	436	Sphere heaven sight	N	state	Qian Lezhi	Jiankang	6.0825	(150.2)	18.2625	(451.1)	†3	†365¼ *du*	5.00	(12.4)
11	440	Small sphere heaven	N	state	Qian Lezhi	Jiankang	2.2000	(54.3)	6.6000	(163.0)	†3	–	2.00	(4.9)
12	< 536	Sphere heaven effigy	N	priv.	Tao Hongjing	Jiankang	> 3	(74.1)	–	–	–	–	–	–
13	< 557	Sphere heaven effigy (wood)	N	state	–	Jiankang	–	–	–	–	–	–	–	–
14	< 562	Sphere sight	–	priv.	Zhang Zixin	Sea isle	–	–	–	–	–	–	–	–
15	< 593	Sphere heaven sight	–	royal	King Jing, Shu	Yizhou	–	–	–	–	–	–	–	–
16	< 600	Sphere heaven sight	–	royal	King Jun, Qin	Chang'an	–	–	–	–	–	–	–	–
17	< 604	Sphere heaven sight	N	royal	Geng Xun	Hengzhou	–	–	–	–	–	–	–	–
18	633	Sphere sight	Y	st. / roy.	Li Chunfeng	Chang'an	–	–	–	–	–	365 *du*	–	–
19	720	Sphere heaven diagram	N	royal	Nangong Yue	Chang'an	–	–	–	–	–	–	–	–
20	720	Sphere heaven diagram	N	state	Nangong Yue	Chang'an	–	–	–	–	–	–	–	–

21	725	Yellow-road displ. sight	Y	state	Yixing and Liang	Chang'an	4.5900	(139.1)	14.6100	(442.7)	†3.18	365¼ *du*	4.00	(12.1)
									14.5900	(442.1)		100 *ke*	14.59	(44.2)
									"	"		360 *ce*	4.05	(12.3)
22	> 725	S.H. look-up diagram	N	royal	Yixing and Liang	Chang'an	–	–	–	–	–	–	–	–
23	979	Sphere sight	N	royal	Zhang Sixun	Bianjing	–	–	–	–	–	365 *du*	–	–
24	995	Bronze watch/sphere sight	Y	state	Han Xianfu	Bianjing	6.1300	(193.7)	18.3900	(581.1)	†3	365 *du*	5.04	(15.9)

* The majority of these instruments are treated in Wu & Quan (2008). Whether or not an instrument is observational (obs) is determined by context (attested use and/or presence of sighting tube). The obelisks (†) indicate values calculated according to the data provided. Note that Item 21 includes the measurements of two separate rings.

and the a priori cosmic centrality of the Zhou capital. It changed not just the *shape* but also the *meaning* of the world: in what was once the domain of omenology, for example, it provided simple physical explanations for lunar eclipses – the earth's shadow – and the moon's capricious equatorial pace. The world was indeed so irrevocably altered that thinkers like Ma Rong began to read the sphere back onto the ancient past, as if unable to comprehend how such a perfect idea could be anything less than divine, or, indeed, how a world without it could have been. *All of this*, to experts like Shen Gua (1031–95), remained irrevocably linked with the instrument:

> *Du* cannot be seen; what *can* be seen are stars, and the course of the sun, moon and five [planets] is replete with stars. Those [stars] that act as demarcations of *du*, they are twenty and eight in total, which we call lodges. Lodges are that by which *du* are measured out, and *du* are that by which numbers are born. *Du* are in heaven; but make a 'device-traverse' (sphere sight), and you have *du* on an apparatus. If you have *du* on an apparatus, then the sun, moon and five [planets] can be modelled (*tuan*) within the apparatus, and heaven will have no play (*yu*). And if heaven has not play, then the things in heaven will not be difficult to know (*Song shi*, 48.954–5).

Most of what we know about the history of cosmology in early imperial China we know through Shen Yue and Li Chunfeng's respective *tianwen* monographs. Their histories frame the matter as a tragically long and pointless debate between the *jia* ('expert lineages' or 'schools') of 'sphere heaven' and 'umbrella heaven'. What did the umbrella heaven world of the *Gnomon of Zhou* and its successors look like? It looked like an umbrella diagram: heaven and earth form parallel planes (or vaults) that meet by either optical illusion or by a tapering off of their edges, the sky etched with concentric grooves allowing the sun and moon to travel up and down it through the year like record needles. What did the sphere heaven world look like? It looked like the sphere effigy: a tilted heaven englobes the (flat) earth within it, carrying the seven luminaries nightly beneath the earth (and the waters upon which it floats) as they march eastward and slantwise along a single (yellow) path. The history of this debate, as Shen and Li tell it, was one of extrapolating dimensions and of reconciling the world-as-instrument with everyday facts, measurements, classical precedent and *yin-yang*, five-agents metaphysics. It was a history in which the sphere had already emerged the victor by the second century CE but struggled for centuries thereafter against detractors who were 'all whimsical and fantastical opinions, *not* people who discussed heaven by plumbing numbers' (*JS* 11.280; cf. *Song shu*, 23.680).[29]

The theme of senselessness and attrition is one that runs throughout the history of the sphere's supposed triumph. Take, for instance, Cai Yong's

[29] On the historiography of cosmology in medieval China, see Section 5.2.1 below.

famous memorial from Shuofang, with which Shen Yue, Li Chunfeng and modern scholars alike begin their histories:

The discourse on heaven's form comprises three expert lineages (*jia*), but the study of expansive night has died out and has no master's method (*shifa*). Both the procedures and numbers (*shu shu*) of the *Gnomon of Zhou* survive, but when examined (*kao*) and verified (*yan*) against the case of heaven, there is much that misses the mark. It is only sphere heaven which completely grasps the true circumstances (*qing*). The observatory bronze sight employed by the Clerk's Office of our day is patterned upon this model (*fa*): a sphere erected eight *chi* [in diameter] and possessed of the forms of heaven and earth by which one aligns the yellow road to observe the release and restrain (of luminary motion), by which one moves the sun and moon, and by which one paces the five [planets] – fine and subtle, profound and mysterious, it is a *dao* that shall remain unchanged for a hundred generations. The officials have the apparatus but not the original book(s), and the previous monographs are, for their part, faulty . . . (*Song shu*, 23.673; *JS* 11.278).[30]

Li Chunfeng cuts off there, but the rest continues:

. . . and do not discuss it. [I] had desired originally to lie beneath the sight [to] contemplate its subtleties and master its numbers according to *du* in order to write it up into a piece; for [my] inexcusable crime, [however], [I] have been banished to the north to be annihilated by dust and broken by rain, and denied access to any route to influence. It would be appropriate [then] to inquire among [Your] ministers, and down all the way to [hermit] grottoes, about someone knowledgeable on the idea of sphere heaven and order [him] to recount its substance (*Song shu*, 23.673).

It is in this plea that we find the germ of the discourse to follow. This, like most early statements on 'sphere heaven', is a *plea*; it is a plea of unwavering confidence, for sure, but it comes from a position of hopelessness and loss. The root of the petitioner's conviction lies in the now familiar values of *li* – in 'tightness' and 'truth' as 'verified' by 'examination' – but it is the form that this 'examination' takes wherein the problem lies. The 'true nature' of the world is so thoroughly entangled with the physical instrument in Cai Yong's mind that, to truly know the one, he must lie beneath, gaze upon and explore the other with his hands. Heaven is tangible. So too is the tragedy of this entanglement, for the loss of one threatens the loss of the other, and Cai's only window into the mysteries of the universe was a sole instrument, commissioned seventy-five years prior, held under lock and key in a city on the brink of ruin.

2.2.2 *The Material Sphere*

Perhaps the most important thing to remember about the Chinese armillary sphere is that it was made of money. Sure, you could use iron, or even wood,

[30] Tr. modified from Ho (1966, 49).

but to do it right you needed bronze, and bronze was the basis of the currency. Being difficult to construct and fast to disappear, it is best to approach the history of the armillary sphere in China in terms of individual instruments, where and when they existed. For the first millennium CE, written sources document a total of seven observational models ('sights'), fourteen non-observational models ('spheres', 'effigies', 'sights' and 'diagrams'), and three 'sights' of unknown function (see Table 2.1). Most of what we know about the number, design and use of these instruments comes down to us through *tianwen* monographs, which tend to focus on government assets. Private instrument building comes up, but it is only ever mentioned in passing. There is the Daoist recluse and *materia medica* author Tao Hongjing (456–536 CE), who, according to his biography, 'also once constructed a sphere heaven effigy . . . saying "It is a necessity of practising the *dao* and not only the Clerk's Office who uses this" ' (*Nan shi*, 76.1898).[31] As to the *observational model*, our only lead is that offered by Li Chunfeng:

At the end of the [Tuoba] Wei, Zhang Zixin (d. 577 CE) of Qinghe, who was broadly conversant in knowledge and skill, and particularly fine at *li* numbers, ensconced himself on a sea isle in order to flee the Ge Rong turmoil (526–8 CE), [and there he] racked up thirty-odd years devoting himself to measuring (*ce*) and watching (*hou*) via sphere sight the numbers of the change in the sun, moon and five [planets'] discrepancies (from their mean motions). Pacing them through calculation, [he] first awakened [to the realisation] that the crossed roads of sun and moon see inside and outside (i.e. latitudinal displacement) and retardation and acceleration (i.e. uneven motion), and that the appearance and hiding (i.e. visibility phenomena) of the five [planets] experience stimulus and calling towards and away (i.e. seasonal variation) (*Sui shu*, 20.561).

Whatever the nature of Zhang Zixin's device, it is, more than anything, a reminder of how peripheral the observational armillary sphere was in the history of astronomy in China. In the first millennium CE, the 'sphere sight' was mostly the work of barbarian courts and Chinese with foreign ties: its very inventor – Luoxia Hong – was a Red Di tribesman from Ba (Liu Changdong 2012). The 323 CE sight was commissioned by a Xiongnu court, and the Xianbei copy of 398 CE was built by a Grand Clerk from Xiangping (modern Liaoyang) and an artisan with a Xianbei surname (Li Hai 1994). In Yixing, moreover, we have a translator–monk with intimate ties to the Indian religious elite of Chang'an, and one wonders whether his '360-rod' ecliptic ring might not have been a concession to the Indian and Nestorian clans running the Tang (618–907 CE) Clerk's Office.[32] Between Luoxia Hong and Yixing, it is worth noting, the only person other than Zhang Zixin attributed with *using* a 'sight' was Jiang Ji (fl. 384 CE), a subject of the Yao Qin court from Tianshui.[33]

[31] On Tao Hongjing and religious instrument building, see Ren Song (2008).

[32] On the 'Western' lineages at the Tang astronomical office, see Jiang & Niu (2001, 105–33) and Lai Swee Fo (2003).

[33] See *JS* 19.566–7; *Sui shu*, 19.513–14.

Cai Yong's plea reminds us that even a device in Chinese hands may not have been at arm's reach. The 103 and 164 CE 'sights' were installed at the Numinous Terrace observatory at the Eastern Han (25–220 CE) capital at Luoyang. Excavated in 1974–5, we know that this site was walled – access to the expensive equipment therein thus being restricted – and that both sights went missing from it at some point in the historical past.[34] It seems a safe bet that the observatory was pillaged between 189 and 311 CE, when the city was sacked by Military Governor Dong Zhuo (d. 192 CE) and by Xiongnu Liu Zhao forces respectively. Destroyed, Luoyang was abandoned as the imperial capital in 196 CE, only to be rebuilt over the Cao Wei (220–65 CE) and lost by the Sima Jin (265–420 CE) in 311 CE, the rebellion of whose barbarian mercenaries forced the Chinese court from the Yellow River heartland in 311 CE.

The Liu Zhao built their 'sight' in occupied Chang'an in 323 CE. This passed (with the city) to the Jie armies of Shi Zhao (319–51 CE) in 329 CE, the Fu armies of Fu Qin (351–94 CE) in 350 CE, and to the Qiang armies of Yao Qin (384–417 CE) in 385 CE. In 417 CE, General Liu Yu (363–422 CE) retook the city for the Sima Jin court, transporting the Xiongnu sight 'to [a] royal palace' in Jiankang (*Yiwen leiju*, 1.6a–b). Over the change of hands to Liu Song (420–79 CE), Xiao Qi (479–502 CE) and Xiao Liang (502–57 CE) rule in Jiankang, the device wound up 'installed before Double Cloud Hall at [the closed imperial park at] Hualin' (*Sui shu*, 17.517), where the Liang emperor held Buddhist lectures and *sūtra*-readings. When Double Cloud Hall caught fire in 559 CE, the Chen emperor 'deeply vilified [General] An Du for leading soldiers and wearing armour into the hall [to help], secretly plotting against him from then on' (*Nan shi*, 66.1612), which suggests that the site was *very* off-limits. Li Chunfeng is quick to criticise He Chengtian and Shen Yue for misidentifying this 'sphere sight' as Zhang Heng's indoor model – '[they] all missed by [a mile]' (*Sui shu*, 19.518) – but no one in fifth-century Jiankang probably got close enough to read the label.[35]

The Tuoba Wei built their version in Pingcheng, three months after moving their capital there in 398 CE. Originally commissioned for the Hall of Heavenly Patterns, in the Western Palace, Li Daoyuan's (d. 527 CE) *Sui jing zhu* places a 'turning mechanism' (*xuanji*) potentially fitting this description in the Pingcheng Bright Hall/Numinous Terrace complex.[36] Whatever came of it after the court's move to Luoyang in 493 CE, it eventually passed with the eastern territories in 534 CE to General Gao Huan (496–547) and the Gao Qi dynasty

[34] For the Luoyang observatory site report, see *Kaogu* 1978.1, pp. 54–7.

[35] *Song shu*, 23.678; cf. *Sui shu*, 19.517–18. Note that Li Chunfeng may not have actually seen said instrument either, his knowledge of the label potentially coming from Emperor Wu of Liang's (r. 502–59) *Zhonglü wei*, as cited in *Sui shu*, 16.408.

[36] *WS* 2.33; *Shui jing zhu shi ding'e*, 13.10b. I thank one of my anonymous reviewers for bringing the latter source to my attention.

(550–77 CE). The Xianbei Yu Zhou (557–81 CE) then captured and transported the sphere back to their capital at Chang'an in 577 CE, which subsequently changed hands to the Chinese Sui (581–618 CE), who, 'upon completion of the new metropolis [in 583 CE], installed it atop the observation terrace' (*Sui shu*, 19.505). At the Chang'an observatory, the Pingcheng 'sphere sight' saw state service through the Tang until replaced in 725 CE. Cast from iron, and slapped together in a matter of months, the Pingcheng 'sphere sight' was a disaster. Li Chunfeng complains that 'its design and construction are loose and rough' (*JTS* 35.1293), while Yixing reports,

> The ring construction is crude and rough, the *du* notches are uneven, the red road does not move – being thus like a [stubborn] glued peg (on a moving-peg zither) – and it is not installed with a yellow road – there being no guideline for [ecliptic] advance/retreat. This, verifying sequence (*li*)-entry fast/slow (true speed correction over anomalistic month) based on red-road lunar travel [results in values] too high, sometimes up to seventeen *du*, and too low, only giving ten *du* – this is insufficient to upwardly examine (*ji*) signs of heaven and reverently grant the seasons of man (*JTS* 35.1295).

For his part, Li Chunfeng convinced Emperor Taizong (r. 627–49 CE) to finance a new 'sphere sight' that he had designed for the observatory in 627 CE. Upon completion in 633 CE, however, Taizong also coveted it for his garden:

> Taizong ordered the sphere instrument that [Li Chunfeng] had constructed installed in the Pavilion of Congealed Light in order to use it for measuring and watching, and though it was right there in the palace, when [later] looked for, [they] had lost track of where it went (*JTS* 35.1293).

It was only in 725 CE that a Chinese observatory once again had access to a bespoke 'sphere sight' with an ecliptic ring, and this time the Clerk's Office held on to it in relative peace.[37] As to access, however, one notes that the Tang Clerk's Office operated in increasing isolation, to the point where, in 836 CE, it was ordered that 'the officers and clerks of the [astronomical] directorate are no longer allowed to communicate or come and go with court officials or any other type of person, the censorate being charged with inspection [of compliance]' (*XTS* 36.1336).[38]

If we want to understand the armillary sphere – where and when there was one (Table 2.2) – we had best lie with Cai Yong beneath its rings to contemplate the subtleties of their numbers. Chapman (1995, 12) argues that in Europe

> it was improved angular measurement, not enhanced visual acuity, that held the key to a whole network of problems from the sixteenth to the early nineteenth centuries, and,

[37] On the Yixing and Liang Lingzan 'sphere sight', see Needham, Wang & Derek (1986, 74–83) and Li Zhichao (2014, 184–90).

[38] On the insulation of the Clerk's Office, see Lai Swee Fo (2003, 332–5).

Table 2.2 *Availability of observational armillary spheres by observatory (104 BCE–907 CE)*[*]

Observatory	Lat.	Long.	Inst.	Const. lat	Political control		Years
Chang'an	34° 21′ N	108° 53′ E	#1	”	???–???	W. Han	?
			n/a		???–323		?
			#8	”	323–329	Liu Zhao	6
			#8	”	329–350	Shi Zhao	21
			#8	”	350–385	Yao Qin	35
			#8	”	385–417	Fu Qin	32
			n/a		417–583		166
	34° 16′ N	108° 48′ E	#9	40° 05′ N	583–618	Sui	35
			#9	40° 05′ N	618–725	Tang	107
			#21	”	725–907	Tang/Zhou	182
Luoyang	34° 45′ N	112° 28′ E	#2	”	103–189	E. Han	85
			†#2	”	189–220	E. Han	31
			†#2	”	220–265	Cao Wei	45
			†#2	”	265–311	Sima Jin	46
			n/a		311–907		596
Pingcheng	40° 05′ N	113° 18′ E	#9	”	398–534	Tuoba Wei	136
			†#9	”	534–550	E. Wei	16
			†#9	”	550–577	Gao Qi	9

* Grey indicates certain absence; obelisks (†) indicate uncertainty. For instrument numbers, see Table 2.1.

with the exception of the discovery of Uranus in 1781, the weight of decisive evidence came from better graduated scales, not better optics.

One could make a similar case for the importance of the embodied *du*. Men like Cai Yong thought about circles differently than we do: not in terms of *angle* – terms they did not possess – but *lü* ratios between diameter and circumference. There were no 'circles' (*yuan*) in heaven, only 'circuits' (*zhou*) – a concept extended beyond the (at first) equatorial and (then) ecliptical 'roads' of the luminaries to the untraveled 'circuit' of the meridian. The only true circles were physical objects of lacquer, wood, bronze and iron. As an 'effigy', the material object stood in proportion to its celestial counterpart, and so any distance measured along its perimeter would correspond to a relative distance in heaven. The *du* was typically thought to correspond to 2,000 *li* (831.6 km [Han]), but who could agree on the exact dimensions of the world?[39]

'If you have *du* on an apparatus', Shen Gua vows, 'then the sun, moon and five [planets] can be modelled . . ., and heaven will have no play' (*Song shi*, 48.955), but we have seen enough evidence for 'play' thus far to take his optimism with a grain of salt. Pan Nai (1989, 271–2) argues that, up until the thirteenth century, the Chinese armillary sphere was graduated only to the integer *du*, the fractions 'less' (*shao* = ¼), 'half' (*ban*), and 'more' (*tai* = ¾) used in observational data being the product of estimation. Pan's argument rests on three points. The first is that, in 1280, Guo Shoujing claims to be the first to *really* empirically measure the twenty-eight lodges down to fractional widths. The second is the degree of precision witnessed in the extant observational record, where 'less–half–more' notation is rough and rare. The third is the documentation of 365-*du* rings in late sources. The first two points are arguments from authority and absence, respectively, and we have dealt with the third in Section 2.1.4, but we have yet to put forward an alternative.

In terms of material evidence, we have two exemplars of circle graduation from opposite ends of the chronological and professional spectrum. The first is the 165 BCE lodge disc, which is graduated in integer *du* at ≈ 1.98 mm (0.86 *fen*) per hole (Figure 2.4). The second is the fifteenth-century reproduction of Guo Shoujing's 'simplified instrument' (*jian yi*) at Purple Mountain Observatory, which features an exterior equator ring of 6.4 *chi* (202.2 cm [Yuan]) diameter graduated to $\frac{1}{36}$ notch at 1.76 mm (0.56 *fen*) and an interior ring of six *chi* diameter (189.6 cm) graduated to $\frac{1}{10}$ *du* at 1.63 mm (0.52 *fen*) (Figure 2.7).[40] As to the terminal fraction, the lodge disc is damaged, making it hard to say, but one can see that the 'simplified instrument' rounds to the unit of graduation,

[39] On the distinctly non-angular nature of the *du*, see Guan Zengjian (1989), Li Guowei (1991) and Huang Yi-long (1992a). On the terrestrial length of the celestial *du* and its implications for measuring the dimensions of the cosmos, see Kalinowski (1990).

[40] On the 'simplified instrument', see Wu & Quan (2008, 449–52) and Sivin (2009, 194–8, 561–6).

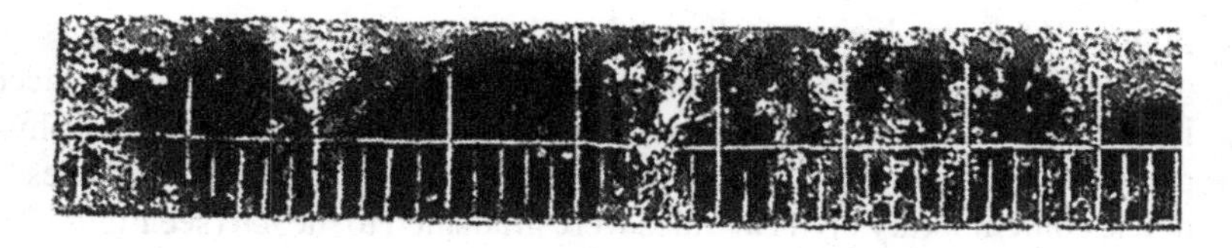

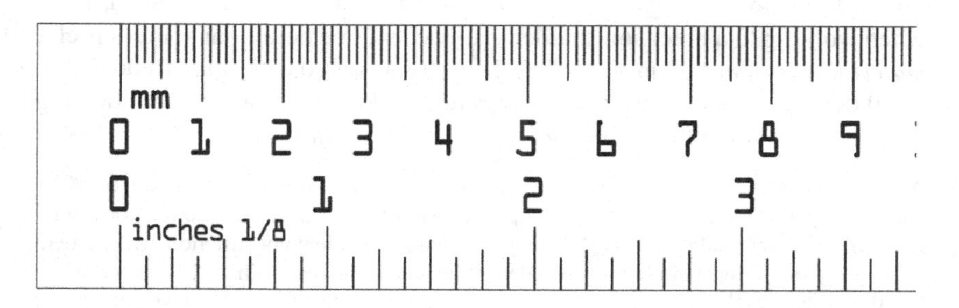

Figure 2.7 Graduation. Top: Eastern Zhou *chi*-rule, Nanjing University Art and Archaeology Museum. Bottom: Guo Shoujing 'simplified instrument' interior equator ring (Pan Nai 2005, 168, fig. 6–4–7). Note the terminal fraction of $\frac{2}{10}$ *du* before prior to Ox.$_{L09}$ at the end of Dipper.$_{L08}$. I would like to thank Zhou Xian, director of the Institute for Advanced Studies in Humanities and Social Sciences, and Shi Mei, vice director of the Art and Archaeology Museum at Nanjing University, for kindly supplying me with the upper image, and Professors Han Qi and Qu Anjing for helping me identify the origin of the lower image.

giving $365\frac{2}{10}$ *du*. Neither tells us how early imperial 'sphere sights' were graduated, but they are nonetheless revealing of the technical limits faced by instrument makers. Namely, over sixteen centuries, the range of physical precision fell between 0.52 and 0.86 *fen* (1.63–1.98 mm). At their finest, one notes, contemporary civil *chi*-rules were graduated to the *fen* (see Figure 2.7).[41]

Turning back to textual evidence, the fact that actors discuss ring construction in terms of *chi* suggests that a *chi* rule of some type was involved. Even when working to a precision smaller than the *fen*, one notes, our sources prefer *material* to *immaterial* units, such as the *li* (0.1^{f}), *hao* (0.01^{f}), *miao* (0.001^{f}) and *hu* (0.0001^{f}) of *suan* mathematical literature. Take, for instance, the following passage from Wang Fan's *Sphere Heaven Effigy Explained*:

> Ancient and old sphere effigies took two *fen* to the *du*, for a total circuit of $7^{ch}3^{c}$ and a half *fen* (171.67 cm). Zhang Heng's redesign took four *fen* to the *du*, for a total circuit of $1^{z}4^{ch}6^{c}1^{f}$ (343.34 cm). [I, Wang] Fan find the ancient construction arrangement to be [too] small, the stars being [too] densely packed together; [Zhang] Heng's device, [on the other hand], was painfully large, making it difficult to turn. [I] have [therefore] redesigned the sphere effigy taking three *fen* to the *du*, for a total circuits of heaven of $1^{z}\text{–}^{ch}9^{c}5^{f}$ and ¾ *fen* (262.98 cm) (cited in *Song shu*, 23.677).

At these rates, the ring circumference works out to 365¼ *du* every time. At four *fen* per *du*, it would be easy enough to graduate to the 'less–half–more' quarter-*du*: cast a ring 1,461 *fen* in circumference, use a *chi*-rule to mark a string with the appropriate number of *fen*, tie the string around the ring's edge, and mark the ring accordingly for incision. The two- and three-*fen du*, on the other hand, raise an interesting problem. How can we believe Wang Fan when he claims to be working to the quarter-*fen* (0.6 mm) or dividing a three-*fen du* in four when we see (Table 2.1) circumferences and diameters given in a *lü* of 3:1 (i.e. $\pi = 3$), smacking of idealised calculation? How realistic is it when Zhang Heng, for that matter, instructs the reader to 'make 182 *du* and ⅝ [*du*]' (*HHS zhi* 3, 3076 (comm.)) on a bamboo strip? It may well be that our sources are speaking in terms of *mathematical precision* as divorced from the sort of *material approximation* that left Guo Shoujing's equator ring with a terminal fraction of $\frac{2}{10}$ *du*. That said, instrument makers may well have found a way around the limitations of the material *fen*, since neither the lodge disc nor simplified instrument are beholden exactly thereto, and since graduating nested rings of different diameters would prove otherwise impossible.[42]

Just because an instrument maker could graduate to a precision of 1.63–1.98 mm does not, of course, mean that he/she could achieve the same degree of accuracy. Yixing, if we remember, says of Artisan Xie Lan's 'sight' of 398 CE

[41] See Qiu Guangming (1992).

[42] See, for example, the diameters of the Yixing and Liang Lingzan rings in Li Zhichao (2014, 185–90).

that 'the ring construction is crude and rough, and the *du* notches are uneven' (*JTS* 35.1295). The example he gives of measured lunar velocity implies a potential error of roughly ± 2½ *du*.[43] One gets a clear idea of what this might look like from the shadow template on the Eastern Han survey gnomon (Figure 2.2): there we see *cun* of 1.91 cm and 2.12 cm flanking those of 2.3–2.4 cm where the normal error for finer *chi*-rules tends to run around one *fen* per *cun* (Qiu Guangming 1992, 42).

2.2.3 The Applied Sphere

Early imperial sources report using the armillary sphere mostly as an object of contemplation. Take, for instance, the opening lines of He Chengtian's argument for the water cycle:

> The meticulous search of previous explanations [shows that] by observing (*guan*) the sphere sight [they] probed and discovered its meaning, some awakening thereupon to [the fact that] heaven's form is a perfect circle, and that water completely [fills] its bottom [half] (*Song shu*, 23.677).

Mr He goes on to suppose that the sun passing nightly through the ocean, west to east, replenishes the rivers with water evaporated over the course of its journey. As is typical of the genre, one notes, he is not *looking through* a 'sight' but *reading* about *looking at* one. Looking *through* also happened, but it has left us with far fewer traces. As ample as is the observational record in the *tianwen* monograph chronicles, the nine monographs prior to the *New Book of Tang* (fin. 1060) preserve only sixty-six phenomena measured in *du*, thirty-eight of which are the positions of solar eclipse (which can only be calculated). As preserved in *lü-li* monographs, the sort of data we see applied to *li* assessment in Chapter 4 is more reliably 'observational', but this data is even rarer. In the end, the best source to which we can turn for lists of 'sight'-derived *du* are star catalogues, and these too fall on either end of the period in question: *Mr Shi's Star Canon* and the Sovereign Protection survey of 1052.

What do these star catalogues tell us about 'sphere sight' observation? As to precision, *Mr Shi's Star Canon* uses 'less–half–more' notation, while the 1052 catalogue is only precise to the 'half-*du*'.[44] As to accuracy, it is more difficult to say because the problem is essentially circular. The first to supply co-ordinates

[43] According to Yixing, 'verifying sequence-entry fast/slow based on red-road lunar travel [results in values] too high, sometimes up to seventeen *du*, and too low, only giving ten *du*' (*JTS* 35.1295). Compare this to Liu Hong's 'slow–fast' speed correction, for example, which gives the moon a daily velocity of between $14\frac{10}{19}$ *du* (perigee) and $12\frac{5}{19}$ *du* (apogee).

[44] By comparison, the first instance of the 'half-*du*' in *tianwen* chronicles occurs in 493 CE (*NQS* 12.205), while 'less–half–more' notation is already at play in the earliest such positional data recorded in a *lü-li* monograph, as cited in the introduction to this chapter.

for *all* the Chinese stars is the *Chongzhen lishu* of 1629–34, which draws from the catalogue of Tycho Brahe (1546–1601);[45] prior, all we are given is the co-ordinates of an asterism's 'reference star'. The first order of business for modern studies is usually to correct Jesuit-era identifications for pre-Jesuit times, but without all the co-ordinates, there is only so much that one can do. Working on the Sovereign Protection survey of 1052, for example, Pan Nai (1989, 190–238) sets ± 1° as the acceptable rate of observational error, identifies fifteen out of 360 'reference star' co-ordinates as outliers by this criterion, and reassigns each to a different star in the same (Jesuit-era) constellation. Nothing has really changed, and, to get there, observational error must be *posited*, rather than measured, and anything beyond the cut-off must be assumed correct. *Mr Shi's Star Canon* poses an even bigger problem, because it is undated, throwing yet another variable into the mix. Once thought to originate from separate surveys centuries apart, Maeyama (1975; 1976; 1977), using Jesuit-era identifications, famously posited a ≈1° misalignment of the instrument's pole that allows dating the entire catalogue to 70 BCE ± 30 years via minimum standard deviation in polar distance. Maeyama's individual identifications are disputed in Sun and Kistemaker (1997), but the polar-misalignment thesis still stands, giving us at least one systematic ≈1° error to which we can point in the history of 'sphere sight' observation.[46]

The *potential* sources for error in early imperial 'sphere sight' operation are, of course, manifold. First, the movement of the Xianbei sphere from Pingcheng (40° 05′ N) to Sui–Tang Chang'an (34° 16′ N) would have introduced a nearly 6° error in polar alignment (in addition to the apparent ± 2½ *du* graduation error) to observatory sight data from 577–725 CE. Second, the alignment of material and celestial lodges via a 'three-chronogram group' would introduce errors in lodge *du* readings commensurate with the degree to which the instrument lodges were rounded (even Guo Shoujing's ring rounds reference stars to integer *du*). Third, in the absence of an ecliptic ring, 'advance–retreat' conversion could introduce an error of up to ½ *du* in longitude (Cullen 2000, 376–9). Fourth, with an ecliptic ring *fixed to the equator* and *graduated in lodge du*, there would be no physical allowance for precession, which would introduce growing errors in ecliptic readings over time.[47] Fifth, without a concept of refraction, the sphere would introduce errors in upwards of ½ *du* in polar

[45] See Pan Nai (1989, 318–57) and Sun & Kistemaker (1997, 34–5). I thank Han Qi for pointing me in the right direction on this topic in a personal communication of 22 September 2016.

[46] One notes that the systematic nature of this error aids, furthermore, in distinguishing it from the sort of textual corruption that might affect only individual data. On the issue of copy errors and textual corruption in 'numbers and procedures' literature, see Morgan (2015) and Mo & Lin (2016).

[47] Yixing was the first to offer a solution to this problem, installing the ecliptic ring (at a constant obliquity) on peg holes bored at one-*du* intervals along the 'three-chronogram group' equator ring; see Wu & Quan (2008, 439–40).

Table 2.3 *Distribution of* 'tianwen' *and* 'li *and mathematics' titles in bibliographic monographs**

Bibliography	Titles				Fascicles			
	TW	*zhan*	I/C	LM	TW	*zhan*	I/C	LM
Book of Han	**22**	21	–	**18**	**419**	402	–	**566**
Book of Sui	**98**	70	12	**108**	**677**	562	15	**265**
Old Book of Tang	**25**	12	8	**56**	**261**	241	9	**124**
New Book of Tang	**41**	26	8	**98**	**481**	451	9	**462**
History of Song	**140**	54	10	**165**	**534**	312	38	**590**

* Bibliographic entries are identified with omen literature (*zhan*) and instrument-cosmos literature (I/C) based on title keywords. Note that the totals for *tianwen* (TW) and *li* and mathematics (LM) entries here are tabulated rather than copied from the totals listed at the end of the monograph headings.

distance as well as readings of RA and longitude taken outside the plane of the meridian. The armillary sphere, in short, is a clumsy, high-maintenance instrument, and when one looks at all the spherical trigonometry that Ptolemy (*c.*90–*c.*168 CE) used to get around the thing, one wonders what use it would have been without (Włodarczyk 1987). If makers *did not* graduate to the ¼ *du*, as Pan Nai argues, it was probably because this level of precision was superfluous vis-à-vis the device's accuracy.

2.3 The Ambiguities of 'Observation'

None of the *tianwen* treated thus far relate to 'observing the signs' in the classical sense; there was another literature for that, a literature that was bigger, older and infinitely more prevalent in contemporary culture. *Tianwen* was a field long rooted in the list-literature of omen-reading (*zhan*) by the time that the instrument/cosmos emerged as a sub-genre in the second century CE. By the seventh century, writings on the instrument/cosmos counted for only 12 per cent of the *tianwen* titles held by the imperial library, and 2 per cent by fascicle (see Table 2.3). Omenology was huge. The standard history *tianwen* annals provide us with the longest, largest and most continuous observational record in human history.[48] Legendarily, the literature extends even further back and farther out, as early catalogues claim to originate from ancient culture heroes, divine revelation or 'the middle of the sea' (*hai zhong*).[49] At the 'modern' (*jin*) end,

[48] The imperial-era observational record is collected in *Zhongguo gudai tianxiang jilu zongji* and studied in Saitō & Ozawa (1992) and Zhuang Weifeng (2009). On astral omens in Shang dynasty (?–1045 BCE) oracle bones, see Keightley (2000, 17–53).

[49] On the 'middle of the sea' omen corpus, see Lü Zifang (1983, vol. 1, 216–26).

omenology was integral to political discourse, to historiography and to the weft and prophecy literature that would find itself a court-ordered mainstay of the gentleman's education.[50] Omens were likewise everywhere in art, be it for the state, the individual, the living or the dead, and so omnipresent were they in contemporary culture that they served as a sort of symbolic vocabulary by which to communicate intentions independent of the sky.[51]

Tianwen omen literature mostly involves 'watching' (*hou*) for strange activities, the role of the observatory 'watchman' mirroring that of his counterpart in wall and border defence (see Chapter 1). In modern terms, a 'watchman of the wall' (*chengmen hou*) is the kind of sentry you find at a checkpoint tower, and a 'watchman of the stars' (*hou xing*) the one you find at an anti-aircraft gun. Be it a kamikaze in the ancient or the modern sense, the foremost task of the 'watchman' is to alert his superior of 'signs' (*xiang*) of 'disturbance' (*bian*), and what little 'measurement' (*ce*) this involves, like the anti-aircraft gun, is done in feet and inches. Take, for instance, the following passage from the third-century CE *Jingzhou Omens*:

If [Mars] travels more than one *chi* (per day?), then in a period of five to fifteen months there is abrogation; more than five *cun*, then in a period of fifteen to twenty-five months there is abrogation; more than three *cun*, then in a period of twenty-five to thirty months there is abrogation; more than one *cun*, then in a period of thirty to fifty months there is abrogation. For the various abrogation periods, the number of months is sometimes given instead as the number of days (*Kaiyuan zhanjing*, 30.4b).

Tianwen chronicles, for their part, document thousands of examples of such measurements taken over the course of the imperial period. For example:

6-I-*jihai*.$_{36}$ (3 March 565): [Venus] trespassed upon [Mars] at a distance of two *cun* (4.94 cm). Omen-reading: 'There are soldiers and death in its field (i.e. the corresponding geographic region), marquises and king are changed and enthroned'.

III-*dingmao*.$_{04}$ (31 March):[52] after sunset, and before the stars had appeared, there was a flowing star white in colour, and as large as a *dou*-measure (two litres), which travelled south through Privy Council (Taiwei; Leo, Com and Vir) with a tail more than one *chi* (24.7 cm) in length. Omen-reading: 'There are soldiers and death' (*Sui shu*, 31.598).

The written record is silent about the practice of *chi–cun* measurement, but the case of 'cubits' and 'fingers' in cuneiform astronomical diaries reminds us that the Chinese watchman was not alone.[53] In terms of scale, Wang Yumin's (2008) analysis reveals the astronomical *chi* to be a measure of altitude and angle of

[50] On weft and prophecy literature, see Lü Zongli (2003) and Xu Xingwu (2003).
[51] See Wu Hung (1989, 73–107), Tseng (2011) and Harper (2007).
[52] Something seems to be wrong with this date, as Zhang Peiyu (1997, 189), has 6-III begin on *guiwei*.$_{20}$.
[53] On linear measures in early Mesopotamian astronomy, see Brown (2000).

separation corresponding to one *du* – *du* being reserved for right ascension, longitude and declination – the value of which remained constant over the centuries-long inflation of the civil *chi*. We have no idea what tool this involved, since it is not one that interested gentlemen like Li Chunfeng. It is possible watchmen used something like a cross-staff, but Liu Ciyuan (1987) concludes that the error distribution points rather to an outstretched human hand. The scale is off, since the ratio of one *chi* to one *du* implies a circumference of 365¼ *chi* (85.8 metres) and radius of 58.1 *chi* (13.7 metres), but Wang Yumin (2008, 40–54) argues that hand measurements were simply projected onto such a circle. Indeed, if 'watching' worked via a 365¼-*chi* reference circle drawn around the observer, this might explain how the pseudo-angular *chi* remained constant in the face of linear inflation; it might also suggest that the *Gnomon of Zhou*'s thought experiment with such a circle in Section 2.1.2 had some (misconceived) basis in observatory practice.

The field of instrument/cosmology emerged in response to completely different questions than those that informed the mainstream of *tianwen*. In military terms, old *tianwen* was concerned with spotting trespassers, and new *tianwen* with the position and speed of the normal guard. The goal was to feed *li* calculation with data that it could use – data in equatorial and longitudinal *du* – and the men involved happen to be the same as those we meet in any *lü-li* monograph. *Li* men brought their questions to *tianwen*, and with those questions came the epistemic values with which they were entangled. When it came to the instrument–cosmos, 'tightness' and its 'verification' took precedence over the impossible antiquity of a conflicting written source; when it came time to *write*, authors identify themselves and assign credit to the real contemporary figures to which it is due. This new, quantitative *tianwen*, for its part, inspired untraditional approaches to 'observing the signs'. Setting aside the old question of what signs *mean*, gentlemen went about deducing the size, shape and mechanics of the elements that constitute the world.

The emergence of quantitative *tianwen* coincides with a transition in omen literature from sponsorship to suppression, and from production to collection. In terms of legality, the private study of omen literature was banned at least nine times between 268 and 617 CE. Issued in the founding years of political dynasties whose own claims to legitimacy were rooted therein, Lü Zongli (2003, 35–81) argues that bans were aimed at denying 'misuse' by any political cause or religious cult with its own thoughts on legitimacy. Omen literature did not disappear overnight, but, from the third century on, imperial library holdings reflect a move from small, miscellaneous pseudepigrapha to multi-volume 'collections' (*ji*) compiled by loyalist intelligentsia. More than half of the Sui library's 677-fascicle *tianwen* holdings, for example, falls to six titles: Chen Zhuo's ten-fascicle *Collection of Tianwen Omen-Readings* (third century), Liu Yan's twenty-fascicle *Jingzhou Omens* (third century), Han Yang's Jin forty-fascicle *Essential Collection of Tianwen* (third century), Zu Geng's thirty-fascicle *Record of Tianwen* (sixth century) and Yu Jicai's 115-fascicle *Secret Garden of the Numinous Terrace* and 148-fascicle *Monograph on the Suspended Signs* (sixth century).[54] Chen, Han, Zu and Yu, it is worth noting, were all Grand Clerks otherwise active in *li*. Criticising sage-time omens as 'old' (*jiu*), 'absurd' (*miu*) and 'misconceived' (*bu de*), *li* men like Li Chunfeng talk about compiling in terms of 'selecting those whose reasoning (*li*) is appropriate and, by deletion and enumeration, writing them into [a] piece' (*Yisi zhan*, 1.10b).[55]

As the sense of what it meant to 'observe the signs' evolved – and as the sage kings themselves evolved to sport the water clocks and the armillary spheres that this entailed – our gentlemen writers become so wrapped up in sphere–umbrella polemics that they neglect to talk about the instrument's *use*. In the words of Monk Yixing:

> Some guard the respective apparatuses of the traditions to which they belong to describe the body of heaven, saying that the primal sphere can be measured by attending to the numbers and that the great signs can be espied by the operation of counting rods – and in the end [we are left] with the theories of six schools (*jia*) alternating in contradictions (*JTS* 35.1307).

The case of omenological practice comes as a rude awakening from the circles in which intellectual spherism was turning. In terms of precision, *chi–cun* measurements appear in one-*cun* (0.1-*du*) increments from the second century on, and Wang Yumin's (2008, 46) analysis of observational data reveals a level of accuracy around the same range – a level surpassing that of the armillary sphere all the way up to Jesuit times. *Chi* are not *du*, of course, but the 'sphere sight' was not the only option for measuring *du*.

It is doubtful that the early imperial 'sphere sight' outperformed the gnomon–water clock combo, whence the *du* arose, by the metric of right ascension or 'advance–retreat' longitude. The fact that the gnomon operated in the plane of the meridian would eliminate issues like refraction and polar alignment. The fact that the water clock 'arrow' was straight, on the other hand, would make it easier to graduate and to break the diurnal 'circuit' into manageable physical units. The Korean *chagyŏngnu* (fifteenth century), for example, featured fifty-notch arrows of 10.2 *chi*, for a total length of 20.4 *chi* (652.8 cm) at 20.4 *fen* (65.2 mm) per notch, and Han Zhongtong's twelfth-century 'leak notch' split the day into *four* – each 'arrow' 3½ *chi* in length for a total of 14 *chi* (442.4 cm) at 14 *fen* (4.4 cm) per notch.[56] These are our earliest sources for professional 'arrow' dimensions, but one doubts that the idea of tall instruments and

[54] See *Sui shu*, 34.1018–21.
[55] On the evolution of omen literature over this period, see Morgan (2016a).
[56] See Needham (1986, 23) and *Tonghu loujian zhidu*, 5a.

multiple arrows took centuries to grasp. If graduated to the civil *fen*, for example, the *chagyŏngnu* and Han Zhongtong 'arrows' would give a precision of $\frac{1}{20.4}$ notch (42.3s = 0.18 *du*) and $\frac{1}{14}$ notch (61.7s = 0.26 *du*), respectively. This is well within the range of precision assumed in earlier *li* literature: $\frac{1}{10}$ notch (1m26.4s = 0.37 *du*) in the solar table of 174 CE, $\frac{1}{100}$ notch (8.64s = 0.04 *du*) in Zhang Zhouxuan's rise/set time of 597 CE, and the $\frac{1}{60}$ *notch* (14.4s or 0.06 *du*) 'arrow' of 938 CE, which Ma Chongji claims to be 'in use since antiquity'.[57] As it happens, $\frac{1}{60}$ notch is also where Hua Tongxu (1991) places the line of diminishing returns as concerns the poly-vascular water clock's error over a twenty-four-hour period. Looking back to Section 2.2, the worst-case scenario for the gnomon–water clock combo beats the best-case scenario for the 'sphere sight' hands down.

'Observing (*kui*) the sun gnomon and dropping the leak notch', Li Chunfeng acknowledges, 'are the root of measuring (*ce*) heaven and earth and correcting (*zheng*) sight and effigy' (*Sui shu*, 19.529), but what gentlemanly writing survives on instrumentation is too absorbed in cosmology and thought experiments to document mundane observational practices such as this. We should not conclude from this that *guan* ('observation') did not matter, of course, because it most certainly did. Recall, for example, that Zhang Heng excuses his mechanical approach to 'advance/retreat', saying 'what one *should* do is *du*-measure these over days and months via a/the bronze sight' (*HHS zhi* 3, 3076 (comm.)). In the same vein, Li Chunfeng insists that '[the effigy] is inferior to the sphere sight' absent a 'traverse tube ... [for] the measure and estimation of sun and moon and the division and pacing of stars and *du*' (*Sui shu*, 19.519). Nor should we insist upon too hard a line between 'real' and contemplative practice, of course, because *guan* is only possible through cosmology and mathematics. Without the *idea* of the sphere, after all, no one could measure along its roads, and without co-ordinates, algorithms and unit conversion, a 'sphere sight' was nothing more than the garden ornament it was seemingly destined to become. Instead, we might ask what it actually means to *guan*.

Thus far we have encountered two distinct practices of *guan* suggestive of modern notions of 'observation' – 'measuring' (*ce*) for *li* data and 'watching' (*hou*) for *tianwen* omens – but this is not the only way that actors *guan*. The confusion we saw between 'sight' and 'effigy', 'diagram' and 'sight', comes down to a confusion of whether the instrument is made to *guan* through, with or at. Actors *guan* the 'sight', 'effigy' and 'diagram' alike, but we also see the Yellow Emperor '*guan* leaking water' (Section 2.1.1) and Sima Qian '*guan* the clerk's records' (Section 1.3.2). The term 'observation' falls further short when one moves to examples of *guan* as used with ideas, propositions, metaphysics and

[57] For the 174 CE solar table, see Section 2.1.2; for Zhang Zhouxuan, see *Sui shu*, 19.528–9; for Ma Chongji, see *Jiu Wudai shi*, 96.1284.

classical commentary, but to understand the full semantic range of this word we must really turn to Buddhism. As Chinese-speakers absorbed Sanskrit ideas, it is telling that the word they chose for *vipaśyanā* and *ārammaṇa* was *guan*. *Vipaśyanā* refers to meditative contemplation, the focus of which is an *ārammaṇa* 'meditation object' by which to, in the words of Zhiyi (538–97 CE), 'pierce through' the phenomenal world to attain 'penetrating *guan*' of that beyond.[58] The *ārammaṇa* was often an effigy of the Buddha (for which you really wanted bronze), but you could also *visualise* it in the idol's absence. Bronze and eternal, the Buddha was made of numbers – the 'thirty-two marks', the 'eighty subsidiary signs', and the 'hundred and forty distinctive characteristics' – the subtleties of which held the mysteries of the universe. In the words of the art historian:

> [*Guan*] means a systematic building-up of visual images, each as complete and precise as possible, in a sequence from the simple toward the complex. In following this step-by-step advance the practitioner was certainly aided by his memories of Buddhist art. The [*guan*] sūtras more than once recommend the man-made icon or statue as a natural first step toward realizing the beauty and glory of divinity. If these mental pictures were correctly formed, on the basis of iconographic rules, they were already approximations of the truth; as so to cross the frontier from reason to ecstasy brought no absolute change, but rather an immense widening of the field of vision, and *seeing* instead of mere visualizing (Soper 1959, 144).

Criticising the implications of the Buddhological rendering of *guan* as 'visualisation' apropos the active/passive nature of the act and the interiority/ontology of what is 'seen' (*jian*), Greene (2012, 194) furthermore insists,

> What the practitioner hopes to achieve is not simply a 'visualization' of some previously seen object, but a vision of something new and different, a vision that arises not by willing it into existence or gradually building it up piece by piece, but by calming the mind through focus on some other object until reaching a mirror-like state from which new and marvelous things were thought to pour forth. Put in simpler terms those practicing meditation were apparently not expected to know what it was that they were eventually hoping to see. Indeed the spontaneity (in this sense) of the visions appears to have been crucial to their validity.[59]

The word that the Buddhologist is looking for would seem to be 'observation', in the very history-of-science sense. Conversely, that needed to reconcile how Cai Yong and many others since his day have 'observed' through, with and at 'sphere heaven' might well be *vipaśyanā*. Either way, *guan*, as an actor's category, makes no such clear distinctions, not that we should expect our subjects to 'see' the same way or the same things as do we.[60]

[58] *Mohe zhi-guan, T* no 1911, 46.21:c5–6. On Zhiyi's *Mohe zhi-guan* and the idea of meditative 'observation' developed therein, see Donner & Stevenson (1993).

[59] I thank Paul Copp for bringing this study to my attention.

[60] For an excellent study on the problem of 'seeing' and ontology in ancient Rome, see Lehoux (2012).

2.4 Conclusion

Writing on 'expansive night' – Cai Yong's third 'school' of cosmology – Joseph Needham saw in China a certain forward-looking freedom to which we should aspire:

> These cosmological views are surely as enlightened as anything that ever came out of Hellas. The vision of infinite space, with celestial bodies at rare intervals floating in it, is far more advanced (and the point is worth emphasizing) than the rigid Aristotelian–Ptolemaic conception of concentric crystalline spheres which fettered European thought for more than a thousand years. Though sinologists have tended to regard the [*xuan ye*] school as uninfluential, it pervaded Chinese thought to a greater extent than would at first sight appear . . . Chinese astronomy is often reproached for its overwhelmingly observational bias, but the lack of theory was an inevitable result of the lack of deductive geometry. It might well be argued that the Greeks had had too much, since the apparent mathematical beauty of 'cycle upon epicycle, orb on orb' came eventually to constitute a strait-jacket posing unnecessary difficulties for a Tycho, a Copernicus and a Galileo (Needham 1959, 220).

Needham is right that the idea of buoyant *qi* would find itself everywhere. As to the 'school', however, intellectuals declared it dead and lost from the time of the second century on, survived by only a single passage of 186 graphs (*Sui shu*, 19.507). If you asked Cai Yong, Shen Yue or Li Chunfeng what happened, ironically, they would triumphantly answer 'the sphere'.

Chinese intellectuals were as much prisoners of their spheres as anyone else, but where Aristotle (384–322 BCE) worked in crystal, theirs were cages of iron, bronze and jade. Dolorous, they rattled at cages that were no more real than Aristotle's, for they were prisoners of the idea of an object locked away in foreign capitals and restricted sites that many had never seen. The closer you got to one, the more you wanted to build another, because the object itself was 'loose and rough' (Li Chunfeng), 'painfully large . . . [and] difficult to turn' (Wang Fan), and 'insufficient to upwardly examine signs of heaven and reverently grant the seasons of man' (Yixing). Cai Yong might well demur if *this* is what he awoke to in the morning, but having never lain with her under the stars, she remained out of sight, and far away, the perfect object of desire. The 'sphere heaven' *was* the world it was built to measure, so it need not have strictly been outside, in the cold and rain, nor even fitted with a tube. Nor was she suited to the outdoors life, as it was in a sealed room where she outshone the sticks and hands and leaky pots, turning in perfect effigy of 'heaven's patterns' and churning '*li* numbers' effortlessly forth. *This* is what *li* men were building, and *this* is whence they contemplated heaven, but this is precisely what *guan* entails, and *guan* is no less important for the history of science than any analogue to modern 'observation'. The meditation object, after all, furnished the cosmology and the numbers necessary to 'observe'.

The sage kings did not always work in right ascension. Before the algorithms and armillaries of the second century CE, they once 'observed the signs' as did their 'watchmen', spontaneous and unmediated but for the human hand. As gentlemen of *li* inserted themselves in this activity, however, they increasingly imposed themselves upon it. In omen literature, after all, the sages insisted upon the 'anomalous' (*yi*) nature of phenomena whose 'constancy' (*chang*) the *li* man could now predict with unfailing 'tightness'. Befuddled, gentlemen set about cleaning house: they gathered sources, weeded out 'absurdities' and systematised contents into proper, authored books. After centuries of intensifying bans, these books were then most of what was left to read. As they shaped the practice of omen reading, so too did *li* men bring their own demands to *tianwen*. It was no longer enough to look for '*li* numbers' in pitch pipes and the *Book of Changes*, as they once did in the Han (Chapter 1 above); it was time that they went straight to the source. To do this, they needed new tools, new methods and more mathematics than *tianwen* had ever seen. Their efforts towards a new, quantitative *tianwen* apparently paid off, judging from the subsequent 'turns' and 'accumulation' in *li* (Chapter 5 below), but the practice of data collection itself was not what captivated the gentleman's imagination in writing about this field. It was not just the 'signs' that one stood to see through/on the 'sphere heaven', it was 'true numbers' revealing even greater truths, and *this*, in certain circles, became what it meant to 'observe the signs'.

3 Granting the Seasons

We often speak about 'the Calendar' in China as if it were a monstrous engine – a sort of terraforming device, the influence of which extends everywhere at once, moulding the fabric of society and the very world that sustains it. Like an engine, it churns, it stutters and it eventually breaks, grinding life to a halt if not tuned and replaced in a timely manner. It is only fitting that this was the very first matter to which Yao attended as king, and so too for any veritable sage to come.

Thearch [Yao] said: 'Oh, you, Xi and He, the year has three hundred, sixty, and six days, [and so] by means of an intercalary month should you fix the four seasons and complete the agricultural year. If you sincerely manage the hundred workers [thereby], [then your] achievements will be numerous and resplendent' (*Shangshu zhushu*, 2.21b).

What marvellous engine is this by which brothers scattered to the valleys of the rising and setting sun may synchronise to their calls the life of one and all in between? Natural *shi* ('seasons/time') unfold at all points under heaven at once, the quarter-moon rising no more contemptuously over the sands of Edsen-gol than over the palaces of Luoyang, but 'the seasons of heaven' (*tian shi*) are beside the point. Heaven is there for all to see; what the world needs a sage to 'respectful grant' it is rather 'the seasons of *man*' (*Shangshu zhushu*, 1.21a). As the stars turn in perfect step around the pole, and as the birds and flowers come and go with the sun, so too is humanity to live and die in spontaneous choreographic rhythm around its thearch – such is the order of things, and such does the order of things depend upon. The *Book of Documents* does not go much beyond the previous quote in terms of *logistics*, mind you, but the way that ancient and modern scholars alike discuss 'the Calendar' in China would seem to make it the *primum movens* of civilisation.

We may not be able to speak for the mysteries of 'the Calendar', but we do have plain old *calendars* from the early imperial period. Humble affairs, preserved in tombs and administrative dumps, these documents have garnered very little attention from modern historians. *Calendars*, admittedly, do not make for as good reading as do the ideological and ceremonial debates of gentlemen scholars, but, as physical traces of how the state actually imposed itself on the lives of its citizenry, they have infinitely more to tell us about the situation on the ground. *Calendars*, moreover, provide us an invaluable link within the networks of knowledge and textual production that connected the intellectual labours of the 'great men' of the history of science to those of the early imperial Everyman. *Calendars*, lastly, give us a meditation object of our own through which to consider how all the abstract ideas debated by *li* men and policy-makers alike translated into physical objects and social reality. Calendars, after all, are supposedly the entire point of *li*.

Given the peculiar use of 'calendar' in Western-language sinology (Section 1.1.3), allow me to define my terms. By 'calendar', I refer to a table of months and days for use in civil timekeeping such as that one might find pre-installed on one's smartphone. More specifically, of the two periods into which scholars divide the history of calendars in China – that of the *liri* ('*li* days') *calendar* and, starting between 630 and 658 CE, that of the *juzhu liri* ('*li* days with annotation') *almanac* – we shall be dealing with the former.[1] This *liri* 'calendar' is a catch-all observer's category for documents featuring a variety of media, layouts and titles (unlike the *juzhu liri* of the seventh century on, not everything that scholars call a *liri* calls itself the same thing). Unlike the *juzhu liri*, furthermore, the *liri* corpus is quite diverse in terms of archaeological provenance as well, providing us exemplars from every corner of the empire and from the world of the living and the dead alike. Fragmentary, diffuse and ever-expanding, we will rely on the corpus collected in Bo Shuren (1993) and Yoshimura (2003), citing photographs and transcriptions in individual site reports published since.

In the previous chapter, we reflected upon the practice of 'observing the signs' as mediated by writing, calculation and material technology, and we shall take here the same approach to 'granting the seasons'. Broadcast over oceans and airwaves from government servers powered by the atom, to us, the calendar is now every bit as ethereal and instantaneous a force as it apparently was in the day of Yao, but such was not always the case. The calendar such as we shall see it here is *handmade*, a physical product of calculation, conversion and tabulation by human hands, and, to arrive everywhere at once, it is also *hand-copied* – 'the one giving birth to two, the two giving birth to three, and the three giving birth to the myriad things' (*Dao de jing*, ch. 42) – and this requires a certain amount of human and material logistics. In this chapter, we are interested to know how this worked, both in practice and in theory.

We begin in Section 3.1 with a survey of excavated calendar types. Interested in manuscripts as physical traces of 'season granting', our approach is codicological, and I shall pay special attention to copy errors as evidence of the processes of textual (re)production. What we find in this section reveals the

[1] On the transition between 'calendars' and 'almanacs', see Jiang Xiaoyuan (1992a) and Deng Wenkuan (2002, 134–44). For the 'annotated calendars' of the Dunhuang cave library, see *Dunhuang tianwen lifa wenxian jijiao* and Kalinowski (2003, 85–211).

extent to which 'the Calendar' was a hybrid of personal and imperial space, as it was up to the individual to both *make* and *make use of* this imperial symbol in daily life. From there, we move in Section 3.2 to the question of the state apparatus for time control. Starting from scriptural precedence for the ceremonies by which the Western Zhou (1045–771 BCE) supposedly went about this, we examine how early imperial courts adapted the *wen* ('cultural patterns') of the classics to the exigencies of their own day. Perhaps surprisingly, we find that the empire was far more successful with the *logistics* of co-ordination than with the *ceremonial*, the latter of which was a morass of intellectual dispute, political deadlock and historical irony. Having examined 'the Calendar' in its concrete individual and institutional manifestations, we then turn in Section 3.3 to the question of how much any of this had to do with *li*. Asking what calendar making meant to each of those involved, we will reconsider the centuries-long debate about 'fixing' the civil month via speed correction. The debate, I argue, pitted the theoreticians' demand for astronomical 'tightness' against the practicalities of calendar making as experienced by computers, copyists and users, the former winning out only after five centuries of self-justification.

3.1 Calendar Types

Previous scholarship on early imperial calendars is mostly concentrated in two areas. One is the reconstruction of the civil calendar and the '*li* numbers' behind it. The archaeological record begins to provide us with calendars and unambiguously dated documents from 246 BCE on, but it is only after 104 BCE that it begins to square with received *li*.[2] Here, scholars are caught in something of a loop, dating fragments via tables, and correcting tables via fragments, cognisant all the while of the potential for human intervention in calculation.[3] Scholars of 'numbers and procedures' (*shu shu*) have also taken a particular interest in calendars for the 'taboos' (*ji*) and 'prohibitions' (*jin*) of 'day selection' (*ze ri*) that they note – 'the teeming taboos and negligent prohibitions' about which we are warned in Section 1.1.2 above.[4] As much as the diffusion of hemerologies like 'establish and eliminate' (*jian-chu*) would seem to argue for *belief*, Huang Yi-long (1999) makes an excellent case from the Yinwan M6 travel diary and prohibitions that people were not particularly rigorous about their application. Here again we find ourselves in a loop: we compare hemerologies in calendars and 'day books' (*rishu*), correcting one against the other, but absent evidence of how *text* translated to *practice*, the question remains an exercise in philology.

[2] For studies of the pre-Qin calendar, see Hirase (1996) and Gassmann (2002). On the Qin and early Han, see Huang Yi-long (2001), Zhang Peiyu (2007) and Li Zhonglin (2010).

[3] See Luo & Guan (2000, 58) and Huang Yi-long (1992b).

[4] See Wang Su (1989), Liu Lexian (2002, 255–72), Deng Wenkuan (2002, 105–22, 296–303) and Arrault (2002).

Rather than trying to resolve incongruities in the true workings of 'the Calendar', or the 'selection of days' therein, I should like to *embrace them* and focus instead on questions of material production. Because our focus is material – because the dimensions, layout and codicological features of the physical support have as much to tell us as the text written thereupon – I have chosen illustration over translation for the manuscript sources considered in this section. For those who do not read Chinese, I promise that visual layout here speaks volumes over what is written, what is written being mostly month numbers and stem–branch dates (see Appendix). As to layout, I have divided the corpus into four typologies based on function and organisational principle: is it a civil calendar, or a table for computing one? Does it list every day of the month, or just the first? Is the sexagenary cycle counted from the month, or is the month counted from the sexagenary cycle?[5] Lastly, to avoid some of the terminological confusion that surrounds the topic in Chinese-language scholarship, I shall refer to titled calendars by their titles, and untitled ones by their archaeological source, transliterated reign period, regnal year, Julian year and type.

3.1.1 Day Tables

The earliest form of calendar with which the Chinese archaeological record has provided us thus far is what I shall call the 'day lookup table' or 'day table'. As of 2017, we possess six manuscripts of this type from southern tombs dating to 220–142 BCE, five of which bear the title 'Zhiri' 質日, following a year number within an implied reign period.[6] These *zhiri* are written on 'bookmats' of bamboo slips (once) bound by silk at their centre and upper/lower margins. Each slip providing a natural one-graph column, the roughly sixty-slip mat is divided into six rows, each row running right to left through the stem–branch dates of two consecutive months. Coming at the beginning and middle of each row, the month is written *above* or *to the right of* the stem–branch date of its inception and highlighted by a heavy block of black ink. Intercalary month IX^2, where it occurs, is appended to the end of the *zhiri*, its stem–branch dates running across multiple registers to save space.

By chance, we happen to have two *zhiri* for the same year: that from Zhoujiatai M30 and, of unknown provenance, the Yuelu Academy 'Sasi nian zhiri' (Figures 3.1 and 3.2). In terms of form, the two are identical but

[5] See also the alternative typologies proposed by Arrault (2002) and Yoshimura (2003).

[6] In addition to the (1) Zhoujiatai M30 and (2) Yuelu Academy Shihuang 34 (213 BCE) calendars illustrated below, there is (3) the Yuelu 'X-qi nian zhiri' (220 BCE), (4) the Yuelu 'Sawu nian si zhiri' (212 BCE), (5) the unpublished Zhangjiashan M136 'Qi nian zhiri' (179 BCE) and (6) the unpublished *zhiri* from Shuihudi M77. For the Yuelu Academy materials, see *Yuelu Qin jian*, vol. 1, 3–9, 19–24, 47–65, 91–106; for the site report for Zhangjiashan M136, see *Wenwu* 1992.9: 1–11; for Shuihudi M77, see *Jiang-Han kaogu* 2008.4: 31–7. One might also add to this (7) the Edsen-gol (Juyan), Mu-durbeljin (Pochengzi) A14 (111·6) Benshi 4 (13 BCE) calendar fragment.

Table 3.1 *Zhoujiatai M30 and Yuelu Academy Shihuang 34 (213* BCE*) calendar annotations (sample)*

Date		Zhoujiatai	Yuelu
X	*wushen*$_{.45}$		Teng occupied clerk of the right.
	bingchen$_{.53}$		Teng went to Anlu.
XI	*jimao*$_{.16}$		Teng came to Anlu (?).
XII	*wuxu*$_{.35}$		Teng returned, rested.
	gengzi$_{.37}$		Teng assumed office.
	bingchen$_{.53}$	Follow Assistant Deng; Clerk Shu transferred.	
	dingsi$_{.54}$	Follow Assistant Deng; Clerk . . . arrived . . .	
	xinyou$_{.58}$	Jiaping [festival].	
	yichou$_{.02}$	Clerk Dan arrested.	
I	*dingmao*$_{.04}$	Jiaping [festival], assumed office.	
	xinsi$_{.18}$		Teng taken into custody by supervisory office.
	dinghai$_{.24}$	Clerk transferred; not sitting at [Accounts] Assistant Section; following his Excellency; lodged in Changdao.	
	wusi$_{.25}$	Lodged in Shangdi, north of Ying Settlement, Zhi.	

for four points: (1) the Yuelu manuscript is titled; (2) the Yuelu slips are 2.5 cm shorter; (3) the Yuelu manuscript omits block highlighting along the centre binding; and (4) the Zhoujiatai manuscript marks intercalary dates with dots. In terms of function, the two record activities by similarly charged local administrators in the Chu region (Table 3.1). The congruity between these (and other) manuscripts suggests that we are looking at a standardised administrative document. Li Ling (2008) understands this document as an archival record of official business, while the appearance of a 'personal' (*si*) *zhiri* in the Yuelu collection has led Su Junlin (2010), on the other hand, to classify the *zhiri* as an aide-mémoire. Whatever their ultimate administrative destiny, the Zhoujiatai and Yuelu manuscripts clearly belong to a single standardised class of document.

Where the Zhoujiatai and Yuelu lookup tables *do* bear the mark of the 'personal' is in the matter of copying – or *miscopying*, to be exact. In one place, for example, the Yuelu copyist writes *wuxu*$_{.35}$ → *jiyou*$_{.46}$ → *gengzi*$_{.37}$ (reg. 6, s. 5–7), and in another, the Zhoujiatai copyist writes *yichou*$_{.02}$ →

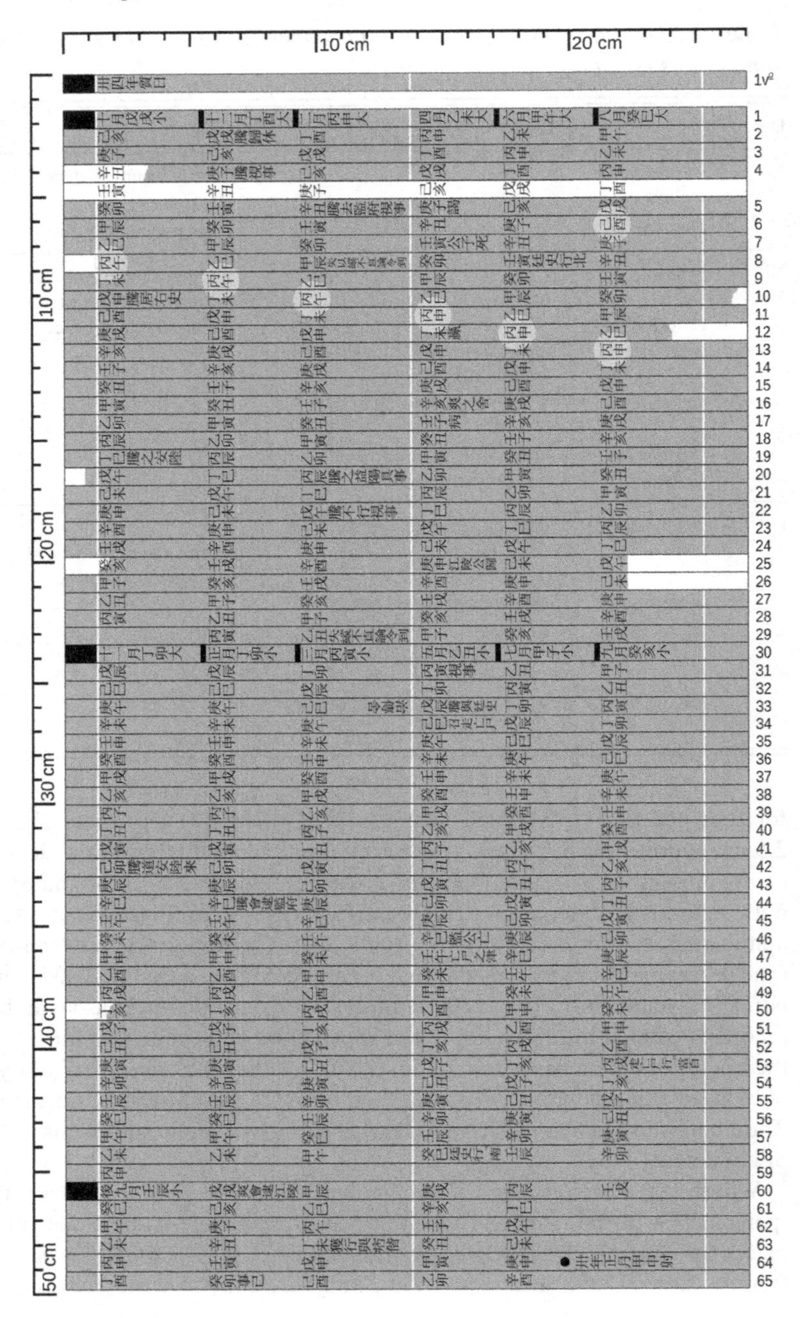

Figure 3.1 Yuelu Academy Shihuang 34 (213 BCE) 'Sasi nian zhiri' bamboo bookmat.

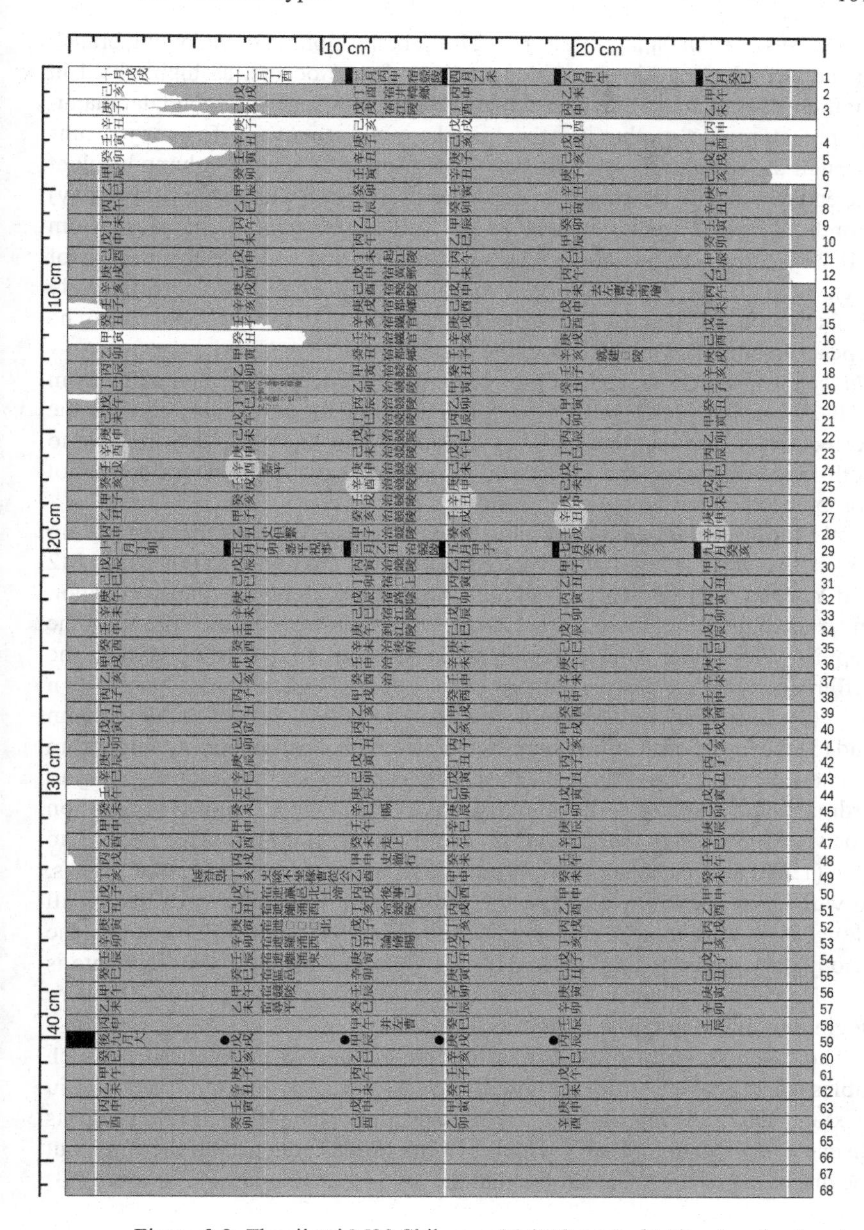

Figure 3.2 Zhoujiatai M30 Shihuang 34 (213 BCE) *zhiri* bamboo bookmat.

bingshen.33 → *dingmao*.04 (reg. 1, s. 27–9), both skipping to the wrong branch in the stem–branch binome. Once such a mistake occurs, as highlighted in Figures 3.1 and 3.2, it cascades down subsequent rows. The Zhoujiatai manuscript features yet another iterative mistake in the same vein. The copyist runs through the sexagenary cycle in row 1, neglecting to leave a column blank at the end of a 'small' (twenty-nine-day) month X. Faced with a 'big' (thirty-day) month in row 2 (and nowhere to put XII-30 *bingyin*.03), he skips from XII-29 *yichou*.02 to I-1 *dingmao*.04 such that the first and subsequent days of each month align. *He then forgets to do this in rows 3–5*, which effectively shifts months III, V and VII one day forward. Coming to VIII-29 in row 6, he repeats an unrelated mistake from the previous row (*jiwei*.56 → *gengshen*.57 → *xinchou*.38), which apparently jogs his memory, because he then skips from VIII-29 *xinchou*.38 (*sic*) to IX-1 *guihai*.60, resetting the sexagenary count to the correct date.[7] The fact that both the Zhoujiatai and Yuelu copyists are able to both repeat mistakes and correct them tells us that the problem here is not comprehension but repetition.

We find a similar type of day table bookmat clustered in northern sites from 134 BCE to 90 CE, two of which are Eastern Seaboard tombs (Yinqueshan M2 and Yinwan M6), and the rest administrative sites along the Gansu corridor. In terms of material, these calendars are written on wood, which provides the copyist with wider slips/columns than bamboo. In terms of layout, the difference with the early, southern *zhiri* is that (1) each month has its own row, (2) stem–branch dates are written horizontally *across* each slip/column and (3) the top of each slip/column is labelled with day numbers counted from the new moon. Interestingly, this last point echoes a shift from sexagenary to ordinal date notation in administrative documents whose conclusion Loewe (1959, 308–14) places in the Eastern Han (25–220 CE).[8] Lao Gan (1970) suggests that the shift represents a concession to popular usages, to which one may now add that the ordinal date was there in calendars all along: it is, after all, the alignment of day-numbers with which we saw the Zhoujiatai copyist struggling in 213 BCE. Whether or not the ordinal date is labelled, it is implicit in the organisation of daily calendars from the very beginning of the imperial period.

The 'Qi nian [li]ri' (134 BCE) from Yinqueshan M2 is one of two such tables that bears a title. Originally transcribed as *liri* ('*li* days'), Liu Lexian (2002, 25) reports that visual inspection of the written title suggests instead *shiri* ('date lookup'), which Li Ling (2008) connects to the *zhiri*, but the discovery of the 'Yuanshi liu nian liri' of 5 CE at Jinguan T23 (Figure 3.3)

[7] For a different explanation, prior to the publication of the Yuelu Academy *zhiri*, see Lin Zhonglin (2009).

[8] Cf. Yu Zhongxin (1994, 15 ff.).

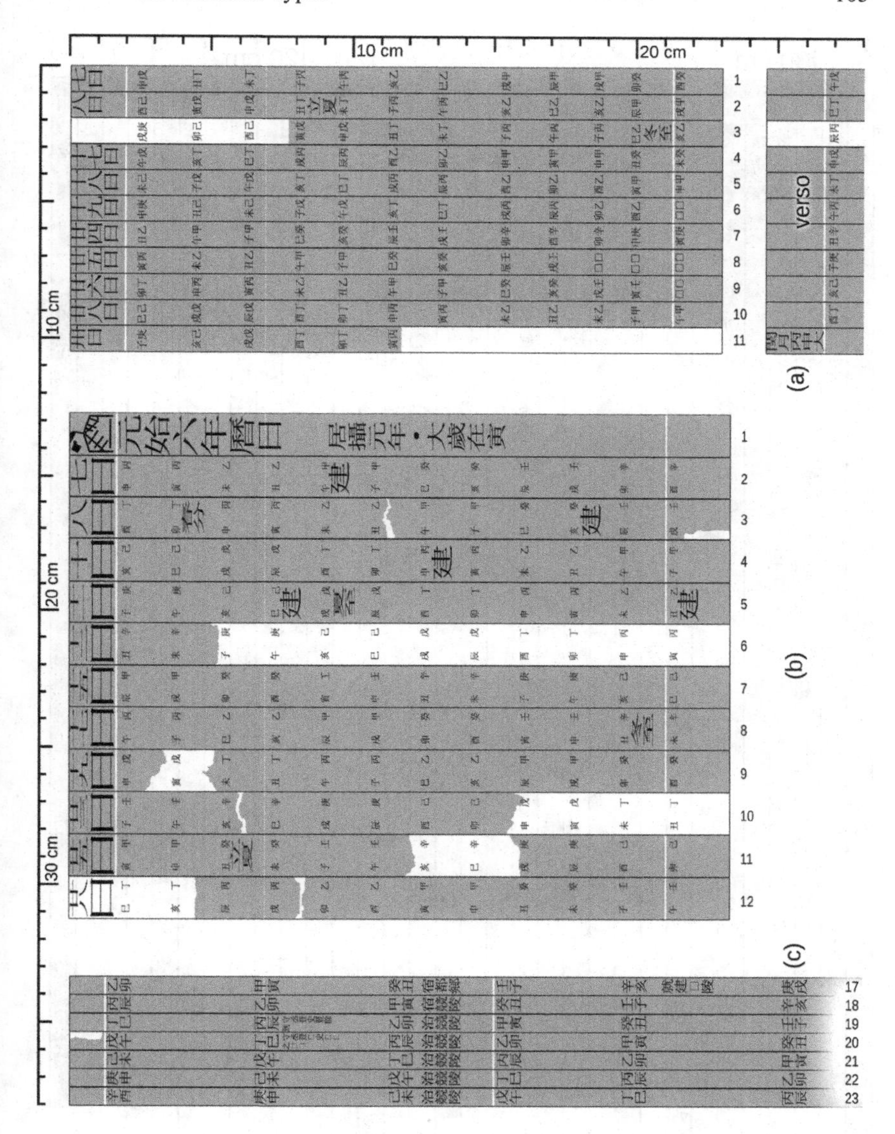

Figure 3.3 Wood bookmat *liri* day tables. Right (a): Dunhuang T6b Shenjue 3 (59 BCE) *liri*. Centre (b): Jinguan T23 'Yuanshi liu nian liri' (5 CE). Left (c): slips from the Zhoujiatai M30 Shihuang 34 (213 BCE) *zhiri* for comparison. Note that the two *liri* are incomplete.

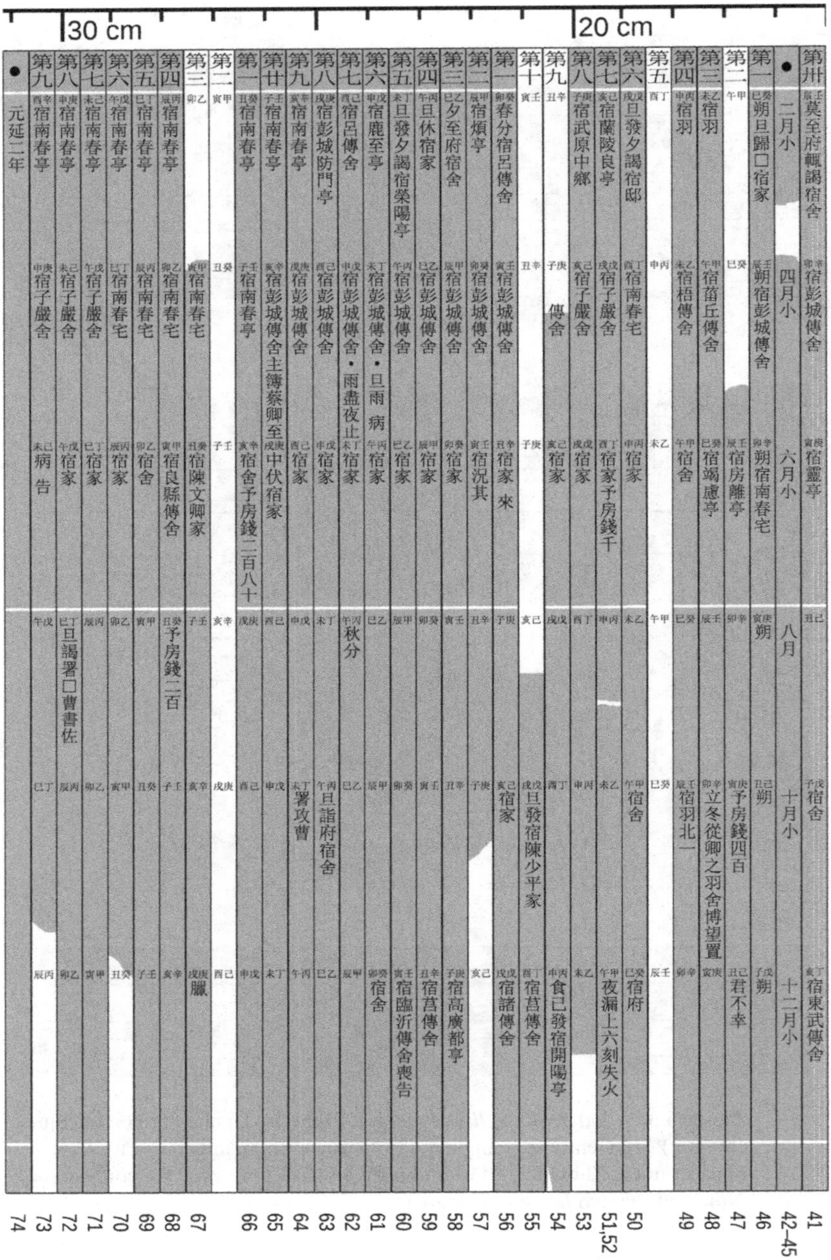

Figure 3.4 Yinwan M6 Yuanyan 2 (11 BCE) 'Yuanyan er nian' bamboo bookmat.

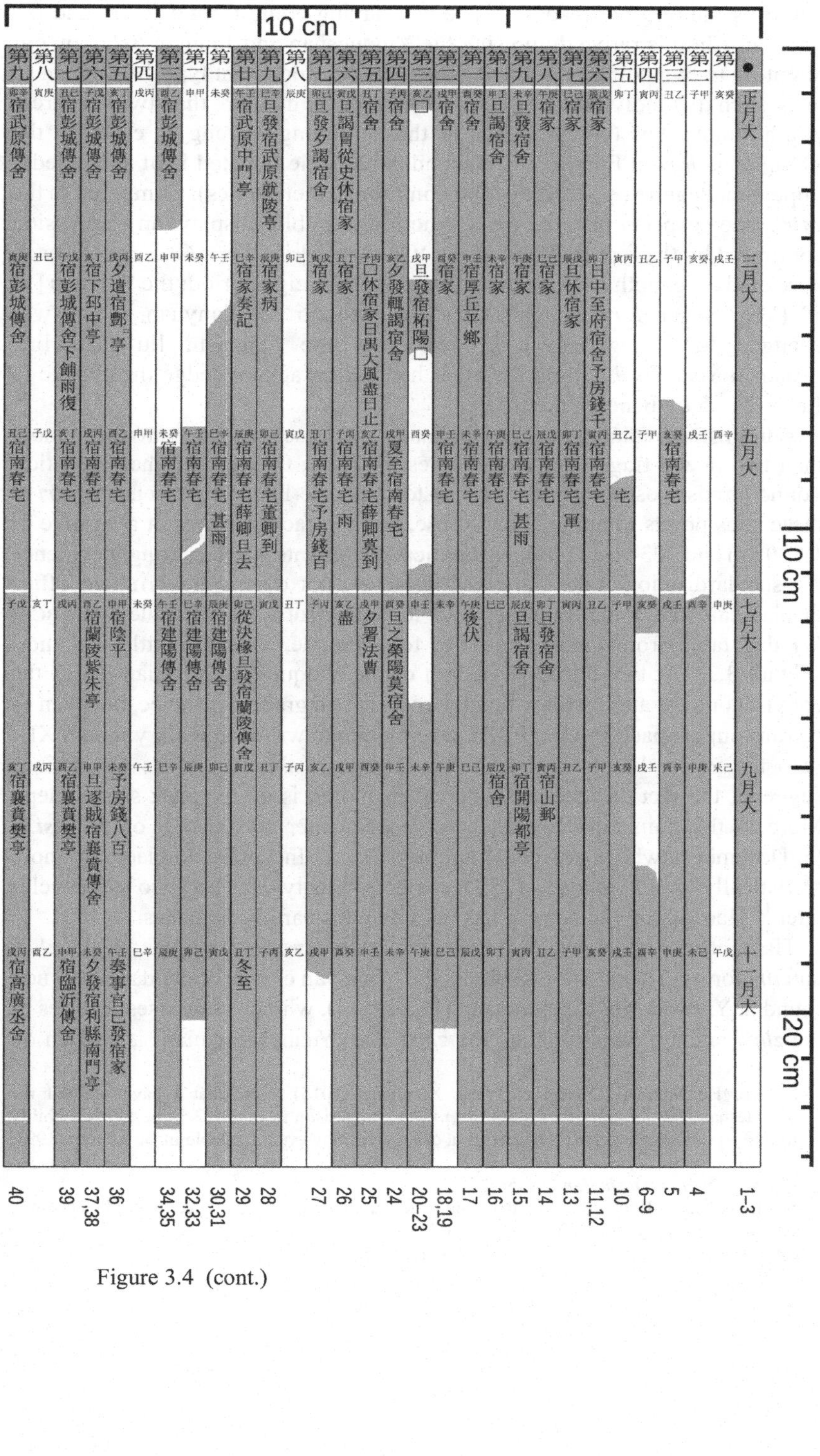

10 cm

20 cm

Strip	Heading	正月大	三月大	五月大	七月大	九月大	十一月大
1–3	●	正月大	三月大	五月大	七月大	九月大	十一月大
	第一	亥癸	戌壬	酉辛	申庚	未己	午戊
4	第二	子甲	亥癸	戌壬	酉辛	申庚	未己
5	第三	丑乙	子甲	亥癸 宿南春宅	戌壬	酉辛	申庚
6–9	第四	寅丙	丑乙	子甲	亥癸	戌壬	酉辛
10	第五	卯丁	寅丙	丑乙 宅	子甲	亥癸	戌壬
11,12	第六	辰戊 宿家	卯丁 日中至府宿舍予房錢千	寅丙 宿南春宅	丑乙	子甲	亥癸
13	第七	巳己 宿家	辰戊 旦休宿家	卯丁 宿南春宅 軍	寅丙	丑乙	子甲
14	第八	午庚 宿家	巳己 宿家	辰戊 宿南春宅	卯丁 旦發宿舍	寅丙 宿山鄉	丑乙
15	第九	未辛 旦發宿舍	午庚 宿家	巳己 宿南春宅 甚雨	辰戊 旦謁宿舍	卯丁 宿開陽都亭	寅丙
16	第十	申壬 旦謁宿舍	未辛 宿家	午庚 宿南春宅	巳己	辰戊 宿舍	卯丁
17	第一	酉癸 宿舍	申壬 宿厚丘平鄉	未辛 宿南春宅	午庚 後伏	巳己	辰戊
18,19	第二	戌甲 宿舍	酉癸 宿家	申壬 宿南春宅	未辛	午庚	巳己
20–23	第三	亥乙 □	戌甲 旦發宿柘陽□	酉癸	申壬	未辛	午庚
24	第四	子丙 宿舍	亥乙 夕發輒謁宿舍	戌甲 夏至宿南春宅	酉癸 旦之榮陽莫宿舍	申壬	未辛
25	第五	丑丁 宿舍	子丙 □休宿家日禺大風盡日止	亥乙 宿南春宅薛卿莫到	戌甲 夕署法曹	酉癸	申壬
26	第六	寅戊 旦謁胃從史休宿家	丑丁 宿家	子丙 宿南春宅 雨	亥乙 盡	戌甲	酉癸
27	第七	卯己 旦發夕謁宿舍	寅戊 宿家	丑丁 宿南春宅予房錢百	子丙	亥乙	戌甲
	第八	辰庚	卯己	寅戊	丑丁	子丙	亥乙
28	第九	巳辛 旦發宿武原就陵亭	辰庚 宿家病	卯己 宿南春宅董卿到	寅戊	丑丁	子丙
29	第廿	午壬 宿武原中門亭	巳辛 宿家奏記	辰庚 宿南春宅薛卿旦去	卯己 從決椽旦發宿蘭陵傳舍	寅戊	丑丁 冬至
30,31	第一	未癸	午壬	巳辛 宿南春宅 甚雨	辰庚 宿建陽傳舍	卯己	寅戊
32,33	第二	申甲	未癸	午壬 宿南春宅	巳辛 宿建陽傳舍	辰庚	卯己
34,35	第三	酉乙 宿彭城傳舍	申甲	未癸 宿南春宅	午壬 宿建陽傳舍	巳辛	辰庚
	第四	戌丙	酉乙	申甲	未癸	午壬	巳辛
36	第五	亥丁 宿彭城傳舍	戌丙 夕遣宿鄞=亭	酉乙 宿南春宅	申甲 宿陰平	未癸 予房錢八百	午壬 奏事官已發宿家
37,38	第六	子戊 宿彭城傳舍	亥丁 宿下邳中亭	戌丙 宿南春宅	酉乙 宿蘭陵紫朱亭	申甲 旦逐賊宿襄賁傳舍	未癸 夕發宿利縣南門亭
39	第七	丑己 宿彭城傳舍	子戊 宿彭城傳舍下餔雨復	亥丁 宿南春宅	戌丙	酉乙 宿襄賁樊亭	申甲 宿臨沂傳舍
	第八	寅庚	丑己	子戊	亥丁	戌丙	酉乙
40	第九	卯辛 宿武原傳舍	寅庚 宿彭城傳舍	丑己 宿南春宅	子戊	亥丁 宿襄賁樊亭	戌丙 宿高廣丞舍

Figure 3.4 (cont.)

confirms the original reading.[9] The one having come from a tomb, and the other an administrative dump site, the Yinqueshan and Jinguan *liri* draw our attention to two points. First, the graphs on this document type are written so large such that individual day cells accommodate no more than two- or three-graph annotations (compare this to those running as long as eight on the Zhoujiatai *zhiri* in Figure 3.3). Second, what little *is* noted is of a decidedly impersonal nature (*qi*, holidays and common hemerologies). Compared to the *zhiri*, one suspects that the *liri*'s function is public display, an impression confirmed by the reported size of the 'Qi nian [li]ri' slips – 68.5 cm![10] On an unrelated note, both *liri* are dated to an obsolete reign period: the 'Qi nian [li] ri' (Year Seven *Liri*) matches the 'seventh year' of Jianyuan, which was changed to Yuanguang 1 in 134 BCE; the 'Yuanshi liu nian liri' (Yuanshi Year Six *Liri*), on the other hand, does acknowledge the change to Jushe 1 in 5 CE beneath the title.

Administrative dump sites in Dunhuang and Edsen-gol have produced more than two dozen fragmentary day tables matching these exact characteristics. Found amidst postal records, duty rosters and the like, the presence of *liri* at these sites points to administrative use, and the fact that sites at a remove of 1,800 km and 223 years produce identical documents is even stronger evidence for standardisation. Like the *zhiri*, however, not even a government office display calendar is totally impervious to copy errors. The Shenjue 3 (59 BCE) *liri* day table from Dunhuang T6b, for example, is just a little too short (Figure 3.3). At two-thirds the length of the Yinqueshan 'Qi nian [li]ri', the copyist compensates by crowding the rows at two graphs' distance, but even so he runs out of space by month XII, forcing him to write intercalary month XII^2 onto the verso. If this is indeed for display, as the size of its annotations suggests, the fact that the last, intercalary month is on the back seems inept. More to the point, the copyist has placed winter solstice.$_{Q22}$ on XI-*yisi*.$_{42}$ (5 December) where the contemporary Grand Inception *li* places it more realistically on XI-*yichou*.$_{02}$ (25 December), twenty days and two stem cycles later.[11] Here again, the copyist has mixed up his earthly branches.

The archaeological record provides us with several variations on the *zhiri* and *liri* forms. There is, for example, the 'Yuanyan er nian' calendar for 11 BCE found at Yinwan M6, Lianyungang (Figure 3.4), which synthesises features of the *zhiri* and *liri* forms.[12] Like the *zhiri*, the 'Yuanyan er nian' is written on

9 Source: He Maohuo (2015); cf. Yang Xiaoliang (2015). Note that I have omitted the Yinqueshan M2 'Qi nian [li]ri' (134 BCE) from comparison in Figure 3.3 due to my inability to find a photograph of said manuscript; a drawing is, however, available in Bo Shuren (1993, 221–3).

10 *Yinqueshan Han mu zhujian (yi)*, 5.

11 See Luo Jianjin (1999, 98). Note that winter solstice that year actually occurred 23 December, 23:33:30 local apparent time (LAT), Chang'an.

12 Source: *Yinwan Han mu jiandu*, 61–7, 139–44.

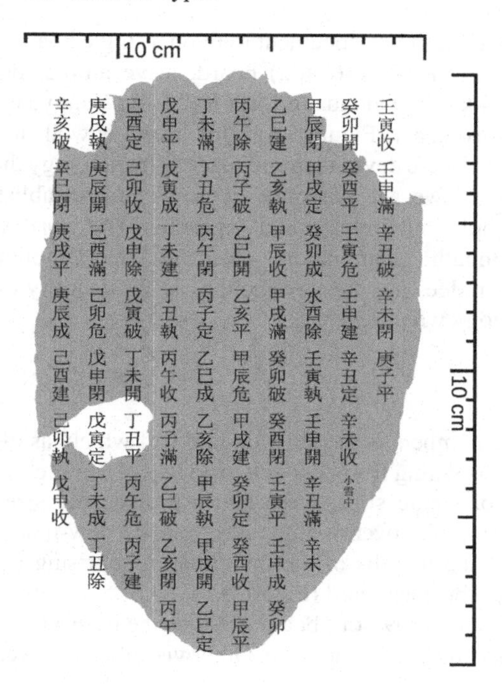

Figure 3.5 Asitana M387 Yanshou 7 (630 CE) paper fragment, Kingdom of Qara-hoja.

bamboo and affords ample space for annotation, noting matters of travel, lodging, business and weather familiar from Table 3.1. Like the *liri*, however, the stem–branch dates are written horizontally, and the top of each slip/column is labelled 'number *x*' (*di x*) in a variation on 'day *x*' (*x ri*). We also have the Yanshou 7 (630 CE) paper calendar fragment from the Kingdom of Gaochang, i.e. Qara-hoja (460–640 CE), excavated from Asitana M387, Turpan (Figure 3.5).[13] Here, it is interesting to note that the latest and furthest-flung calendar prior to the age of the *juzhu liri* is a variation on the day table. Where this document departs from the *zhiri* and *liri*, however, is that it leaves no space for annotation beyond the 'establish-and-eliminate' sequence and the twenty-four *qi* nestled therein.

We also find wooden-board day tables for individual months.[14] Given in ordinal, *then* sexagenary, notation, the one-month table arranges dates into

[13] Source: Deng Wenkuan (2002, 228–40).

[14] Note that register 3 of the Dunhuang T15A (D1968) Yongyuan 6-XII (94 CE) calendar contains dates for a month VII, but month VII of another year; see Luo Jianjin (1999, 94–5).

multiple registers, allowing the entire document to fit on a single board. Some, like the Yinwan M6 Yuanyan 3-V (10 BCE) board, leave ample room for annotation, while others, like the Dunhuang D1968 (T15A) Yongyuan 6-XII (94 CE) board, are far too compact (Figure 3.6). This layout is well suited for annotation, the two examples here devoting more space to hemerology than any other pre-almanac calendar: we see not only 'reverse' and 'establish-and-eliminate' days, for example, but 'release and expansion' (*jie-yan*), 'return' (*fu*), 'pitfall' (*xian*), 'month examination' (*yuexing*) and 'month kill' (*yuesha*).[15] That said, form does not dictate function, as the majority of these tables feature no hemerology whatsoever.[16]

3.1.2 Lunation Tables

Next in terms of absolute numbers is the lunation table, of which, as of 2017, we possess seven exemplars dating from 209 BCE to 451 CE.[17] Simple, compact lists on wooden boards or single slips, these tables provide the size ('big' versus 'small', thirty versus twenty-nine days) and new-moon date (stem–branch binome) of the months of the civil year, comprising no more text than that could fit on a business card (Figure 3.7).

The Zhoujiatai M30 Ershi 1 (209 BCE) board gives us the basics of size and new moon, filling less than one-half of the recto. Otherwise blank, the verso has been used as a notepad:

Jiaping [Festival] on XII-*wuxu*.$_{35}$ (9 February 209 BCE), four days out from the end of the month.

XII-*jimao*.$_{16}$ (21 January 209 BCE) . . . arrived at court tax office a total of twenty mat cushions.

The following registers enumerate the stem–branch binomes for month XII. Given how the verso counts around the Jiaping festival in terms of ordinal and sexagenary dates, it would seem as if its owner had written out the latter in order to plan how to solve his mat-cushion problem before going on holiday. If so, the fact that he wrote out *all of them* suggests that stem–branch binomes may not have come to him as naturally as numbers.

The Zhangjiashan M247 Gaozu 5–Gaohou 2 (202–186 BCE) bookmat is even more compact, fitting seventeen years into as many slips. The layout here is less

[15] On these hemerologies, see Liu Lexian (2002, 255–72).

[16] See, for example, the Edsen-gol, Mu-durbeljin P9 (457·19) Benshi 2 (72 BCE) table; the Jinguan A33 (179·10) Shenjue 1 (61 BCE) table; and the Edsen-gol, Mu-durbeljin A21 (290·11A) Jushe 1 (6 CE) table in Yoshimura (2003, 500–7).

[17] In addition to the four examined here, there is the Dunhuang D38 (81.D38:59–61) Yuanfeng 3 (78 BCE) table, the Dunhuang D21 (D565) Yangshuo 1 (24 BCE) and the Huaguoshan LHM1 Yuanshou 2–3 (1 BCE – 1 CE) table; see Dunhuang-xian wenhuaguan (1984), Yoshimura (2003, 464–6) and Li Hongfu (1982).

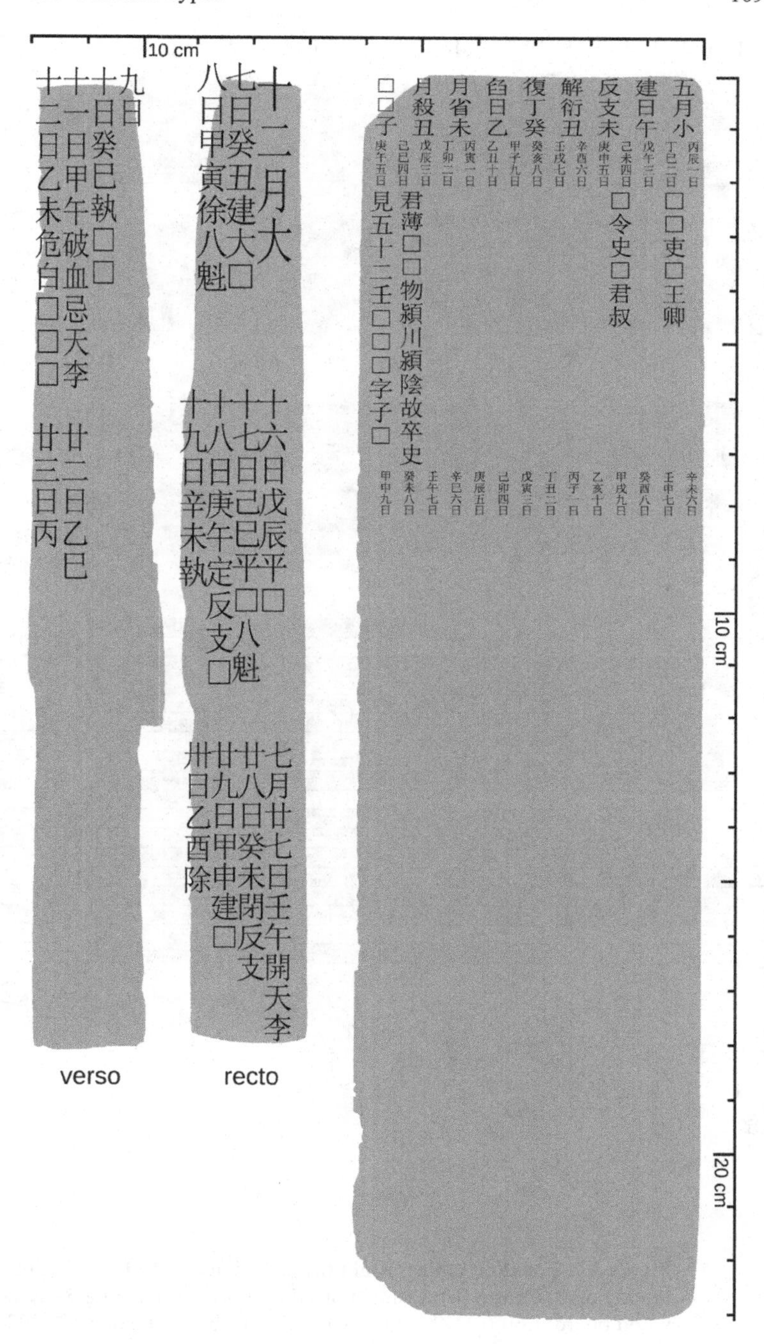

Figure 3.6 Monthly day tables. Right: Yinwan M6 Yuanyan 3-V (10 BCE) wood board. Left: Dunhuang D1968 (T15A) Yongyuan 6-XII (94 CE) wood board.

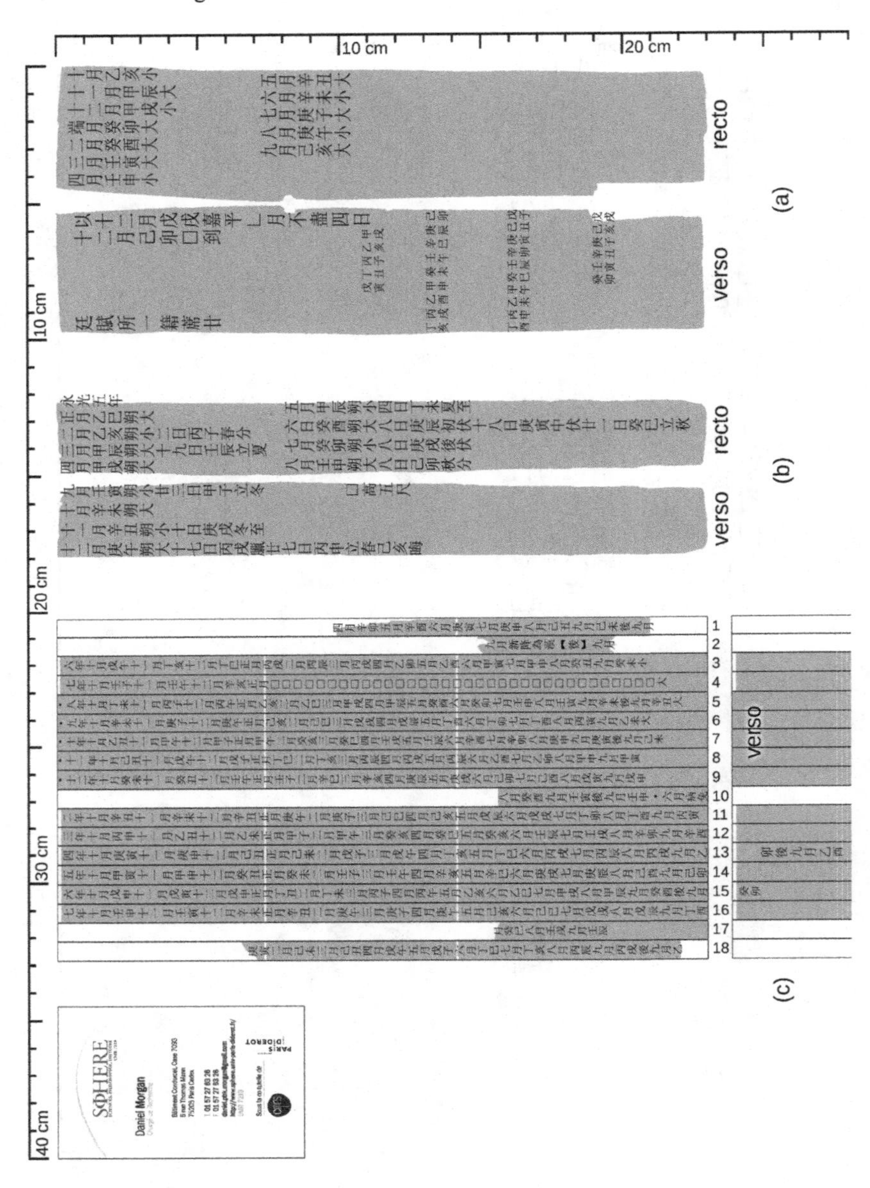

Figure 3.7 Lunation tables. Right (a): Zhoujiatai M30 Ershi 1 (209 BCE) wood board. Centre (b): Dunhuang T5 (1560B) Yongguang 5 (39 BCE) wood board. Left (c): Zhangjiashan M247 Gaozu 5–Gaohou 2 (202–186 BCE) bamboo bookmat.

strictly tabular than the other examples in Figure 3.7, each slip running through the new-moon dates of a single year without space or punctuation, the unevenly sized graphs creating columns of different lengths, two of which run onto the verso. This table was clearly made for reference, rather than note keeping. More specifically, it is made for *past* reference, as reign years cannot be counted into the future. However cramped the surface, the copyist does squeeze in a note at the end of two different years:

[Gaozu 5 (201 BCE):] Month IX, newly surrendered to the Han (s. 2).
[Hui 1 (195 BCE):] Month VI, illness, relieved [of office] (s. 10).

The presence in the same tomb of a dove-head staff, an honour awarded in one's seventies, would seem to comport with the occupant's retirement.[18]

Some lunation tables tell the reader a little more about what happens from one new moon to the next. The Dunhuang T5 (1560B) Yongguang 5 (39 BCE) board, for example, appends the ordinal and sexagenary dates of holidays and principal *qi* (Figure 3.7).[19] Taking this one step further is the 'Taiping zhenjun shiyi nian liri' (450 CE) and 'Taiping zhenjun shier nian liri' (451 CE) paper lunation table from Dunhuang.[20] In addition to holidays and *qi*, this Tuoba Wei (386–535 CE) lunation table notes hemerologies like Taisui, Taiyin and Dajiangjun, as well as a change of reign periods – 'Taiping zhenjun 12 *liri*, this year was changed to Zhengping 1'. It also marks the date of lunar eclipses predicted for 12-II-16 (2 April 451) and 12-VII-16 (27 September 451), making it the only extant calendar of this period to do so.[21]

3.1.3 Sexagenary Rounds

By far the rarest type of calendar recovered from the early imperial period is the sexagenary round, which arranges the new moons around a single sexagenary cycle like letters on a decoder ring. As of 2017, we possess four manuscripts that function on this principle. Within these four exemplars, furthermore, we count two distinct layouts: sixty-slip bookmats, where the sexagenary cycle is written linearly across the upper margin, and unique boards, where it is written in a circle around its edges.

The earliest sexagenary rounds are those of the linear variety found in the Zhoujiatai M30 Shihuang 36–7 (211–210 BCE) 'Saliu nian ri' and the untitled

[18] On the dove-head staff, see Loewe (1965). [19] Source: Bo Shuren (1993, 227, 238).
[20] The manuscript in question found its way into the hands of a private Japanese collector, who, in 1997, anonymously donated it to the Dunhuang Academy, where it is held under the accession number 0368V; see Deng Wenkuan (2002, 188). The transcription is available in *Dunhuang tianwen lifa wenxian jijiao*, 101–10), but, as of 2017, no photograph of the manuscript has yet been published.
[21] See Deng Wenkuan (2002, 189–200) and Martzloff (2009, 267–79).

Kongjiapo M8 Jingdi Houyuan 2 (142 BCE) calendar.[22] The two manuscripts feature identical structures, so let us focus on the former in Figure 3.8. With the

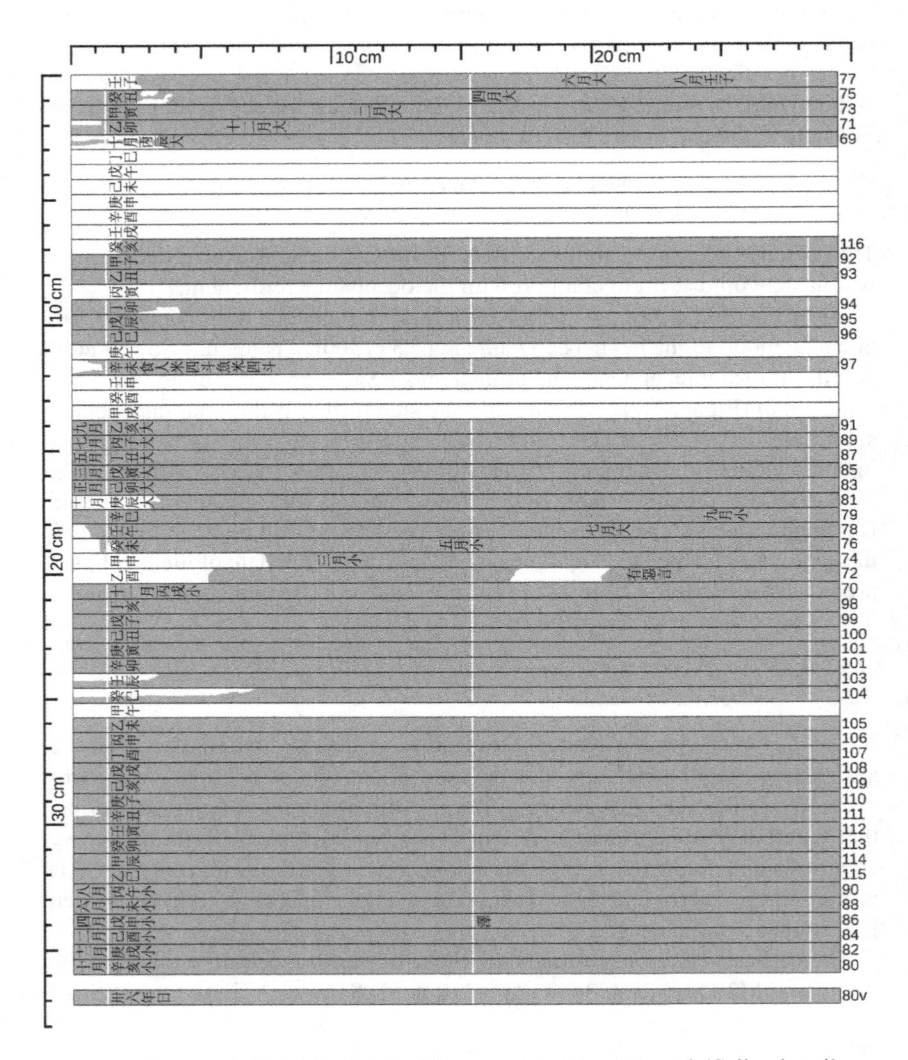

Figure 3.8 Zhoujiatai M30 Shihuang 36–7 (211–210 BCE) 'Saliu nian ri' bamboo bookmat.

[22] Source: *Guanju Qin-Han mu jiandu*, 18–24, 99–102; *Suizhou Kongjiapo Han mu jiandu*, 117–22, 191–4. Note that I follow the arrangement of the Zhoujiatai slips in Liu Guosheng (2009).

sexagenary cycle running across the top of the bookmat, the months are placed in separate registers beneath the sexagenary dates of their respective inception, running up and to the left like stairs. This layout leaves the writing surface comparatively free, though neither exemplar adds much by way of annotation. The Zhoujiatai round, for example, records a total of three entries:

Ate four *dou* (eight litres) of others' refined grain, four *dou* of fish refined grain (reg. 2, s. 97).
Marsh (reg. 5, s. 88).
There were vicious remarks (reg. 6, s. 72).

The advantage of this layout, it seems, was its expandability, seeing that the Zhoujiatai manuscript owner adds the months of the following year (Shihuang 37, or 210 BCE) to the upper margin. The *disadvantage* of this layout, however, is also its expandability: having nested two years in one sixty-day cycle, it is unclear, for example, if its owner ate all that grain on day *xinwei*$_{.08}$ of 36-X or any one of the even-numbered months of the following year.

Hardly suited for official business, the archaeological context of the Zhoujiatai 'Saliu nian ri' points to private use. Not only were its slips buried in a private tomb, but also Xia De'an (2007) argues from the fact that they were rolled within the slips of the 'daybook' from the same tomb that two were probably once bound into a single bookmat for use together (portions of the 'daybook' likewise reference the year Shihuang 36). Given that the Kongjiapo M8 sexagenary round was also interred with the owner of a 'daybook', one can assume that it was intended for hemerological use. As to what we call these documents, the title of the Zhoujiatai manuscript is sufficiently vague – 'Year 36 Days' – and, without more examples, it is difficult to surmise how common such an appellation may have been among private calendars of this type.

The other type of sexagenary round we find in the Yinwan M6 Yuanyan 1 (12 BCE) 'Yuanyan yuan nian' and the Edsen-gol Wufeng 3 (55 BCE) boards (Figure 3.9).[23] Winding the sexagenary cycle around the edges and confining the months to two ends, this layout uses about one-tenth the surface area of the bookmat variety – the 'Yuanyan yuan nian' comes in at 149.5 cm^2, for example, versus the 1060.2 cm^2 of the 'Saliu nian ri'. This leaves considerably less space for annotation, seeing especially as annotation must run towards the centre, one line potentially blocking others radiating from different directions. Like some of the lunation tables we saw in Section 3.1.2 above, the rectos of the Yinwan and Edsen-gol sexagenary rounds mark only the typical selection of *qi* and holidays, while the versos are left blank, supplying the user with

[23] Source: *Yinwan Han mu jiandu*, 21, 127; and *Zhongguo gudai tianwen wenwu tuji*, 38, Plate 36.

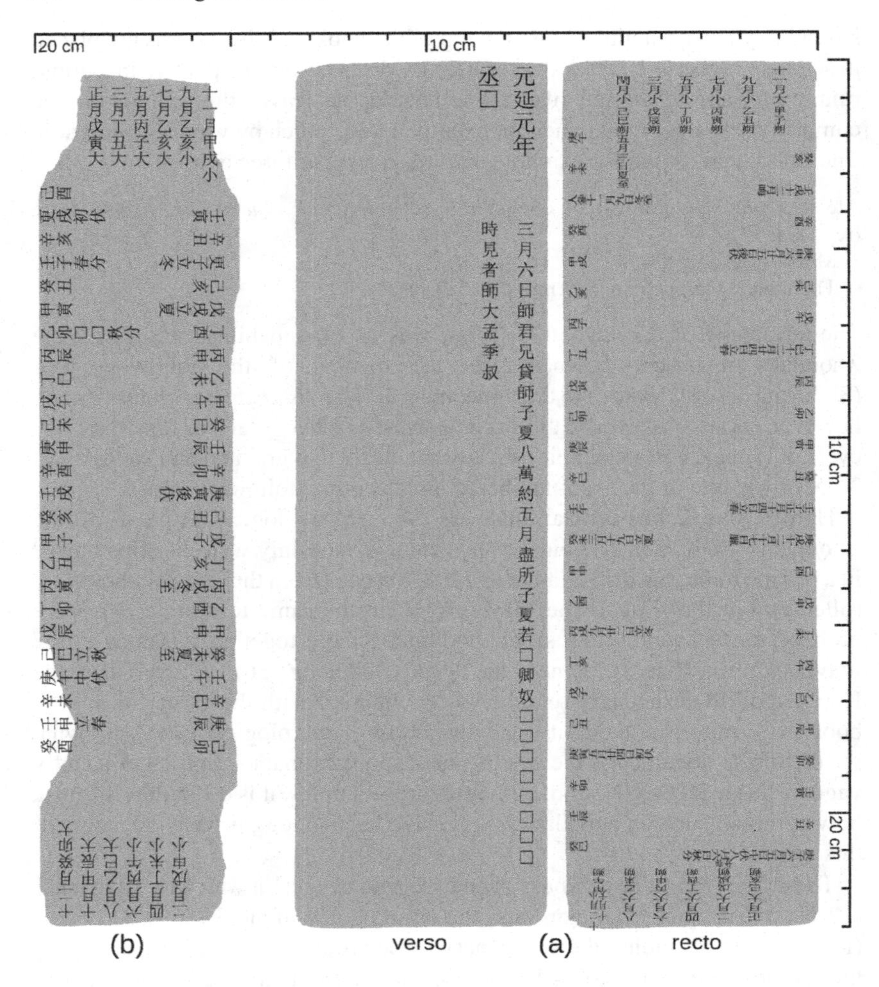

Figure 3.9 Sexagenary rounds. Right (a): Yinwan M6 Yuanyan 1 (12 BCE) 'Yuanyan yuan nian' wood board. Left (b): Jinguan Wufeng 3 (55 BCE) wood board.

a writing surface to do with what he likes, which, in the case of the former, the tomb occupant Shi Rao used to record a contract:

Yuanyan 1-III-6, Shi Junxiong (i.e. Shi Rao) borrows 80,000 from Shi Zixia in agreement that by the end of month V, if Zixia . . . chamberlain (?) slaves . . . assistant . . . those who witness [this contract] at the time [of its swearing] are Shi Dameng and Ji Zishu.[24]

[24] See Cai Wanjin (2006).

3.2 The Ritual and Administrative Apparatus

The sort of time control that the classics ascribe to the sage kings is not a matter of calendars alone, for the calendar is but one cog in a larger machine. Administratively, the Clerk's Office was a subsidiary of the Ministry of Rites from the (imagined) time of the *Offices of Zhou* to the late sixth century CE. The *Book of Later Han* 'Hundred Offices Monograph' describes the function of the parent ministry thus:

> [The minister] controls ritual ceremonies and sacrificial rites. [He] petitions on the [correct] ceremonial in advance of every sacrifice, and when it comes to the performance of [ritual] affairs, [he offers] constant assistance to the son of heaven. [He] petitions on the acceptability of erudites at every selection exam, and [he] petitions on the [correct] ceremonial in all cases of big shoots, elderly care and great mourning. [He also] makes an inspection trip of the necropoles and temples on the [last day] of every month (*HHS zhi* 25, 3571).

The role of *li* was to provide a metronome, in place of heaven, by which to choreograph the ceremonies of state, and it is thus in *ritual*, properly timed, that lies the miraculous engine of utopia. The classics say that this is what the sages did, and they say that it must be done again, criticising those who failed and giving us a glimpse of what lies in store:

> The practice of the great *dao* [created] a world-under-heaven for the public [good]. [They] selected the worthy and engaged the capable, [they] taught trust and cultivated harmony, and thus did man not treat his parents alone as parents, nor his sons alone as sons, [but] made it so that the old saw fulfilment, the strong saw employment, the young saw nurture, and [moreover] pitied the widowed, orphaned, lonely, disabled and ill [such that] all of them saw the care [they needed]. The men [all] had a [professional] part, the women [all] had a [home to] return [to], and [they] hated to throw commodities away on the ground, but neither did they [feel the] need to collect them for personal [gain], and [they] hated not to exert their own strength [in labour], but neither did they [feel the] need to [labour] for personal [gain], and this is why scheming and secrecy never arose, and [why] robbery, theft, rebellion and banditry never came up, and so were [their] outside doors [never ever] shut – [we] call this the 'Great Uniformity' (*Liji zhushu*, 21.413a).

The classics hint at how this was accomplished under the Western Zhou. Foremost among the early imperial classicist's sources in this regard was the *Spring and Autumn Annals* – an annals of the ducal court of Lu (722–481 BCE) supposedly redacted by Confucius – and the three 'traditions' (*zhuan*) devoted to expounding its 'subtle words' (*wei yan*). This, specifically, is what the *Annals* has to say:

> Wen 6 (winter, 621 BCE): Intercalary month, no announcement of the month (*gao yue*), [duke] still holding court at temple.

Wen 15-VI-1, *xinchou*.$_{38}$ (28 April 612 BCE): Sun eclipsed, drumming and animal sacrifice at altar to the soil.

Wen 16-V (summer, 611 BCE): Four times the duke did not sight the new moon (*shi yue*).

The *Zuo Tradition* explains that 'upon sighting the new moon, the duke ascended the observation terrace to look out upon it and write it down – this is the [correct] ritual' (*Zuo zhuan*, Xi 5). The scene is reminiscent of the Islamic calendar except that *shuo* ('new moon') is counted from *heshuo* ('syzygy'), when there is nothing to see. To this end, the *Guliang Tradition* explains that the duke was working here from an announcement written in advance: 'The son of heaven announces the new moon(s) to the marquises, and the marquises receive them at the temples to their fathers – this is the [correct] ritual' (*Guliang zhuan*, Wen 16). The passivity of the performer's role in this is apparent from the parallel royal ceremony. According to the *Record of Rites*, the king 'wears his dark cap-robes . . . to hear the new moon outside the southern gate, where, in the case of an intercalary month, [he] has the left leaf of the gate closed and stands at its centre . . . every new moon it is a big pen [that is sacrificed]' (*Liji zhushu*, 29.543a).

Co-ordinated with the royal 'hearing', the ducal ceremony likewise makes special provisions for intercalary months:

> 'Intercalary month, no announcement of the month, [duke] still holding court at temple'. Q: What does it mean to 'not announce the month'? A: It means not announcing the new moon. Q: Why not announce the new moon? A: Because there is no such month in heaven (*Gongyang zhuan*, Wen 6).

As concerns the duke, the *Gongyang Tradition* equates the announcement of the 'month' (*gao yue*) with the 'new moon' (*gao shuo*), and it would seem to have the ducal 'announcement' and 'sighting' coincide. *Analects* III.17 attests to the presence of a 'sacrificial sheep of the announcement of the new moon' on this occasion – a scaled-down version of the king's 'big pen' (an ox, a sheep and a pig).

Co-ordination is key, and the task of co-ordinating royal and regional courts with the moon falls to the Great Clerk (*dashi*), whose duty the *Offices of Zhou* describes thus:

> The Great Clerk . . . rectifies the agricultural and civil years (*sui nian*) so as to put (ritual) affairs in order, disseminating it to offices and storehouses from the metropolis [to its] peripheries, and [also] disseminating the announcement of new moon(s) to [vassal] states and domains (*Zhouli zhushu*, 26.401b).

For this to work out, one would need to allow time for delivery. The classics are silent on this point, but He Xiu's (129–82 CE) commentary to the previous catechism from the *Gongyang Tradition* reads between the lines: 'The [correct] ritual is that the marquises receive the new moons and policies *of the twelve months* from the son of heaven and store them in [their] great ancestral temples' (*Gongyang zhuan*, 13.168a–b (comm.)). In other words, they received something like a lunation table at the beginning of every year.

One of the reasons for the aforementioned ceremony was preparedness. New moon is the one day when one might expect a solar eclipse, and it was important that everyone act together in the face of this challenge. According to the *Zuo Tradition*,

> When there is an eclipse of the sun, the son of heaven does not offer sacrifice, but has drums beaten at the altar to the soil, while the marquises sacrifice offering cloth at [their] altars to the soil and have drums beaten at [their] courts. [They do this] to illustriously serve the spirits, train the commoners to serve their lords and show that there are distinct ranks of authority – this is the *dao* of antiquity (*Zuo zhuan*, Wen 15).

Our interest here is not so much the history of scholarship per se, nor what archaeology has revealed of real Zhou ceremonial, but how early imperial courts went about 'granting the seasons' on the ground. Real-world institutions like the Clerk's Office were obviously conceived in response to contemporary exigencies. The Shi Zhao court (319–51 CE) at Chang'an, for example, 'established a female Grand Clerk at the Numinous Terrace ... to examine the veracity of the external Grand Clerk' (*JS* 106.2765), and by the Sui, the Clerk's Office had been moved to the Palace Library – closer to the records with which they worked, to the historiographical projects with which they were involved, and to a throne apprehensive of their independence (Deane 1989, 138–9). The most important exigency, however, was legitimacy, and *that* lay in the sort of cultural patterns (*wen*) seen here. The question is what sort of balance was achieved between practical and antiquarian considerations. The written record is not always forthcoming as concerns the former, so I will attempt to fill in the gaps here as best I can with excavated sources and by working backwards from better-documented periods.

3.2.1 Calendar Distribution

In fitting with the *Offices of Zhou*, the *Tang liu dian* compendium of state regulations, compiled in 738 CE, defines the Bureau of the Grand Clerk's duty as 'each year to create in advance a *li* of the upcoming year and promulgate it through the subcelestial realm' (*Tang liu dian*, 10.13a). Between the two, the 'Hundred Officials' genre is silent on 'promulgation', but it is safe to assume that earlier Clerk's Offices shared this basic directive.

How was one office to 'promulgate' a time-sensitive document throughout the known world? The Yuan (1271–1368) astronomical office housed its own printer and printing staff for this purpose (Sivin 2009, 169). It is unclear what sort of text-production facilities earlier Clerk's Offices may have enjoyed, but we do know that printing caught on very early. Mentioned in Feng Su's 'Memorial on the Banning of the Woodblock Printing of Almanacs' (Jin banyin shixianshu zou) of 835 CE, the first printing ordinance in human history was indeed aimed at excesses in the Chinese calendar industry:

> We have by imperial decree prohibited the woodblock printing of calendars. Throughout the markets of Jiannan's Two Rivers [region] (Sichuan) and Huainan Circuit there are woodblock-print *liri* for sale. Every year, before the [Clerk's Office] has even submitted the new *li* for approval and promulgated it [around], these printed *li* have already flooded the subcelestial realm. This is a perversion of the *dao* of 'reverently granting [the seasons]' (*Quan Tang wen*, 624.6301a).

According to the *Tang yulin*, the problem seems to have persisted for decades, especially in the provinces and in times of political turmoil:

> When Xizong (r. 862–88 CE) entered Shu (modern Sichuan) (amidst the Huang Chao Rebellion of 881 CE), the Grand Clerk's edition *li* was not reaching River East (south of the Yangtze). There were prints for sale at market, but they invariably confused the last and first days of the month. Each merchant maintained the veracity of his [solar cycle] and, thus, they wrangled with one another. They were detained by the men of the quarter and delivered to the authorities. The functionary in charge responded: 'You're fighting over [nothing more than the arrangement of] the big and small months? If concurrent orders are [off by] a day or a half a day, then it is really no big deal'. He then shouted [in exasperation] and walked out, ignorant of the fact that the *yin-yang li*, from which auspicious and inauspicious [days] are chosen, had introduced numerous errors to the masses (*Tang yulin jiaozheng*, 7.671).[25]

The introduction of printing presented the court with a powerful tool for standardising the production, branding and consumption of civil time, but its potential was not realised overnight. Early calendar printing continued to operate on a cultural model based upon – and still fully intertwined with – manuscript production.[26] The *Jixian zhuji*, for example, offers this to say about early Tang (619–907 CE) distribution:

> Ever since the establishment of the [Jixian] Academy (in 725 CE), the academy has been ordered to make 120 copies of the new *li* to disseminate to the kings and princesses of royal blood as well as the Grand Councillor, Excellencies, Ministers, etc. within month XI of every year. All of these are ordered to be distributed in red and black ink with annotations concerning the sequence of stars (*li xing*) so that they may be passed around for copying. [The master copies] are referred to as the Jixian Academy edition (cited in *Yuhai*, 55.43b).

[25] Tr. modified from Arrault (2003, 94–5). For more on censorship and calendar-printing, see Whitfield (1998).

[26] On this point, I take particular inspiration from the study of early European print culture in Johns (1998).

The Tiansheng Decrees of 976 CE confirm that this model persisted well into the printing age:

> Every year, the [Clerk's Office] is to create in advance a *liri* for the upcoming year, giving one copy each to the three capitals and various prefectures. [They] are to be dispatched sequentially, based on the measured distance of the journey required. The Privy Council disseminates [them] and are furthermore ordered [to do so such that they] arrive before the beginning of the civil year (*nian*).[27]

Given the physical numbers, weight and distances involved, it is reasonable to assume that the 'pass-around-for-copying' model goes back much earlier than 725 CE, by which time paper had displaced bamboo and wood as the writing support of choice. This model put production in the hands of the *individual* – be it the end user or the professional copyist – the results, however, were far from *individualistic*. In Section 3.1 above, we noted a 'great uniformity' over time and space as concerns layout, which evidences the acceptance, if not enforcement, of national standards.

One thing that *does* change around the mid- to late Tang is the language of product delivery. Where eighth- and ninth-century sources speak in terms of 'promulgating the *li*' (*ban li*), earlier sources persist in the classical idiom of new moons.[28] This is still very much the norm in the *Book of Sui* (656 CE), for example:

> At the beginning, the [Xiao] Liang followed the [Xiao] Qi (479–502 CE) in using the [Liu] Song Epochal Excellence *li* . . . In [510 CE] the *Jiazi*.$_{01}$-origin *li* constructed by Zu Chongzhi (429–500 CE) was used to promulgate the new moons (*Sui shu*, 17.416).

'The new moons' may well be a synecdoche for 'calendar'. The coincidence of this shift with printing and the 'calendar'/'almanac' divide in archaeology, however, suggests there is more to it than that. I suspect that we should read 'the promulgation of the new moons' as exactly that – the distribution of lunation tables. Of the official document types we know to have existed, the lunation table certainly fits the description. Logistically, it would make sense to deliver 'the Calendar' in its lightest, most compact form, knowing that it could be expanded into the necessary *zhiri* and *liri* at the point of delivery.

How well did this work? One would expect the worst disruptions to occur at border outposts, and scholars have been quick to point out where this is the case. Loewe (1959, 316–19) lists nine documents at Edsen-gol that are dated to obsolete reign periods, due, potentially, to 'the delay in the promulgation and

[27] *Tianyige cang Ming chaoben Tiansheng ling jiaozheng (fu Tang ling fuyuan yanjiu)*, 734. For more on these edicts and the 'pass-around-for-copying' transmission model, see Chen Hao (2007).

[28] The *locus classicus* of *ban li* appears to be in 597 CE, in *Sui shu*, 2.41.

receipt of the new order'.[29] Chen Hao (2007, 218) points to an even clearer example from a document at Turpan from 754 CE:

> Because the *liri* for month II had not arrived by month I, [we took the liberty of] setting it on a [stem–]branch date befitting a small month. Later, when the calendar arrived, [we] discovered that it was a double big month, and so [we] counted two days' feed. On II-13 of this year, the *die* invoice was dispatched to Director of the Granaries Section Chong He to purchase the full amount of grain (*Tulufan chutu wenshu*, vol. 4, 487).

Seeing that an imperial edict could take two to four months to arrive at these places, it is not surprising to think that the calendar might show up late.[30] What *is* surprising is how quickly the vast majority of documents at these sites reflect the vicissitudes of capital calendrics. In months VII–IX of 8 CE, for example, Wang Mang (*c*.45 BCE–23 CE) changed Jushe 3 to Chushi 1, then, only a matter of weeks later, he changed Chushi 1-XII to Shijianguo 1-I.[31] At the Shule river, we find a document dated 'Jushe 3-XI' (292), unaware of what was going on, but at Edsen-gol, amazingly, we see documents make the *first transition* between 'Jushe 3-IX-*bingchen*.$_{53}$' (T59:101) and 'Chushi 1-XI-*renzi*.$_{49}$' (312.6), and *the second* by at least 'Shijianguo 1-V' (T17:3).[32]

All in all, the situation at north-western border outposts such as these speaks pretty highly for the early empire's control over civil time. The throne had to rely on 'Chinese whispers', manuscript calendars passing from one man's copy to the next, transforming to meet their needs, and this diffusion opened the door to human error. Copyists made mistakes, postmen got lost and fraudsters peddled their own calendars to turn a quick buck, but the fact that the same document from Lianyungang could be mistaken for one from Edsen-gol suggests that 'Chinese whispers' did the job.

3.2.2 State Ritual

The logistics of maintaining distribution networks and scribal norms from Pyongyang to Kashgar were, it turns out, the easy part. The hard part was getting a handful of capital intellectuals to agree on the proper ceremony to cap it off. *This*, the real science of 'granting the seasons', experienced centuries of deadlock broken by paroxysms of questionable action. The standard histories, for example, record nothing of 'hearing', 'sighting' or 'announcement' as having been practised at imperial court, the discourse around these ceremonies being limited to exegesis and intellectual debate. There is, perhaps, no better

[29] See also Yu Zhongxin (1994) and Arrault (2003, 93).
[30] See Hsing I-t'ien (1998, 25–7) and Sanft (2008–9, 137n16).
[31] See *HS* 99.4095, 99.4113. This episode is treated in Section 1.2.1.
[32] See Yu Zhongxin (1994, 145–6). Loewe (1959, 316) reminds us that we do not always know the exact month when new reign names took place, nor can we eliminate the possibility of scribal error or the advanced preparation of documents as concerns the use of obsolete reign names.

illustration of this disconnect than Director of Rites, Erudite Pilü Renxu's response to Wu Zetian's (r. 690–705 CE) edict of 698 CE ordering that 'the announcement of the new moon ceremony be held at the Bright Hall on the first of every month':

Careful examination of the *wen* of the classics and histories reveals that there is no such thing as the son of heaven's monthly announcement of the new moon. The only thing is the *Record of Rites*, 'Yu zao', which says that the son of heaven 'hears the new moon outside the southern gate'. The *Rites of Zhou*, 'Celestial Offices', Great Steward [entry reads]: 'On the first day of the first month, policies are disseminated to the states, cities and settlements'. Gan Bao's (d. 336 CE) commentary [explains]: 'Zhou month I . . . is the day of the announcement of the new moon(s)'. *This* is the 'hearing the new moon(s)' in the 'Yu zao'.

Nowadays, on New Year's Day of every year, court is held, the seasonal ordinances (*shi ling*) are read, and policies are disseminated at the Tongtian Palace, while capital officials of grade nine and above, territorial representatives from the prefectures, etc. all line up in tiers in the courtyard – it is at this that the ceremony of hearing the new moon(s) is finished, and this accords with the *wen* of the *Rites of Zhou* and 'Yu zao'. Zheng Xuan's (127–200 CE) commentary to the 'Yu zao', however, posits the Qin-instituted monthly ordinances with sacrifices to the Five Thearchs and Five Officials and then says: 'The hearing of the new moon necessitates the sacrifice of a single ox, so that one may announce oneself to the thearch and spirit of that season as well as pair oneself with King Wen (r. 1099–1050 BCE) and King Wu (r. 1049–1043 BCE)'. This is where the Zheng commentary is wrong (*wu*), and that is why, from the Han and [Cao] Wei (220–65 CE) all the way to today, no one has ever done this (*JTS* 22.868–9).

The disconnect here is double. On the one hand, there is a divide between the world of scholarship and state policy, where there is 'no such thing' and 'no one has ever done this'. On the other hand, the basic function of the temple-based 'hearing', 'sighting' and 'announcement' ceremony, as seen in the *Spring and Autumn Annals*, has been supplanted by the Bright Hall (Mingtang) and 'monthly ordinances' (*yue ling*).

The Bright Hall only appears in classics antedating Confucius' supposed redaction of the *Documents, Odes, Changes* and *Spring and Autumn Annals*. It may have slipped the master's mind, but the Han-era (206 BCE–220 CE) *Record of Rites* and Dai the Elder's first-century recension thereof are clear that *this* was the cornerstone of Zhou ceremonial, devoting a total of four chapters to the subject. They do not agree on *what it was*, exactly, Dai the Elder offering us two distinct possibilities:

The Bright Hall once existed in antiquity . . . The Bright Hall was that by which the nobility and baseness (hierarchy) of the marquises were clarified . . .. The Bright Hall [is where] monthly ordinances [are promulgated] . . . The top is round above, and the bottom is square. [It has] nine chambers and twelve halls, each chamber [with] four doors, and each door [with] two windows. The palace is a square three hundred *bu* to a side. [It] is located in the nearby suburbs, within thirty *li* [of the capital]. Some believe that the Bright Hall was the ancestral temple to King Wen ... This was the son of heaven's road chamber, no one [of] unequal [status being allowed to] dwell under its roof. Audience was awaited in the southern palace, and audience was sent off, exiting the southern gate (*Da Dai Liji*, 8.6a–11a).

It was there, in the case of the former, that the Zhou king supposedly made a monthly progression through its rooms, changing numbers, colours, ordinances and prohibitions as he moved from one space to another. Reciting precise sidereal, climatological, natural, agricultural and ritual events for which his subjects should prepare, as per the 'Monthly Ordinances' chapter of the *Record of Rites*, the son of heaven guaranteed continued harmony between heaven, earth and man.

The whole thing was delightfully modern for an ancient rite. Redolent with five-agents metaphysics, the ceremony gave concrete religious form to the political philosophy of legitimate conquest/succession that had emerged all the rage in the last century prior to unification – an ideological complex pioneered by the hated Qin Shihuang (r. 221–210 BCE). As he cycled through positions on a schematic representation of the cosmos issuing prohibitions, so too did the king act out the role of hemerological spirit, hemerology being likewise all the rage. Whatever its origins in antiquity, it was the Qin (221–206 BCE), as Pilü Renxu notes, that got the ball rolling in imperial times, and when exegetes like Zheng Xuan and Kong Yingda (574–648 CE) looked back at the *Spring and Autumn Annals*, it is increasingly the Bright Hall that they saw there.[33]

When it came time to *build* a Bright Hall, things could get pretty dark. In 140 BCE, Emperor Wu (r. 141–87 BCE) set about planning one for the southern suburbs, withdrawing his support later amid allegations of corruption and resistance from the empress's clique. He revisited the idea in 110 BCE, this time deciding that it should go at Mount Tai.[34] And so the first historical Bright Hall was built 800 km east of Chang'an as a support structure for the Qin *feng-shan* complex, where, like Qin Shihuang before him, Emperor Wu would make quinquennial treks to offer sacrifice to heaven and earth. The site was abandoned after his death, seeing desultory expeditions in 56, 85 and 124 CE. In 4 CE, Wang Mang oversaw the construction of a second Bright Hall in the southern suburbs. Debated in a fatalist millenarian atmosphere, commissioned by the regent and completed a year before the child emperor was poisoned and replaced, it was built to herald a new beginning. And so, as 'regenting emperor', Wang Mang held New Year's Audience there in 6 CE as per Zhou

[33] For Zheng Xuan's equation of the 'sighting', 'hearing' and 'announcement' ceremonies with the Bright Hall, see *Liji zhushu*, 29.543a–545a (comm.).

[34] For the *li* reform coinciding with Emperor Wu's newfound interest in 'granting the seasons', see Section 1.2.1 above.

ceremonial, making him – the regent – the first to perform this royal rite. As emperor, in 9 CE, Wang Mang then repurposed the complex as a temple to Shun, because *Han* : *Yao*, and *Yao* → *Shun* (see Section 1.1.2 above). The temple was subsequently destroyed with the sack of Chang'an in 23 CE. It is only really in 59 CE, upon completion of a *third* version for the new capital at Luoyang, that the Bright Hall began to see the use for which it was intended.[35]

Aside from the New Year's audience (Wang Mang) and sacrifices to the Five Thearchs (Qin Shihuang), the one other ceremony we know to have been performed at the Bright Hall was 'the reading of seasonal ordinances' (*du shi ling*). One gathers from historical monographs and compendia that the latter was regularly held since at least the time of the 56 CE Bright Hall:

> According to the stipulations of the Later Han, the Grand Clerk memorialises the annual *li* every year, [after which] it is customary to read the ordinances of the five seasons just prior to establishment of spring.$_{Q01}$, establishment of summer.$_{Q07}$, greater heat.$_{Q12}$, establishment of autumn.$_{Q13}$ and establishment of winter.$_{Q19}$. The [vestments] worn by the emperor [on these occasions] should be of a colour appropriate to each of the five seasons. When the emperor ascends to his throne, the prefect of the Masters of Writing descends to take his mat (seat). The Masters of Writing, the Three Excellencies and the gentlemen of the palace then place the ordinances upon a desk, which is carried in and presented with both hands. They then proceed to their mats, where they genuflect until the reading is finished, at which point they are bestowed a goblet of wine (*Tongdian*, 70.1922).

The ordinances themselves do not go back much further than this. According to Hsing I-t'ien (1998, 51), 'the monthly ordinances implemented in the Han dynasty were not in strict accordance with any one Ruist classic or system, they were instead the [product] of Han experts formed from an admixture of exigency and "ancestral precedent" through a process of eclectic cobbling and ceaseless revision and adjustment'. We see annual moratoriums on forestry, fishing, hunting and construction, for example, in early legal manuscripts, but it is not until the first century BCE, Hsing I-t'ien argues, that the court began to codify these practices and speak of them under a single theoretical rubric.[36]

When the court *does* begin to talk about the seasonal/monthly ordinances, the central theme is that things are bad, that the laziness and ignorance of imperial subjects are to blame, and that the solution is that the throne micromanage their daily activities based on a single correct timetable taken from

[35] On the history of the Bright Hall, see Maspero (1951), Soothill (1951), Forte (1988), Lewis (2005, 265 ff.), Tseng (2011, 17–88) and Pankenier (2013, 342–50).

[36] For examples of such ordinances as seen in early legal manuscripts, see Sanft (2008–9, 171–2) and Hulsewé (1985, esp. 21–6).

ancient literature. One of the first edicts on the subject was that issued in the spring of 23 BCE:

> In the past, when Thearch Yao established the offices of Xi and He, [he] commanded [them] to adhere to the matters of the four seasons such that [they] not get out of order. Thus it was, the *Book of Documents* says, that 'the masses were there and then prosperous and concordant' (2.20b) – a clear indication of the foundational importance of *yin* and *yang*. Today, some Excellencies, Ministers and Grandees do not believe in *yin* and *yang*, dissembling and belittling it, and many of the petitions that they have memorialised violate seasonal policies. Passed around out of ignorance, they circulate the subcelestial realm, but if the desire is to harmonize *yin* and *yang*, then is this not absurd?! May the monthly ordinances of the four seasons be duteously observed (*HS* 10.312).

Wang Mang, for one, took the matter very seriously, sending in 15 CE 'eleven grandees of the first order to spread out [across the countryside] to exhort agriculture and sericulture and promulgate seasonal ordinances' (*HS* 99.4140). We see physical evidence of this campaign in the 'Edict on the Fifty Monthly Ordinances for the Four Seasons, Supervised and Inspected by Emissary Hezhong' (Shizhe Hezhong suo ducha zhaoshu sishi yueling wushi tiao) of Yuanshi 5-V-*dingchou*.$_{14}$ (14 June 4 CE), found painted on the wall of a government office in Xuanquanzhi, Dunhuang.[37] Distilled from the *Record of Rites* and *Mr Lü's Spring and Autumn Annals* (239 BCE), the edict is a list of classical building and agricultural prohibitions provided with item-by-item vernacular explanation. Seeing that the prohibitions are divorced from the sidereal, climatological and ritual events that give them meaning, the intended audience would seem to be the peasantry (Sanft 2008–9, 141). That said, the ordinances are framed at either end by a dialogue pregnant with classical allusions between Wang Mang and the Empress Dowager (his maternal aunt). The edict begins thus:

> The Empress Dowager said: 'In the past, *yin* and *yang* were not in accord, and wind and rain were not in time; indolent farmers were self-satisfied and did not diligently rise to their labours – this is why [they] endured numerous disasters and suffered grievously therefrom. It is the sagacious emperor and perspicacious king who never fails to embody the *li* numbers of heaven, faithfully upholding the balance between them, respectfully obeying *yin* and *yang*, reverently granting the people the seasons, and . . . [text damaged] ingly exhorting ploughing and seeding to ensure a bounteous harvest [text damaged] . . . all of this is out of weighty consideration of the hundred surnames' fate. Thus [have We] created [the office of] Xi-He and established [the offices of] the four masters [to fix the four] seasons to complete the year (*sui*) and deliver happiness unto [text damaged] . . . may every year they split up and travel through each commandery under their respective jurisdictions' (lines 1–4).

When Wu Zetian ordered an 'announcement' ceremony in 698 CE, it was in the context of a state ritual apparatus long since dominated by Bright Halls and

[37] See *Dunhuang Xuanquan yueling zhaotiao*, cf. Sanft (2008–9).

monthly ordinances.[38] Why is *this*, of all things, what came to the fore in the ritual sciences? Tseng (2011, 21–36) argues that it was because post-Qin dynasts were most concerned about establishing legitimacy – most of them having risen from anonymity to wrestle the throne from an ancient bloodline – and it was the Bright Hall that best met their needs. On the one hand, the monthly ordinance complex tapped into contemporary intellectual currents in metaphysics, placing it at the cutting edge of the modern sciences. Its pretensions to antiquity also lent the practice a pedigree rooted in Zhou ceremonial (Wang Mang) and/or rites going back as far as the Yellow Emperor (Emperor Wu). More importantly, classical precedence, be it as late as the *Record of Rites*, provided the perfect cover for continuing extant imperial rites pioneered by the likes of Qin Shihuang and Wang Mang. The whole point of the Bright Hall, Dai the Elder reminds us, is to 'clarify hierarchy', and that is precisely what it did, from the 'sagacious emperor' all the way down to the ignorant shepherds of Dunhuang.

In practice, what is good for the Zhou heartland, for which the monthly ordinances were composed, is not always good for the soil and climate of the whole empire. This was not a problem at first, seeing that court ceremonial languished a good two centuries behind the logistical infrastructure that it was it was meant to justify. It took charismatic leaders like Emperor Wu and Wang Mang to finally push through the appropriate facilities, which they did out of personal ambition and in contravention of every possible classical precedent. It was only around Wang Mang's regency that the ritual apparatus was finally in place to begin 'granting the seasons' for real. On the ground, the results took the form of public edicts and education campaigns, like the sort we see at Xuanquanzhi, and when these failed, it then turned to mass arrests and forced resettlement, recalcitrant peasants marched to the mountains of Xihai to rethink who knows best how to farm.[39] Like the Great Leap Forward, Wang Mang's agricultural ordinances were followed by years of unexplained drought, natural disaster and famine, pushing subjects across the empire from their homes to scavenge, steal and kill for food. Make what you will of the winner's account, but this is what Emperor Guangwu's forces report upon entering the capital in 25 CE:

> The Red Eyebrows [rebels] had torched the palaces, halls, markets and neighbourhoods of Chang'an to damage Gengshi (r. 23–5 CE). The populace was starving from famine and eating one another, the dead [numbering] several hundred thousand. Chang'an was empty, there was no one walking [the streets] within the city. The ancestral temples, parks and necropoles had all been dug up, only [the two at] Ba and Du being [left] intact (*HS* 99.B.4193).

[38] On the innovations of Wu Zetian's Bright Hall, see Forte (1988).
[39] *Dunhuang Xuanquan yueling zhaotiao*, 45.

3.3 Calendar Making

As products of '*li*-calculation', the physical documents presented in Section 3.1 above offer us a valuable perspective on how *li* was put into action. Preserved in the *li* and *lü-li* monographs of the standard histories, procedure texts give us the exact 'numbers and procedures' used to calculate the state calendar from 104 BCE on. This, in fact, is where we get date-conversion tables like Zhang Peiyu (1997, 2), which are 'calculated from the *lifa* that saw actual use through successive dynasties and are, to the extent that is possible, checked against sources both from these new [archaeological] discoveries and transmitted almanacs and calendars'. 'Checking' is crucial, as many dates in excavated and received sources conflict – some due to textual corruption, some due to ambiguities of when and how a given *li* was implemented, and some due to active human intervention.[40] That said, most extant calendar dates check out, and Martzloff (2009) demonstrates how you can use procedure texts to reproduce them exactly as they are written.

With texts on both the front and the back end of 'granting the seasons', we are in an excellent position to speak about matters of theory versus practice. The most striking discrepancy between procedure text and calendar is how little of the former actually goes into the latter. *Li* procedure texts are typically divided into three sections – civil time, eclipses and planets – each of which comprises 'numbers' (*shu*), 'procedures' (*shu*), 'sequence-tables' (*li*) and, in the case of the planets, '*du*-travelled' (*xingdu*) models for motion and visibility. To give the reader a sense of the contents and structure of a typical procedure text, I offer an itemised table of contents for the received Supernal Icon *li* in Table 3.2. Of the forty-one procedures in Table 3.2, only the first five enter into the sort of calendars we have seen thus far; the rest, you might say, were beyond the scope of 'season granting'.

Planetary astronomy, for one, has nothing to do with civil time, but the case of eclipse prediction is more ambiguous. The modelling of lunar 'slow–fast' to this end initiated a centuries-long debate about 'fixing' (*ding*) the (mean) civil lunation to the speed-corrected (true) moon of eclipse theory (Section 1.2.2 above). This position won out in the seventh century CE, since which time its advantages have appeared self-evident, and its opponents short-sighted. Chen Meidong (2003, 262), for example, dismisses the counterargument of 443 CE as 'an

[40] There are sufficient conflicts and ambiguities in civil dates to merit monograph-length treatment in studies like Yu Zhongxin (1994) and Liu Ciyuan (2015), to which I direct the reader for examples of textual corruption. As to the question of when and how effectively '*li* reform' (*gai li*) is implemented, I offer the aforementioned 'Taiping zhenjun shiyi nian liri' (450 CE) and 'Taiping zhenjun shier nian liri' (451 CE) as a case in point: *WS* 107A.2659 clearly states that 'when Shizu (r. 423–52 CE) pacified Liang soil (in 439 CE), he obtained the Epochal Beginning *li* devised by Zhao Fei, and since the latter was tighter, it was used to replace the Luminous Inception [*li*]', and yet, as Martzloff (2009, 267–79), for example, shows, the later manuscript calendar was calculated by the Luminous Inception *li*. As to human intervention, see, for example, Huang Yi-long (1992b) on how Li Chunfeng's (602–70 CE) Unicorn Virtue *li* was simplified and manipulated as concerns the civil calendar over the course of its official tenure from 665 to 728 CE.

Table 3.2 *Supernal Icon* li *procedure text table of contents*[*]

Section 1: [Mean lunisolar/calendrical elements]
[Numbers:] *Lü* resonance periods for solar year, lunation and eclipse month
Procedures: (1) 'Calculate era-entry [years]', (2) 'Calculate new moon [dates]', (3) 'Calculate winter solstice [date]', (4) 'Find the twenty-four *qi* [dates]', (5) 'Calculate the intercalary month', (6) 'Calculate quarter and full moon [dates]', (7) 'Calculate disappearances', (8) 'Calculate solar *du* position (for midnight)', (9) 'Calculate lunar *du* position (for midnight)', (10) 'Calculate syzygy *du* position', (11) 'Calculate [lunation of] lunar eclipse', (12) 'Calculate hexagram domination dates', (13) 'Calculate five-agent domination [dates]', (14) 'Calculate added hour (for lunar phases)', (15) 'Calculate the water clock notches (for lunar phases)'.
Section 2: 'Procedure(s) for the three roads of lunar motion'
Table: 'Slow–fast sequence' (pseudo-equation of centre)
[Numbers:] Elements for sidereal and anomalistic month
Procedures: (1) 'Calculate syzygy sequence entry', (2) 'Calculate fixed [date and time] of quarter and full moon', (3) 'Calculate added hour and fixed *du* position of new, quarter and full moon', (4) 'Calculate lunar motion midnight sequence entry', (5) 'Find the fixed *du* position of moon at midnight', (6) 'Method for finding [interpolation]', (7) 'Find subsequent sequence [entries]', (8) 'Find the fixed *du* position at midnight for subsequent day', (9) 'Find pseudo-equation of centre at midnight for subsequent day', (10) 'Find lunar *du* position at dusk and dawn', (11) 'Find fast–slow of lunar motion'.
Table: '*Yin-yang* sequence' (latitude)
[Numbers:] Elements for nodical month
Procedures: (1) 'Calculate *yin-yang* sequence entry for new moon', (2) 'Find subsequent month', (3) 'Find fixed numbers for new and full moon', (4) 'Calculate sequence entry at midnight', (5) 'Find fixed date at midnight', (6) 'Find dusk and dawn numbers', (7) 'Find polar *du* distance of moon'.
Section 3: 'Calculate the five stars (planets)'
[Numbers:] *Lü* resonance periods for planetary conjunction and visibility
Procedures: (1) 'Calculate month of planetary conjunction', (2) 'Calculate month-entry days', (3) 'Calculate *du* position of planetary conjunction', (4) 'Find month of subsequent conjunction', (5) 'Find syzygy date of [month of] subsequent conjunction', (6) 'Procedure to find subsequent month-entry days', (7) 'Find subsequent *du* position', (8) 'Procedure for planetary *li* pacing'.
Motion-*du* [models]: (templates for motion and visibility between conjunctions)

[*] For a full translation of the Supernal Icon *li*, which uses different translations and procedure numbers than those given here, see Cullen (2017, 235–355).

unreasonable reason', but it is a reason that stood, nonetheless, from the second century CE on. There is a tension here, centuries-long, that polarised educated men, and I would like to explore that tension with the intent to further decompose 'the Calendar' into the human elements behind its operation.

3.3.1 Computer

To get a sense of the back end of calendar making, let us consider the computer and the sort of calculations that the procedure text instruct him/her to make.

I offer here an example from the Supernal Icon *li* (*JS* 17.504–31) as computed for Xiping 3 (174 CE).

The Supernal Icon *li* opens with a list of the fundamental values with which the user will be calculating (I list only those relevant to our example):

From high origin, [year] *jichou*.$_{26}$, to Jian'an 11, year *bingxu*.$_{23}$, the year has accumulated 7,378 years.

Supernal divisor:	1,178
Era divisor:	589
Communication divisor:	43,026
Day divisor:	1,457
Rule years:	19
Rule months:	235

The one other datum required is 'the [number of years from] high origin to the year sought', which, in the case of Xiping 3, is 7,345.

From there, the Supernal Icon *li* moves to 'procedures', the first of which is 'Calculate Era entry' (find the author's worked examples in parentheses, and note that I translate as directly from the primary source as possible to convey its ambiguity):

Set out the [number of years elapsed from] high origin to the civil year sought.
(answer: 7,345 years)
Eliminate (*chu*) it by the supernal divisor.
(answer: 7,345 mod 1,178 = 278)
[That which] does not fill the supernal divisor, eliminate (*chu*) it by the era divisor. Any remainder that does not fill the era divisor is the years entered into inner era *jiazi*.$_{01}$, [while any remainder which] fills the divisor, remove (*qu*) it, [and that] is the years entered into outer era *jiawu*.$_{31}$.
(answer: 278 < 589, thus 'inner era jiazi.$_{01}$*')*

With that answer in hand, we then proceed to 'Calculate New Moons':

Set out the era-entry years, excluding that sought.
(answer: 278 − 1 = 277)
Mount (*cheng*) it by the rule months, the rule years then one (*er yi*) – what is obtained makes the fixed accumulated months, and what is not exhausted makes the intercalary remainder.
(answer: $277 \times \frac{235}{19} = 3{,}426\frac{1}{19}$ *or, in big–small notation, 3426;1)*
As to the intercalary remainder, if 12 or above, the months of the year [experience] intercalation.
(answer: 1 < 12, thus no intercalary month this year).
Mount (*cheng*) the fixed accumulated months by the communication divisor for the temporary accumulated days.
(answer: 3,426 × 43,026 = 147,407,076)

Table 3.3. *Lunation table calculated for Xiping 3 (174* CE*)*

	Supernal Icon *li*		Quarter-remainder *li*	
Month	New moon	Size	New moon	Size
XI	*yihai*.$_{12}$	big	*yihai*.$_{12}$	big
XII	*yisi*.$_{42}$	small	*yisi*.$_{42}$	big
I	*jiaxu*.$_{11}$	big	*yihai*.$_{12}$	small
II	*jiachen*.$_{41}$	small	*jiachen*.$_{41}$	big
III	*guiyou*.$_{10}$	big	*jiaxu*.$_{11}$	small
IV	*guimao*.$_{40}$	small	*guimao*.$_{40}$	big
V	*renshen*.$_{09}$	big	*guiyou*.$_{10}$	small
VI	*renyin*.$_{39}$	small	*renyin*.$_{39}$	big
VII	*xinwei*.$_{08}$	big	*renshen*.$_{09}$	small
VIII	*xinchou*.$_{38}$	small	*xinchou*.$_{38}$	big
IX	*gengwu*.$_{07}$	big	*xinwei*.$_{08}$	small
X	*gengzi*.$_{37}$	big	*gengzi*.$_{37}$	big

Fill (*man*) the day divisor for the fixed accumulated days, and [that which] is not exhausted makes for the small remainder.

(answer: 147,407,076 ÷ 1,457 = $101{,}171\frac{929}{1457}$ *or 101,171;929)*

Remove (*qu*) the accumulated days by 60 for the big remainder.

(answer: 101,171 mod 60 = 11)

Name off [the sexagenary day] according to the era entered, counting exclusively, and [that] is the new moon day of heaven I, [civil] month XI, for the year sought.

(answer: jiazi.$_{01}$ → *yichou.*$_{02}$ → *bingyin.*$_{03}$ *... jiaxu.*$_{11}$ → *yihai.*$_{12}$*)*

To find the subsequent month(s), add a big remainder of 29 and small remainder of 773.

(answer: 11;929 + 29;773 = 40;1702)

[If] the small remainder fills the day divisor, [then it] follows the big remainder.

(answer: 1,702 > 1,457, thus carry the 0;1457 for 41;245)

[If] the small remainder is 684 or above, [then] the month is big.

(answer: 245 < 648, thus the month is small)

In 167 graphs, these two procedures give you the lunation table above in Table 3.3, and that, for the purposes of calendar making, is all you need.

Who is reading this text? It is the current fashion in early China studies to assume the minimal level of literacy and comprehension for any and all involved in the textual and material traces to have come down to us as a corrective for the twentieth-century academic's particular experience with literacy.[41] It is in this vein that Sivin (2009, 21) labels the genre 'step-by-step instructions, worked out

[41] For a sample of recent scholarship attempting to nuance our current understanding of (il)literacy and *(il)literacies* in early China, see Li & Branner (2011), Meyer (2012) and Richter (2013).

so that a minor functionary with limited mathematical skills could calculate the annual ephemeris', thus returning us by way of the illiteracy and incomprehension of the Chinese subject to the 'practicality' of his scientific writing. Sivin offers no evidence in support of these characterisations, but their axiomatic status in Chinese studies is such that one would not normally think to demand it. The problem in this case is that we know, administratively, exactly who these hypothetical 'minor functionaries' were: they were called '*li*-workers' (*zhi li*), and they were 'expectant appointees' (*daizhao*) specially recruited for their mathematical skills (see Section 1.3 above). Furthermore, once you set aside the symbolic algebra into which the procedure text genre is so often (and so loosely) translated, its *actual* contents and language give you a pretty good idea as to why, particularly in the light of comparison.

Li procedure texts present us with a genre of mathematics a world apart from the sort of *suan* 'calculation' literature one finds in early tombs and later classics. *Li* literature uses different numbers, switching between cardinal and ordinal numbers and operating on compound fractions as 'big and small remainder' integer pairs. *Li* literature uses unique, cyclic measuring units (without explanation), like *du*, 'notches' and 'added hours'. *Li* literature features operations unique to that context, like *modulo* (*chu* 除, *qu* 去 or *chuqu* 除去) and series subtraction (*ci-chu* 次除, *ci-qu* 次去, *ci-chuqu* 次除去 or *zhuan-jian* 轉減). *Li* literature uses unique expressions for common operational terminology, giving 'fill *x* for y', where in *suan* you see 'per *x* then obtain one *y*' or '*x* then one, *y*'. *Li* literature uses common terms *differently*, employing 'eliminate' (*chu*) for *modulo* where in *suan* it can only mean division or subtraction. *Li* literature strings procedures into an unbroken chain, where *suan* organises procedures into discrete question-and-answer problems.[42] *Li* literature, more importantly, never gives worked examples.[43]

Let us take the previous instructions from the Supernal Icon *li*, for example. In 'Calculate Era entry', the first instance of 'eliminate' signals *modulo*, while the second instance is not to be acted upon, otherwise all years give era *jiazi*$_{.01}$, and there can be nothing to 'remove' (*qu*). Here, 'remove' signals subtraction, but in 'Calculate New Moons' it signals *modulo*. To further confuse matters, the Supernal Icon *li* offers the following definition near the very end, under 'Calculate the Five Stars':

All talk of 'as per' (*ru*), 'overflowing' (*ying*), 'simplify' (*yue*) and 'filling' (*man*) is elimination (*chu*) to find the dividend (i.e. division); 'remove' (*qu*) as well as

42 This holds for early *suan* base texts; on the deployment of *shu* 'procedures' in *suan* commentary, see Chemla (1997; 2009).

43 On the computational practices of division and the evolution of the operation *chu* 'eliminate' in *suan* 'mathematics', see Chemla (2013; 2014). The findings presented here, as concerns *li*, are the product of a collaborative effort with Karine Chemla, the results of which are forthcoming in M. Husson, K. Chemla, A. Keller and J. Steele, eds., *Mathematical Practices in Relation to Astral Sciences*.

'eliminate it' (*chu zhi*) are elimination (*chu*) to take the exhausted (i.e. *modulo*) (*JS* 17.528).

This would be useful to know in advance, one imagines, if the Supernal Icon *li* actually respected its own terms. If the reader makes the mistake of *using* this definition, however, procedures like 'Calculate the Solar *Du* position' return non-sense, as *there* 'eliminate' signals division, subtraction and sequence subtraction.[44]

The reader must intuit what sort of 'elimination' to perform at every step along the way, knowing that his/her errors will cascade, undermining the results of dozens if not hundreds of operations to come. Speaking from personal experience, the only way to produce meaningful results from these texts is to know a priori what you should get. The modern historian is aided to this end by works like Liu Hongtao (2003) and Martzloff (2009), which provide worked examples. That may be why early imperial *li* men wrote *commentaries*, but such commentaries, as we see in Chapter 4, are often written in the process of 'reception' (*shou*). *Li*, incidentally, is one of the few fields of gentlemanly knowledge that *no one* is mentioned as having learned alone from a book.[45] Knowing full well what to expect from a procedure, working from the text alone still involves trial and error, which is why Cullen (2005, 337) promotes the use of spreadsheet automation 'to see whether my understanding of the text led to calculations that actually worked' without having to recalculate everything by hand. The early imperial practitioner, of course, did not have such tools to fall back upon, only training – his 'master('s) method' (*shi fa*) – and hours of long division.

3.3.2 Copyist

The '*li*-worker' works with numbers. More specifically, he works with *lü*, quantities and co-ordinates. *Lü*, once again, are numbers with no meaning except as they relate (proportionally) to one another – numbers, in *li*, used via the rule of three to convert quantities from one unit to another, e.g. from years into months. Quantities like 'the high origin to the year sought' (7,345) and the 'big and small remainder' (11;929) are the numbers that the reader operates upon, by addition, subtraction, multiplication, division and *modulo*. Quantities, in turn, are converted into co-ordinates of position (i.e. 'lodge *du*') and civil time (e.g. 'stem–branch' dates), which, in the case of the latter, have no further role in calculation. The sexagenary days, it merits emphasis, are 'named off' (*ming*), but they are never 'counted' (*shu*) – let alone 'multiplied' (*cheng*) or 'subtracted' (*jian/chu/qu*) – because *they are not numbers*. The product, as we see in the lunation table

[44] See Liu Hongtao (2003, 116).

[45] For more on the role of orality in the teaching and transmission of mathematical astronomy in this period, see Morgan (2015).

computed for Xiping 3 (Table 3.3), is completely devoid of '*li* numbers'.[46] *This* is what the copyist has to work with.

How did the copyist proceed with calendar making? Arguing from a sample of variora in *Book of Odes* citations appearing in early manuscripts (written in different scripts), Kern (2002; 2005) has forwarded the influential thesis that memory, dictation and oral transmission, as he argues is evidenced by these variora, would have necessarily played such a role in early Chinese textual culture as to make the assumption of unbroken chains of visual copying unthinkable for texts intended for transmission. Though the evidential basis of his analysis has been generally rejected, speculations about the role of 'orality' in early Chinese manuscript production abound, leading us to ask the question.[47] There is unequivocal evidence for the role of 'orality' in the practice and transmission of *astronomy*, as is on display in Chapter 4 below,[48] but *calendars* are something else. The nature of the variants (mistakes) in the Zhoujiatai M30 and Yuelu Academy Shihuang 34 *zhiri* provide positive evidence for visual copying.[49] Specifically, the pattern of error reveals that the two copyists were simply copying the stem–branch binomes from one row to the next, moving each row one binome over to accommodate the difference between the sexagenary cycle and the two-month, fifty-nine-day row. Visual copying, of course, raises the question of literacy, but the way that these manuscripts correct their mistakes points, again, to the copyists' comprehension and reflection upon the text. The Zhoujiatai M30 Ershi 1 board speaks likewise to the *user's* comprehension of calendar making, as the user references the lunation table on the recto to make his own day table on the verso.

In my own experience, while producing the figures for the present chapter I found myself falling back upon a number of rote patterns that changed the way that I thought about calendar making. First, the alternation of 'big' (thirty-day) and 'small' (twenty-nine-day) months creates a predictable pattern in stem–branch binomes, where, from 'big' to 'small', the branch moves to the opposite position, and, from 'small' to 'big', both stem and branch retreat one from the previous 'big'. In numerical terms:

b03s03 →	b03s09 →	b02s02 →	b02s08 →	b01s01 →	b01s07 →	b10s12 →
big	small	big	small	big	small	big

[46] On *lü*, see Section 1.1.3 above. Once again, this breakdown of the elements of *li* calculation derives from collaborative work with Karine Chemla developing out of Chemla (2013; 2014), the results of which are forthcoming in M. Husson, K. Chemla, A. Keller and J. Steele, eds., *Mathematical Practices in Relation to Astral Sciences*.

[47] See, for example, Schaberg (2001, 315–26), Meyer (2012) and Richter (2013, esp. 4–5, 101, 107, 108, 158, 172–3).

[48] See also Morgan (2015).

[49] For further examples of visual copy errors in technical literature, see Mo Zihan (2011) and Mo & Lin (2016).

Second, *everything* comes down to the placement of the 'double big', which, as is clear in Table 3.3, interrupts this progression with a return to the same stem–branch, i.e. *yihai*.$_{12}$ → *yisi*.$_{42}$ → *yihai*.$_{12}$. Third, the month size, while not technically necessary for *using* a day or lunation table, is extremely helpful for *copying* one (as a sort of fail-safe). Fourth, the arrangement of *zhiri*- and *liri*-style day tables takes advantage of these patterns, facilitating the sort of copying (and mistakes) seen in the Zhoujiatai M30 and Yuelu Academy Shihuang 34 calendars. Lastly, in the process of copying and pasting in Photoshop, I myself made *many more* of the *exact same type* of copy errors as my early imperial counterpart, forcing me to set aside any presumptions I may have previously harboured about his literacy, comprehension or diligence.

We cannot know how the Everyman thought of the 'the Calendar', but the way he wrote (and wrote upon) his calendar gives us something of an idea. Scholars were previously at a loss to explain why the Zhoujiatai Shihuang 34 *zhiri* 'has at least four new moon stem–branches that are wrong' (Zhang Peiyu 2007, 69); when we consider the logic of the layout and the 'double big', however, we can trace every variant to a single mistake. When we examine the pattern by which new moon stem–branch dates articulate, we can appreciate the ingenuity by which the sexagenary round can compress the days of the year into a single 23-cm board. When we see how the Zhoujiatai lunation table is used to write a date table, lastly, we can see how the former offers a matrix for constructing the latter. Copyists were clever about stems and branches in a way that is quite distinct from how the computer went about them, but they were no means independent of, or more intuitive than, good old numbers. The mistakes we find are, after all, exclusive to the sexagenary cycle, and we can see the copyist of the Zhoujiatai lunation table, for one, struggling with planning via stem–branch dates in a way that no one struggles with ordinal day numbers.[50] In the end, officialdom moved away from sexagenary dates in official documents, but that is not the only concession that men of stems and branches, 'bigs' and 'smalls', made to men of numbers.

3.3.3 Theoretician

The classics frame the sage's task of 'granting the seasons' around intercalation – aligning the seasons (*shi*) of the 365¼-day agricultural year (*sui*) to the lunations (*yue*) of the ≈354½-day civil year (*nian*). When asked for a definition, this is how scholars have presented the science of *li* to our very day, where,

[50] The fact that sexagenary counting was a *learned* practice is suggested, for one thing, by the sort of sexagenary flashcards found at the Warring States tomb at Dongxiaojing, Hebei (*Wenwu* 1990.6: 67–71).

fittingly, the word now means 'calendar'. Reading *li* as 'calendrics', we must understand the sort of lunisolar calendars in Section 3.1 above to be the sole purpose of the centuries-long pursuit of 'tightness' and 'truth' conducted in this name. If this seems at all inevitable, remember that Egypt and Rome had actively abandoned the lunisolar calendar by the time of most of the events in this study. A lunisolar calendar is a lot of trouble: it requires you to 'reconcile the irreconcilable' (Needham 1959, 390), and the manifest nature of lunar phenomena guarantees you a public trial for your efforts. In the best of times, the result is a calendar where no year or month is ever reliably the same. In the Mediterranean, people moved towards simple 365-day affairs modelled upon the agricultural year and divided into solar 'months'. United as their world was by sailing, experts and governments alike decided that the moon had no place in 'granting the seasons' (Neugebauer 1942).

The Chinese persevered. As to *why*, a proponent for a solar calendar like Shen Gua (1031–95) might point to mindless tradition: 'Though they are connected with things like tides and gestation, the waxing and waning of the moon have nothing to do with the year and season, nor the rhythm of cold and heat, [so] it would suffice to [simply] lodge these things within [my solar] *li*' (*Mengxi bitan jiaozheng*, entry 545).[51] The average *li* man, however, would probably insist that it was their devotion to something like scientific empiricism that kept him going. Neither would have been wrong. On the one hand, the operation of a rigorous lunisolar calendar is where calendrics and astronomy coalesce – unlike 'the West', the *li* man might complain, where precise astronomical periods are substituted with thirty- and thirty-one-day 'months'. On the other hand, every calendar is, in the end, a compromise between astronomical reality and cultural convention. The difference is simply *which* compromise early imperial courts made, and how that compromise was negotiated.

Looking at Table 3.3, it is somewhat difficult to get excited about the revolution – the 'final turn towards the fine and tight' (*JS* 17.498) – that later sources ascribe to Liu Hong's Supernal Icon *li*, but that is because its genius lies in its 'slow–fast' lunar model, and *that* does not go into the civil calendar. It could, of course, and Zhang Heng (78–139 CE) and Zhou Xing had suggested as much long before Liu Hong's day. The problem in 123 CE – the problem that lost Zhang and Zhou the deliberation by forty to two – was that 'using the nine roads, the [civil] month would see three bigs and two smalls in a row', which, in the judgement of the dissenters, 'is all loose and off' (*HHS zhi* 2, 3034). The problem is that the 'slow–fast sequence' (*chi-ji li*) 'fixed' (*ding*) the time of true syzygy up to 7h15m before or 10h35m after mean syzygy, which could push the former past the imaginary midnight line separating one civil day from the next. This is not a problem for eclipse prediction, of course, because 'fixing'

[51] On Shen Gua's advocacy of a solar calendar, see Morgan (2016b).

places the moon where it (more or less) actually is vis-à-vis the sun, and Liu Hong was celebrated for the accuracy of such predictions.[52] The problem, rather, is that moving the *calendar* moon forward and back destroys the neat 'big–small' sequence of civil months.

Whether or not this was an 'unreasonable reason' (Chen Meidong 2003, 262), Liu Hong, for one, was content to keep the two moons separate, as were the other great *li* men of the third and fourth centuries CE. It was not until 'Director of the Watches for the Heir Apparent He Chengtian privately wrote (*si zhuan*) a new method (*fa*)' and submitted it to an emperor 'deeply fond of *li* numbers' in 443 CE Jiankang that the issue was officially reconsidered (*Song shu*, 12.260). The specific argument to which Director He appeals for this change is that 'having syzygy and lunar eclipse *not* on new and full moon is counter to the very idea of *li*' (ibid., 12.262). Upon review, Prefect Grand Clerk Qian Lezhi and his assistant, Yan Can, wrote a glowing recommendation of Mr He's Epochal Excellence *li* to the throne with but one modest proviso:

> Re Chengtian's method: the new, full and quarter moons of each month all fix the big and small remainder, which, while prudent for calculating the time notch of crossing coincidence (eclipse) – for which surplus–shrinkage (i.e. speed correction) is always used – it does [result in] the [civil] month experiencing three bigs and two smalls in a row, which is markedly different from the old method (*jiu fa*). In the old [way], solar eclipse does not occur on new moon alone but also on the last and second day of the month, which is what the *Gongyang Tradition* refers to as 'some miss ahead of the mark, some miss behind' (Yin 3). In our humble opinion, it is advisable that [we] persist in the old [way] on this one point (*Song shu*, 12.264).

To this, Supernumerary Gentleman of the Cavalier Attendants Pi Yanzong adds:

> If the last and first day of the month fix the big and small remainder [such that] era head falls on a surplus (a positive speed correction), then [one] must fall back one day and take the last day of the previous year (*sui*) as the head of the new era (ibid., 12.264).

Faced with practical, scriptural and theoretical objections from Clerk's Office leadership, He Chengtian gave in, 'amending his new method according to the old procedure, no longer fixing the big and small remainder for each [civil] month' (ibid., 12.264).

The tide began to turn by the following century. In 544 CE, the Xiao Liang court (502–57 CE) at Jiankang 'issued an edict for the creation of a new *li* . . . that fixed the small remainder of new moon via slow–fast [so as] to have three bigs and two smalls', but the matter was tabled during Hou Jing's military revolt of 548 CE (*Sui shu*, 17.417). In 608 CE, 'fixists' went on the offensive at

[52] According to Xu Yue in 226 CE, for example, Liu Hong issued such a prediction sometime between 176 and 178 CE, and 'after the matter had been inspected, it was found to have happened just as Hong said; [everyone] within the oceans recognised the truth of it, and there was no one who did not hear of it' (*JS* 17.500).

the Sui court (581–618 CE) at Chang'an. There, Liu Xiaosun and Cultivated Talent Liu Zhuo of Yizhou (544–610 CE) denounced the authors of the official *li* of 584 CE for 'only knowing to add a greater remainder of 29 for the new moon without understanding to take the conjunction-coincidence (*hehui*) of sun and moon as the standard for fixing [it]' (*Sui shu*, 17.424). In a long memorial, the two Lius recount the history of lunar speed correction for eclipse modelling as developed by luminaries like Yang Wei (fl. 226–37 CE), He Chengtian and Long Yidi (fifth century):

> These three men – the *li* experts (*shan li*) of previous dynasties – were all of this intention but had yet to correct their writings. What matters in *li* numbers are new moons and *qi*. The [first] new moon crowns [the calendar of] court audiences, and the [first] *qi* is the beginning of the growth of life; in the new moon is invested the cultural pattern (*wen*) of declarations and food offerings, and in the *qi* is invested the ceremonies of the suburban sacrifice and reception – thus it was that Confucius (551–479 BCE) mandated *li* and fixed (*ding*) the new moon and winter solstice as a model for the future [in the *Spring and Autumn Annals*]. Today, Xiaosun's *li* method (*li fa*) accords with both the manifest and patterns (i.e. experience and classical precedent): it fixes (*ding*) the syzygy by means of slow–fast lunar motion, desiring to make [solar] eclipse necessarily fall on new moon, not on the last or second day of the month. Even if it concatenates the months 'one small, three big', [it does so to] get at the integrity (*tong*) of heaven (*Sui shu*, 17.424).

Now, theoreticians were acknowledging the practical concerns of their opponents and appealing to their own scriptural support for modernisation, enlisting Confucius himself as a 'fixist'.[53] Neither made any political headway, their ally, Grand Clerk Zhang Zhouxuan (d. *c.*613 CE), pushing through his own *li* in 597 CE while the two Lius struggled with political enemies at court. It was only with the institution of Li Chunfeng's (602–70 CE) Unicorn Virtue *li* in 665 CE that the civil month was finally 'fixed'.

Experts have been quick to judge objections to 'fixing' civil time since at least the sixth century. If anything, modern critiques are much softer than those that men like Liu Xiaosun and Liu Zhuo hurl at Pi Yanzong, who 'raised objection by decorating wrongness' (*shi fei zhi nan*) and 'had no idea in the first place' (*benlai bu zhi*).[54] In the words of Martzloff (2009, 41), 'l'idée fondamentale et sans cesse répétée, est que les calculs doivent tendre à se conformer aux apparences célestes', and anyone who questioned that idea was clearly standing in the way of progress. But what does it mean to have an astronomically accurate calendar? To us moderns, for whom the flow of time has come unmoored from the revolutions of an increasingly opaque sky, the premodern quest for calendrical 'tightness' seems fatuous. Beyond simple matters of intercalation, around which the legend of the sage kings revolves,

[53] Confucius, of course, only 'fixed' the new moon in the sense that the 'subtle words' of the *Annals* were read as criticising the Lu ducal court for getting the new moon day wrong.
[54] *Sui shu*, 17.424 and *JTS* 79.2714.

astronomical precision in civil calendrics offers no real scientific advantage for agriculture.[55] That much is certain, but none of our theoreticians are talking about agriculture per se – that ship sailed with Wang Mang – they are talking about 'declarations and food offerings', 'the suburban sacrifice and reception'. *That* is where you need to be scientific, and in an age where mistiming meant misgovernment, and misgovernment meant misfortune, the ritual sciences created a demand for astronomical precision every bit as serious as that of the New York Stock Exchange.

As much as we should take our subjects' convictions here seriously, we cannot take them at face value, since calendrical 'tightness' is a paradox. 'What matters in *li* numbers are new moons and *qi*', say the Lius Xiaosun and Zhuo, but *neither* can be directly observed, except in the event of an eclipse, and *qi* are mostly artefacts of interpolation (see Section 2.1.2 above). You can submit the calendar to astronomical testing, of course, but the mean moon too is but a mathematical construct, and whether or not the 'fixed' lunation falls on the right day depends entirely on one's definition of the civil day. Is a *li* that places syzygy on the right calendar day, twelve hours ahead of eclipse, really 'tighter' than one placing it twelve minutes behind on the next? In terms of calendrics, it is, which is why both sides of the debate can appeal to 'tightness' and declare the other 'loose and off' – 'the very idea of *li*' is open to interpretation, and both sides have scriptural support for their positions. One rather poignant example of the circularity of calendrical empiricism is 104 BCE, when Emperor Wu commissioned a 'Grand Inception *li*' to begin from a coincidence of the new moon, winter solstice.Q22 and midnight *as determined by the failing system it was designed to replace* (Section 1.2.1 above).

If the idea of calendrical 'tightness' is circular and open to interpretation, perhaps we should seriously entertain the other side of the debate. The defenders of the 'old method', after all, included men like Qian Lezhi, who ran the Liu Song (420–79 CE) Clerk's Office and made a name for himself in instrument building and tono-metrics. Qian makes it perfectly clear that no one is against 'fixing' the moon for purposes of *astronomy*, 'for which surplus–shrinkage is *always* used', but they are against the introduction of 'three bigs and two smalls in a row' into *the calendar*. Outside *li* men from Liu Xiaosun's day to Chen Meidong's have been at a loss as to how this could outweigh accuracy at a theoretical level, but the problem, I suspect, is that Clerk's Office insiders had to think about the calendar at other levels, respecting the entire chain of calendar production and official use. The computer could probably adapt, but the laborious extra steps of 'fixing' would more than double his workload and the chance of computational error. The real problem, however, was probably the copyist and the ordinary user, for whom the calendar had

55 Jiang Xiaoyuan (1991, 115–23); cf. Sivin (2009, 41).

been a well-ordered sequence of stem–branch pairs, alternating reliably 'big–small–big–small' since the beginning of time. Change *that*, one imagines, and you could guarantee 'looseness' and confusion rippling through every level of society.

Lest the inertia of such conventions fail to impress themselves upon the reader, let us consider an example of 'scientific' calendar reform more close to home: the French Republican Calendar. In a drive to replace the trappings of the *Ancien Régime* with institutions founded on transnational scientific principles, the French Republic introduced a host of social, legal and metrological reforms. Now known as the metric system, the latter sought to replace a confusing array of anthropometric weights and measures that varied from locale to locale with a single universal standard – a *decimal* standard, based on the circumference of the Earth, that would at once facilitate easy conversion between units and transcend both the body and the body politic. Taking decimalisation one step further, Charles-Gilbert Romme's (1750–95) reform commission proposed a new calendar on the same grounds: a calendar of twelve thirty-day months (plus five or six days placed at the end of the year) beginning on the day of the spring equinox, each month divided into ten-day 'decades' with ten hours to the day and 100 minutes to the hour. Instituted in 1793, the Republic disseminated its new calendar via printed almanacs and conversion guides. Though the state promoted calendar reform with all the vigour that it had the metric system, it abandoned the calendar a mere fourteen years later in 1806 amid religious resistance and popular confusion.[56]

3.4 Conclusion

There was never any one such thing in early imperial China as 'the Calendar', let alone as a historical agent that acted alone to move mountains and empty seas. 'The Calendar' was an idea, an idea that translated into physical objects of wood and paper, which depended upon a continent-wide network of computers, postmen and copyists to get where they needed to be. Nothing could be more central to his sagely function, and yet the son of heaven depended upon the Everyman to see it through. The diffusion of calendar production left room for corruption, both textual and economic. The forces of corruption posed a perennial challenge for the central court, but, all in all, the system worked: lunation tables got to where they needed to go, and local offices produced standardised documents from them with relatively few mistakes. 'The Calendar' was never quite the same from one man to another, of course, neither in form, nor in essence. To the computer, it was a problem of *lü* ratios and mathematical procedures – where to subtract for ordinal numbers and what

[56] On the French Republican Calendar, see Shaw (2011).

'elimination' to perform – while to the copyist, it was a sequence of stems and branches, 'bigs' and 'smalls', reset by a 'double big'. To the theoretician, it was a problem of pure astronomy at which you must throw everything you have, while to the Clerk's Office, it was a question of human logistics that needed time to adapt.

'The Calendar', in a way, belonged to everyone, less a monolith of ideology than a stage for its negotiation. People brought different ideas of *li* to that stage, and different aspirations for what it could be. There was the emperor, of course, but most emperors were content to leave matters of 'season granting' to deliberation (*yi*), where they would be settled by popular vote. Others took a more activist role. To ambitious dynasts like Emperor Wu and Wang Mang, 'granting the seasons' provided a venue for pursuing immortality and power, the latter extending imperial time control into the very minutiae of local agricultural practices. To peasants 'transplanted in numbers of thousands and myriads' (*HS* 99A.4077–8) for violating his ordinances, 'the Calendar' was probably felt as a violent imposition on their way of life. In better times, it was a rather neutral document, absent any emperor's name or overt ambitions. Indeed, more than half of the calendars extant from the Qin persist in using the word *zheng* for month I, despite the decades-long taboo on this, Emperor Shihuang's given name.[57] To most, 'the Calendar' was no more than a personal possession by which to record business, plan deliveries and record amounts of money and 'fish grain' owed.

As to the gentleman of *li*, he saw in the civil calendar the potential to be something more – something 'tight'. One would be right to criticise the blindness of his empiricism, as did Pi Yanzong and Qian Lezhi in their time – even compare it to the sort of imposition posed by monthly ordinances – for the calendar is and always has been a practical amalgam of astronomical reality and cultural convention. How 'scientific' does the calendar need to be? We ask ourselves this question time and again in human history, but the dilemma is of our own making, and so too is its solution. Calendrics, after all, is not eclipse prediction, where '[only the actual] appearance of an eclipse should be counted as correct, there being no "distant" or "close"' (*HHS zhi* 2, 3041), so the point is completely moot. *Li* had been up to something all this time – something wholly unrelated to the calendar – and its practitioners fought for five centuries that the calendar look more like it. Whether by force of demonstration or reason or scripture or public spectacle, they would stop at nothing until the calendar was really 'tight', whatever it is that that means.

[57] Those calendars that *do* avoid this taboo by replacing *zheng* 正 with *duan* 端 are as follows: (1) the Yuelu Academy 'X-qi nian zhiri' Shihuang 27 (220 BCE) daily calendar and (2) the Zhoujiatai M30 Ershi 1 (209 BCE) lunation table (Figure 3.7).

4 Reverent Accordance with Prodigious Heaven

As much as the sage kings are celebrated for having left us the *wen* of human civilisation – for having taught us, among other things, to 'observe the signs' and 'grant the seasons' – they also left us, more than that, with a saga patterned by their failure. Zhuanxu was the first to realise that one needs 'rectors' (*zheng*) to make things 'right' (*zheng*) in heaven, as on earth, and so too was he the first to hire the wrong men for the job. Selected from tribes with unruly pasts, Chong and Li performed their duties for a time, but when the Sanmiao rose in turn against the throne, 'the Sanmiao corrupted their virtue, and both offices were abandoned' (*HS* 21A.973). What ensued was chaos, and it was in response to dereliction that Yao made it his first duty to 're-educate Chong and Li's descendants' (ibid.) and assign the Xi and He clans to the task. And they too, for a time, were able to right the world without, but not the world within. And so, some generations later, 'the Xi and He [brothers] deserted their posts, losing themselves in alcohol in their [respective] fiefs' (*Shangshu zhushu*, 7.102a), until they too would be punished and replaced. And so it goes, time and again, in the stories that *li* men tell about their past. There are patterns that you can discern in the vastness of time and space, and so too in the pettiness of man.

The solution is always *mi* ('tightness'); the problem, of course, is that everyone knows the magic word. *Mi* is invoked so universally in our sources that it is what our subjects would call 'empty' (*xu*), and semantic vacuums accommodate crackpots, drunks and visionaries alike. Like 'humanness' (*ren*), 'propriety' (*yi*) and 'ritual/politeness' (*li*), however, there is something empirical (or intersubjective) to this virtue: to celebrate in mourning proves you 'inhumane', and an eclipse that's early proves you 'loose'. The gentleman of *li* needs something 'solid' (*shi*), and the fruits of 'observation' (*guan*) and ancient records, for one, give him what he needs to 'verify' (*yan*) that what he says is 'true' (*zhen*). 'Circumspection' (*shen*) requires that things be 'referenced' (*can*), 'compared' (*jiao*), 'examined' (*jian*), 'investigated' (*kao*) and 'checked' (*cha*), but, here again, the expert is not the only one who knows these words. In 179 CE, for example, the Grand Clerk reported that the eclipse predictions of one Wang Han surpassed the Quarter-remainder *li* twenty-nine to two, prompting the throne to summon our hero:

[Liu] Hong (fl. 167–206 CE) shall go forthwith to [Prefect Grand Clerk] Xiu to compare (*can*) against Han, calculating origins and testing (*ke*) parts, examining and collating (*kao jiao*) lunar eclipses. [If] upon circumspection (*shen*) [Hong finds Han's] *jisi*.06 origin tight and close (*mi jin*) and that [Han] has a master('s) method (*shifa*), then Hong is to receive it from Han. If unable to do so, [please] respond (*HHS, zhi 2*, 3042).

In his response, Liu Hong speaks of 'comparison' (*jiao*), 'reference' (*can*) and 'proving' (*xiao*), but *he* concludes that 'though [Wang Han] has a master('s) method it is the same as if he had none; the test results (*ke*), moreover, were not close and tight (*jin mi*)' (*HHS, zhi 2*, 3043). If it's one man's word against another, which words are we to believe?

Experts disagree, and as fertile as such disagreement could one day prove, the sage king, in the here and now, needs consensus about what will work. At the level of policy reform, therefore, the pursuit of 'tightness' is much less about the precision of instruments and the originality of procedures than about the 'solidity' of words, and *that* is much more difficult to measure. The solution, echoing an untitled Qin (221–207 BCE) manuscript from the Peking University collection, is that we remove words as much as possible from the equation:

Lu Jiuci inquired about numbers from Chen Qi, saying, '[I], Jiuci, read speech (*yu*) and count numbers (*shu*) but cannot fathom the two at once. [I] want to fathom [at least] one [of these] things, [so] which thing is [the most] urgent?' Chen Qi replied to him, saying, 'If my good sir cannot fathom both at once, then abandon speech and fathom numbers, [for] numbers can speak, [but] speech cannot number' (Han Wei 2015, 30).

If the empire could issue standards of weight and measure, and it could fix the price of grain, it could surely institute some sort of metric for *li*.

In this chapter, we will examine how early imperial actors use 'speech' and 'numbers' to settle opposing claims of astronomical accuracy. My approach is that of Cullen (2007a), whose close reading of the Eastern Han (25–220 CE) *li* debates preserved in the *Book of Later Han* provides a thick description of the institutions, motives, participants, tools and rhetoric at play in the history of *li* policy. Where I think it is important to augment this study is in the area of sources: the *Book of Later Han lü-li* monograph, while one of the richest of the genre, provides very little of the 'numbers' involved in such debate. Here, we will focus instead on a debate of 226 CE at the Cao Wei (220–65 CE) court at Luoyang, as recorded (primarily) in Li Chunfeng's (602–70 CE) *lü-li* monograph in the *Book of Jin*.[1] Leaving questions of Li's authorial intention to Chapter 5, the reason for choosing this debate, quite simply, is that it is the first to come to us with the actual data under discussion.

[1] On Li Chunfeng's authorship of the *Book of Jin* 'Lü-li zhi', see *JTS* 79.2718.

Our goal is to make these numbers speak – to speak *in dialogue* with what is said around them. The intention is not to generalise about 'Chinese astronomy' (or 'Chinese thought') but to go into sufficient detail, with one particular event, to discern the thoughts and actions of individuals who vehemently disagree. In Section 4.1, we will begin with an outline of the events surrounding the policy deliberation of 226 CE. Juxtaposing Li Chunfeng with what we can cull from other sources, we will come away with a broader view of the imperial and interpersonal politics involved in one man's adjustment of mean lunisolar parameters. In Section 4.2, I will then present the text of the deliberation. With an idea of what each speaker *wants* – to innovate, to preserve, to praise a master and to destroy – we will consider what they say to get it. The picture of 'scientific debate' that emerges from this event is one that is deeply agonistic, as *non sequiturs*, attacks and surprise revelations derail the conversation, but one, nonetheless, where numbers speak the loudest. In Section 4.3, we will then look at the numbers – five eclipses, fourteen planetary visibility phenomena – to understand what it was that day that proved one man 'loose', and the other 'tight'. The 'numbers', as I will show, are every bit as arbitrary as the 'speech', it being rather the dignity of the experiment – *the contest* – that lent them the status of indisputable, 'solid', fact.

4.1 Cause for Deliberation

Before coming to the debate itself, in Section 4.2, it is important to supply the events of 226 CE with some context. Li Chunfeng's chronicle is a good place to begin, but the chronicle tells us very little beyond the technicians and technical issues involved. Dredging through memorials, biographies, annals and monographs on related subjects, studies like Hasebe (1991; 1993) and Cullen (1993; 2007a) show that *li* reform usually entails a whole lot more than *li*. In Section 1.2, we saw how Emperor Wu of Han (r. 141–87 BCE) was persuaded into action out of political symbolism and personal aspirations for immortality, Cullen (1993, 201) warning us that 'a failure to take account of the real motivations involved . . . will lead to a radical falsification of our account of this aspect of scientific change in early China'. In Section 1.2, we also saw how a voice like Liu Hong's might fail to carry in a political vacuum. Fascicle 2 of Li Chunfeng's *Book of Jin* monograph is the story of the latter:

> From the Yellow Inception period (220–6 CE) on, the reform and creation of *li* procedures all pored over the Supernal Icon's Dipper parts reduction, syzygy remainder, and *yin-yang* and slow–fast [sequences] of lunar motion in search of compromise. Hong's procedures were the exemplar (*shibiao*) of calculation for the subsequent age, and thus [do I] lay them out as follows (*JS* 17.503).

The story of 226 CE begins where we left off in Section 1.2 above, and to tell it we must turn to sources beyond Li Chunfeng: Chen Shou's (233–97 CE) *Monograph of Wei*, Shen Yue's (441–513 CE) 'Monograph on Ritual', and the biographies, where they exist, of the actors in question. In terms of secondary sources, I rely on Goodman (1998) and Hasebe (1991), whose analyses of the politics behind the period's technical policies leave few stones unturned. Most importantly, lest we lose sight of the context that the historian himself thought to provide, I shall frame what we otherwise know of the events surrounding 226 CE in juxtaposition with Li Chunfeng's minimalist, technical account.

4.1.1 *Turns of Fortune, 168–196* CE

> In the time of Emperor Ling of Han (168–89 CE), chief commandant of Kuaiji Eastern Regiment, Liu Hong, examined the Clerk's Office's *li* notes [all the way] from antiquity to his day. [He] traced the motions of advance and retreat, and [he] verified (*yan*) the risings and settings; [he] watched the goings and comings, and [he] *du*-measured the ends and beginnings, and [he] first awoke (*shi wu*) [to the realisation] that the Quarter-remainder [*li*] was loose and wide (*shu kuo*) of heaven and that it was all because the Dipper parts was too large. [He] changed the era divisor to 589, and the Dipper parts to 145, and he created the Supernal Icon method, with the winter solstice.$_{\text{Q22}}$ sun at Dipper.$_{\text{L08}}$ 22 *du*, seeking the motion of the sun, moon and five [planets] via procedures. [He] calculated, and above [it] matched with antiquity, and below [it] responded to the present.
>
> What he did was establish numbers based on the *Changes* [such that] they called out to one another in hidden motion and sought each other out from secret parts – and [at this he] named it the 'Supernal Icon *li*'. Also, [he] created the solar/daily motion slow–fast (speed correction) while concurrently investigating (*kao*) lunar motion, [concluding that] *yin* and *yang* (i.e. negative and positive latitude) cross inside and outside the yellow road, and that the sun travels on the yellow road, experiencing advance and retreat in terms of red-road lodge *du* (see Section 2.1.4) – and only with this was there a turn (*zhuan*) towards the fine and tight (*jing mi*) relative to prior methods.
>
> In [196 CE], Zheng Xuan (127–200 CE) received his method, and considering it to plumb all that is recondite and crown all that is subtle, he further added commentary and explanation to it (*JS* 17.498).

Liu Hong's 'awakening' was not the only 'turn' of the late second century. Taking office amid a succession of young emperors and powerful women, Liu Hong looked to heaven while things on earth unwound. Purges rippled through the capital, eunuchs, officials and family cliques decimating one another's ranks while the provinces fell to revolts and millenarian cults. No longer able to staff its own offices (or collect the taxes for their upkeep), the court resorted to their *sale*, packing central offices with unlikely characters and divesting regional power onto private armies. His body failing, the thirty-four-year-old Emperor Ling summoned Liu Hong in 189 CE with the thought, perhaps, that *li* could stop the haemorrhaging. Halfway out, the emperor succumbed to the ravages of disease, and the capital to those of Dong Zhuo's armies. Called to intercede against the eunuchs, Dong's forces sacked the capital, leaving the palace in flames and the thirteen-year-old heir lost. Recovering the heir, Dong Zhuo moved him to Chang'an, where, in the following year, he would have him poisoned, putting his nine-year-old brother on the throne.

Liu Hong turned back for Kuaiji, where he was relatively insulated from what was happening. In the north, the empire was devolving into warlord coalitions and, soon, upon Dong Zhuo's assassination in 192 CE, total war. Liu was recalled to the Shandong peninsula. It was there, as governor of Shanyang, that he would meet Zheng Xuan. The premier scholar and educator of his day, Zheng Xuan's journey began with mathematics, and while passing through the area in 196 CE, it also brought him to Liu Hong's door.[2] The emperor had a special guest that year too. Released destitute to the ruins of Luoyang amidst rebellion in the ranks, the warlord Cao Cao (155–220 CE) came to take him to his power base at Xu. With no real emperor to petition, Liu Hong was happy to share his work on the peninsula, so he taught Zheng Xuan, he taught Xu Yue, and it was by private exchange that the Supernal Icon *li* began to circulate.

Under the aegis of the fifteen-year-old Emperor Xian (r. 189–220 CE), General Cao Cao launched a prodigious, decades-long drive to reconsolidate imperial lands. Liu Hong would never see the end of it, dying at his post some ten years later, but neither, for that matter, would Cao Cao. Dying in 220 CE, he passed his new title – the King of Wei – and a beleaguered half-empire to his son Pi (187–226 CE).

The fall of Han had been prophesied off and on since the days of Wang Mang (*c*.45 BCE–23 CE), the chatter of omens, prodigies and '*li* numbers' reaching fever pitch by the time of Cao Cao's death on 15 March. The emperor gave in, abdicating to the King qua Emperor of Wei on 11 December. Succession requires a discourse, and Goodman (1998) details how one arose around Cao Pi (187–226 CE) over five weeks of consensus building – a consensus negotiated between parties of scholars, diviners and civil and military officials using omenology, prophecy, classical learning and metaphysics. All in all, it was agreed that the mandate had transferred by force of earth (*Han : fire*, and *fire* → *earth*) and Shun (*Han : Yao, Yao* → *Shun*), who had himself taken the throne through abdication.

[2] According to his biography, having learned the Triple Concordance *li* and *Nine Chapters of Mathematical Procedures* at the Imperial Academy, Zheng Xuan was singled out for Ma Rong's (79–166 CE) inner circle of disciples because he was 'good at maths' (*HHS* 35.1207).

Upon ascension as Emperor Wen (r. 220–6 CE), Cao Pi inaugurated the Yellow Inception period (*yellow : earth*) and ordered the Three Excellencies to deliberate (*yi*) upon the regalia appropriate to his virtue – uniforms, sacrifices, first months and appellations.[3] At the advice of a palace attendant, the emperor ordered that the first month – and the first month alone – be left as it was, adducing two items of classical precedence in his support:[4]

Implement the seasons of Xia, ride the carriage of Yin and wear the cap of Zhou, while, for music, it's the *Shao and Wu* (*Analects* XV.11).

The Xia numbers get heaven (*Zuo zhuan*, Shao 17).

The Prefect of the Masters of Writing objected, begging that three-concordance metaphysics be taken seriously, but the emperor was content to be inconsistent.[5] It was one thing in theory, but Cao Pi decided against switching all the months around after the unmentionable fate of Wang Mang in this regard.

4.1.2 *The Case for Reform, 223* CE

Midway through the Yellow Inception period (220–6 CE) . . . Prefect Grand Clerk Gaotang Long (d. *c.*240 CE) made repeated and detailed deliberation (*yi*) on *li* numbers and [the need for] reform. Assistant to the Grand Clerk Han Yi thought that the Supernal Icon *li* went too far in reducing the Dipper parts and that it would later slip ahead of heaven, so [he] constructed the Yellow Inception *li*, using an era divisor of 4,883, and a Dipper parts of 1,205.

Later, Prefect of the Masters of Writing Chen Qun (d. 237 CE) submitted a petition, which went, '*Li* numbers are difficult to comprehend and have been a point of much contention among the expert scholars (*tong ru*) of the previous age. At the inauguration of the Yellow Inception reign, the Quarter-remainder *li* had long been far-off (*yuan*), loose (*shu*) and wide (*kuo*). The Great Wei has received the mandate, and it is time to reform the *li* and illuminate the seasons. Han Yi was the first to establish [a new *li*], afraid that [it] was not [sufficiently] circumspect (*shen*), [however], [he] thus referenced and compared (*can jiao*) [it] against the Supernal Icon [*li*]. What [he] compared was the motion-*du* of sun and moon and the quarter, full, new and dark moon (i.e. lunar phases). Over the course of three years, [people] exchanged [accusations] of right and wrong, which were at no time [ever] resolved. According to the Three Excellencies' deliberation (*yi*), everyone gathered everything together and put it to a hearing to arrive by different paths at the same conclusion: that [they] desired that it be proven (*xiao*) at the rotating mechanism (*xuanji*, i.e. armillary sphere) – each to the utmost of their methods (*fa*) – such that in one year's time there should be sufficient [grounds] to determine their respective success/failure (*de shi*)'.

The petition was approved (*JS* 17.498–9).

[3] *SGZ* 2.75 (comm.).

[4] For Emperor Wen's edict, see *Song shu*, 14.328; for Palace Attendant Xin Pi's opinion, see *SGZ* 25.696.

[5] *Song shu*, 14.328.

The emperor had spoken out on first-month reform (*gai zheng*), but not on *li*, leaving the court with the lagging Quarter-remainder *li* of 85/86 CE. The way that Chen Qun (and Li Chunfeng) tell it, it would seem that the matter arose quite spontaneously, but it is not a coincidence that the question *c*.223 CE was not *whether* to implement the Supernal Icon *li*, but *how to best it*. The year 223 CE, it just so happens, is when Sun Quan (182–252 CE), the newly declared King of Wu, inaugurated his Yellow Martial reign by 'reforming the Quarter-remainder to use the Supernal Icon *li*' (*SGZ* 47.1129). This was now a space race, and the Cao Wei were losing.

It was Cao Pi's very legitimacy that was at stake, and the figures at the forefront of this initiative were, not surprisingly, those who had helped construct it three years earlier. Gaotang Long (d. *c*.240 CE), for example, tutor to Cao Cao's sons Hui (d. 243 CE) and Rui (204–39 CE), had thrown his expertise in omens, ritual and metaphysics into justifying Cao Pi's ascension and had been awarded, for his part, the position of Prefect Grand Clerk. Whether or not Gaotang was competent in *li* is another matter – his biographer comments that '[his] intentions surpassed his expertise' (*SGZ* 25.719) – but the Grand Clerk is an administrator, like any other, with an office of experts in his employ. Gaotang's thing was the ceremonial – an expert of the Zheng Xuan school, he would go on to compile the dynasty's *feng-shan* rites and pass the torch as a national treasure to thirty disciples chosen by imperial edict.[6] Chen Qun, for his part, was one of the eight 'legitimation experts' (Goodman 1998, 88) at the core of the transition. Born into a well-connected family in the Caos' home town of Yingchuan, Chen was broadly learned, memorialising the throne on topics of spending and ritual, and was otherwise renowned for his knowledge of seals and the classics. Like Liu Hong, Chen worked his way from the military to civil administrative posts in the capital and the provinces; Chen, however, continued up the ladder to master of writing, the prefect thereof and, later, Minister of Works. Chen, like Gaotang, was something of a national treasure, going on to revise the Wei criminal code in 230 CE and propose the famous *jiu pin* system of bureaucratic recruitment.[7]

Neither Gaotang Long nor Chen Qun was an expert, as such, but they knew that something had to be done. Han Yi frames the solution in terms of 'Dipper parts' (*Dou fen*). The 'Dipper parts', in short, is the fractional part of the circuits of heaven that is placed, for the sake of convenience, just prior to winter solstice.$_{Q22}$ in Dipper.$_{L08}$ (see Appendix, Figure 1). Han Yi increased this from Liu Hong's $\frac{145}{589}$ (= 0.2462) to $\frac{1205}{4883}$ (= 0.2468). The number of *du* in a circuit is the number of days in a year, and it is the latter which this adjustment concerns. We could understand Han Yi's fear to be that the *qi*, specifically, 'would later

6 For Gaotang Long's biography, see *SGZ* 25.708–19.

7 For Chen Qun's biography, see *SGZ* 22.633–8; cf. Goodman (1998, 92–5).

slip ahead of heaven', but this would be difficult to detect. The problem is that no one yet saw the 'difference' between tropical and sidereal 'years' (*sui cha*) – of which the Supernal Icon 'year' fell 5m35s long and 14m40s short, respectively – and *that* resulted in 'subtle errors' which no one in 223 CE was equipped to resolve (see Chapter 5). The real problem was the synodic month. Pegged to the year in a 235:19 *lü* ratio – the 'rule' (*zhang*) – any change to the Dipper parts spilled over into the lunar parameters, and *those* were readily detectable, as we shall see in the debate to come.[8]

After three years of 'reference', 'comparison' and 'exchange', the emperor, as per standard protocol, put the matter up to the Three Excellencies for deliberation, and they, as per standard protocol, kicked it down to the Clerk's Office for further 'proving'. This was 'inappropriate' (*bu yi*), Xu Yue and Yang Wei will criticise, on two separate counts. First, the Three Excellencies demanded the use of an armillary sphere – probably the 'Grand Clerk's Bronze Sight' of 103 CE – and *that*, as discussed in Chapter 2, is the last thing you want. Second, testing was in this case handed *back* to the self-same agency. If three years' testing there was not enough to secure a consensus, what would one more year hope to prove?

4.1.3 The Tabling of Reform, 226 CE

It is at this point in the chronicle that Li Chunfeng inserts the transcript of the deliberation, to which we will turn in Section 4.2 below. 'Deliberation' (*yi*), it is true, can take oral and written forms, but the rapid back-and-forth of this exchange points to the sort of public debate studied in Cullen (2007a). All said, the debate goes nowhere, as 'before comparison and deliberation had been settled, the emperor died, and [the matter] was laid to rest' (*JS* 17.503).[9]

Speaking at this deliberation are seven men. We know nothing of (1) Han Yi and (2) Gentleman of the Interior Li En beyond these events, but the rest are familiar faces. (3) Grand Clerk Xu Zhi, for example, was an expert in prophecy and occult knowledge and authored one of the more masterful assertions of Wei legitimacy in this regard in 220 CE – an act, like that of Gaotang Long, that won him a position in the Clerk's Office.[10] (4) Sun Qin, libationer of the erudites, likewise appears in the historical record as concerns the ceremonial

[8] On the derivation of lunisolar constants, see Chen Meidong (1995, 211–77).

[9] Wei Shou (506–72 CE) reports that 'in [178–84] there was a change to the Supernal Icon, and in [220–6 CE they] used that fixed by Han Yi' (*WS* 107A.2659), but he seems to be confounding 'creation' (*zao*) with 'reform' (*gai*) on both accounts. See Section 1.2.2 above on the history of the Supernal Icon *li*.

[10] For Xu Zhi's memorial, see Goodman (1998, 100–5). Xu Zhi appears in 220 CE as Assistant to the Grand Clerk, then, in 226 CE and the Green Dragon reign (227–33 CE), as Prefect Grand Clerk (*SGZ* 2.62 (comm.); *Song shu*, 34.1011; *JS* 12.338, 17.499).

arrangements for Cao Pi's ascension.[11] (5) Dong Ba, erudite and 'serving within the palace', was an expert on ritual and metaphysics known for compiling the 'Carriages and Robes' and 'Five Agents' monograph for the Eastern Han as well as co-authoring a memorial in 220 CE in support of Xu Zhi's assertions.[12] (6) Xu Yue, the outlier, is mentioned only by his place of origin – Donglai – near where Liu Hong was stationed in his final years. Xu is known for two things: his association with Liu and his authorship of *Records for Posterity on Numbers and Procedures* (*Shushu jiyi*), the latter being largely about the former.

> Trekking all through hill and mount, the woods and gullies that need be passed, [I] finally [arrived] at Mount Tai. [There I] saw Liu Kuaiji, broadly knowledgeable and well informed, everywhere in numbers and procedures (*shu shu*). I thereupon received [my] patrimony, rather infected by whence it came. I at one time asked, 'Is there any [point at which] numbers are exhausted?' Kuaiji said, 'I once was roaming in the Eye of Heaven Mountains and saw that there was a recluse [living] there. No one in our day knows his [real] name, nicknaming him "Eye of Heaven Sensei" (Tianmu Xiansheng). I thought to ask him the same question, and Sensei said [to me], "The people of this day and age say that a three's no match for two, speaking about [the games] cut-boredom and four corners" (*Shushu jiyi*, 1b).

Liu Hong, whom Xu names here after his former command, goes on to recount the mysteries of 'number' in riddle and rhyme, turning around games and divination. Whatever the nature of their exchange, Li Chunfeng's chronicle names Xu as a key figure in the Supernal Icon *li*'s transmission:

> The House of Liu were in Shu (221–63 CE), which kept the Han Quarter-remainder *li*. The [Sun] Wu prefect of the palace writers, Kan Ze (d. 243 CE), received Liu Hong's Supernal Icon method from Xu Yue of Donglai and added exegesis and commentary to it. Regular Palace Attendant Wang Fan considered Hong's procedures fine and marvellous, and [he] used them to extrapolate the principles of sphere heaven to build a sight [and/as] effigy as well as a discursive. Thus it was that the Sun Clan used the Supernal Icon *li* until the demise of Wu (222–80 CE).

The crisis at the Cao Wei court *c.*223–6 CE is that the Sun Wu court beat them to the Supernal Icon *li*, and it was *this man* whom Li Chunfeng, at least, names as the vector.[13] Why is he invited to the deliberation? Why is it *Assistant Han Yi's* loyalty and credibility that is impugned? Seeing that the state only began to insulate the Clerk's Office and ban private practice in the seventh century, it is probably safe to assume that, in the third century, there was simply no expectation of secrecy – no more so than for any other gentlemanly art.

[11] See *Tongdian*, 72.1970.
[12] See Mansvelt Beck (1990, 147–9, 242–6) and Goodman (1998 108–9).
[13] For more on the southern transmission of the Supernal Icon *li*, see Morgan (2015).

(7) Yang Wei, lastly, emerges as a *li* expert amongst the Masters of Writing in the following year. After his character assassination of Han Yi in 226 CE (Section 4.2.1 below), the story becomes his.

4.1.4 New Emperor, New Regrets, 227–239 CE

> In Emperor Ming, Luminous Inception 1 (237 CE), Gentleman of the Masters of Writing Yang Wei constructed the Luminous Inception *li* and submitted it to the throne. The emperor thereupon reformed the first month and implemented [Yang] Wei's *li*. [He] took the month established at *chou*.$_{B02}$ as the first month, changing month III of that year to the first month of summer (month IV). Though the first, middle and last months of the seasons were different from those [counted from] the Xia first month, when it came to the suburban sacrifices, the spring and winter hunts, and the proclamation of seasonal ordinances, all of these [were done] according to a first month established at *yin*.$_{B03}$ (as per the old Xia calendar). In year 3 (239 CE) the emperor died, and [the court] reverted to the Xia first month (*JS* 17.503).

Upon taking the throne in 226, Cao Pi's brother Rui – Emperor Ming (r. 226–39 CE) – issued an impassioned call to start over:

> Since the Yellow Inception, Ruists have all been discussing the first month, some considering it advisable to reform, some considering it right to not reform, and, with the intention of selecting from difference and dissent, [the matter] has gone unresolved to our day. [We] had heard of this in [Our] time at the Eastern Palace (as heir apparent), being always of the opinion that the master (Confucius) had created the *Spring and Autumn Annals* and circulated three-concordance [metaphysics] as a model for later kings – that the first month of each should follow [its] colour [correlate], which differs in accordance with inheritance. Since the [time of] Five Thearchs and Three Kings, [this] sometimes passed between father and son, [who were] of the same body but different virtues, sometimes (referring to Shun) by being 'sent to the great plains at the foot of the mountains' and 'receiving [the former ruler's] retirement in the temple of the Cultured Ancestor' (*Shangshu zhushu*, 3.34b–35b), and sometimes by taking up shields and dagger-axes to administer punishment at the bidding of heaven. Though [they] took different steps in different times, there has never been a case of not changing the first month, using robe-colour or elucidating of cultured (ritual) objects to display the tokens of having received the mandate. Speaking about it from this [perspective], why mustn't reforming [the first month] be correct? (*Song shu*, 14.328–9).

The Excellencies, Ministers and their subordinates were thereupon called to a 'broad deliberation' (*bo yi*), where Gaotang Long came out in defence of his former pupil, offering a barrage of classical precedents from the *Changes, Documents, Odes, Record of Rites, Spring and Autumn Annals*, and their respective weft texts. Opinions remained divided along party lines, Wang Su's (195–256 CE) anti-Zheng Xuan ritualists refusing to budge (Table 4.1).

Table 4.1 *Opinions for and against first-month reform in 227* CE *(*Song shu, *14.3330)*

for	Palace Attendant	Gaotang Long
	Grand Commandant	Sima Yi
	Supervisor of the Masters of Writing	Wei Zhen
	Master of Writing	Xue Ti
	Supervisor of the Palace Writers	Liu Fang
	Gentleman-in-Attendance of the Palace Writers	Diao Gan
	Erudite	Qin Jing
	[Erudite]	Zhao Yi
	Captain of the Capital Region by direct appointment	Ji Qi
against	Palace Attendant	Miao Xi
	Regular Cavalier Attendant	Wang Su
	Gentleman of the Masters of Writing	Wei Heng
	Member of the Suite of the Heir Apparent	Huang Shisi

Nothing happened with first-month reform, but some time later in the Grand Harmony period (227–33 CE) the Grand Clerk memorialised an eponymous *li* addressing the inaccuracies in the Quarter-remainder *li* (the Supernal Icon and Yellow Inception *li* being off the table). The emperor ordered his former tutor to work with Master of Writing Yang Wei and Expectant Appointee Luo Lu to perform the necessary tests by merit of Gaotang's 'finesse (*jing*) with *tianwen*'.[14] Yang Wei was less impressed with his collaborator, reporting that, in *li*, 'Lu got the solar eclipses, but not before the dark moon (last day of the month) was out; Long did not get the solar eclipses, though [he did place syzygy] at the end of the dark moon'. Unable to agree, they were ordered to 'follow the Grand Clerk'.[15] Unfortunately, we do not know to whom this refers, nor what came of the Grand Harmony *li*.[16]

Sixteen years and two emperors in, things were at an impasse, but everything changed on 7 April 237, when Shanchi County reported the appearance of a yellow dragon in a replay of events prophesying the new dynasty.[17] This set

[14] *Wei lüe*, cited in *SGZ* 25.708 (comm.).

[15] Ibid.; cf. Hasebe (1991, 10–11). The solar eclipse to which Yang Wei refers is probably that of 9 January 232 (max. 23% at Luoyang) or 25 June 233 (max. 15%).

[16] In 233–7 CE, *SGZ* 25.709 tells us that Gaotang was 'transferred to palace attendant while kept on as Prefect Grand Clerk', but the *Wei lüe*, cited in *SGZ* 25.708 (comm.), states that '[Yang] Wei and [Luo] Lu were Grand Clerks' in the same period. Add to this that Xu Zhi also occurs as Grand Clerk in concurrence with Gaotang Long in *JS* 17.498–9 and one gets the impression that there was more than one Grand Clerk at any given time, the way that the Shi Zhao court (319–51 CE) appointed a 'female Grand Clerk' to supervise her male counterpart (*JS* 106.2765). For whatever it's worth, the *Wei lüe*, cited in *SGZ* 3.108 (comm.), identifies the Grand Harmony *li* with Yang Wei's later Luminous Inception *li*.

[17] *SGZ* 3.108. Earlier, *SGZ* 2.58 tells us that a yellow dragon appeared in 176 CE and was prophesied to return again in less than fifty years' time to mark the ascendancy of the virtue earth, which it did in 220 CE.

things in motion. Emperor Ming (again) called a 'broad deliberation' – this one summoning every pay grade from 600 *shi* up – and when this (again) failed to deliver a consensus in the following month, he decided to take matters into his own hands.[18] All at once, Emperor Ming declared the opening of the Luminous Inception period (237–9 CE) and 'changed month III of that year to the first month of summer (month IV)'. This was a lot to digest, so the emperor left additional matters of regalia to deliberation; he also decided to keep the schedule of state rituals as it was, because the spirit world, apparently, still ran on 'Xia' time.[19] Opportunity at hand, Yang Wei memorialised a Luminous Inception *li from within the secretariat*, which was received, read and implemented without discussion.[20]

On New Year's Day, Luminous Inception 3-I-1 *dinghai*.$_{24}$ (22 January 239), the emperor died. Li Chunfeng notes only that the court 'reverted to the Xia first month'. Whatever the debate – whatever the effects, real or perceived, of running separate civil and ritual calendars – Cao Fang (r. 239–54 CE) was faced with an impossible dilemma by the winter of 239 CE: as emperor, he would be holding New Year's audience with his ministers, feasting and merry-making, on the one-year anniversary of his father's death. He could push the ceremony back to the second, or maybe the sixth, but of all the solutions proposed to him he went with Master of Writing Lu Yu's idea to simply change New Year's Day back to where it always was.

Upon reflection of the matter of [your] memorials, [We feel as if] Our five organs are severed. What to do? [Oh], what to do? [Our] meritorious ancestor, Emperor Ming, abandoned the subcelestial realm on New Year's Day, and every time [We] think about the approach of this day with the Empress Dowager [We feel as if] our hearts are being peeled and torn. [We] cannot hold court for the mass of lords on this day, for that is to receive praise and felicitation, and meeting on the second is furthermore against [all] precedence. [We] have heard that it is right to return to the Xia first month, and though [We] violate the former emperor's understanding of the three concordances, this is, for its part, [an expression of His] descendants' grievous misery and eternal remembrance. Moreover, since the Xia first month 'gets the numbers of heaven' (*Zuo zhuan*, Shao 17), may [this] the month established at *yin*.$_{B03}$ be made the beginning of the year (*Song shu*, 14.332–3).

18 See Wang Yin's *Wei shu*, cited in *SGZ* 3.108 (comm.).

19 *Song shu*, 14.330–2; cf. Hasebe (1991, 11–12). Note that Wang Yin's *Book of Wei*, cited in *SGZ* 3.108 (comm.), quotes an edict to the same effect as that recorded in the *Book of Song*, but in different language.

20 *SGZ* 3.108 and *JS* 17.503 suggest that *li* reform was ordered together with the first-month reform, but Yang Wei's preface to said *li* points to the fact that it came only later: 'Now that the era has been changed to Luminous Inception, it is apt that [this] be called the Luminous Inception *li*' (*JS* 18.536).

4.1.5 *Collapse, 239–317* CE

> When Emperor Wu [of Jin] (r. 265–90 CE) ascended the throne in Grand Beginning 1 (265 CE), [he] inherited the Luminous Inception *li* of the [Cao] Wei but changed its name to the 'Grand Beginning *li*'. Yang Wei's [procedures for] calculating the five [planets] were particularly loose and wide (*shu kuo*), thus [this part of Yang] Wei's *li* was swapped out for the Supernal Icon's five[-planet] method after Emperor Yuan (r. 317–22 CE) crossed to the left bank of the Yangtze (in 317 CE) (*JS* 17.503).

In 265 CE, General Sima Yan (236–90 CE) deposed the last Cao Wei emperor and established the Jin dynasty (265–420 CE). In the new round of consensus building, it was decided that the Simas had come to power by force of metal (*Wei : earth, earth → metal*), but as eager as Sima Yi (179–251 CE) once was to change the first month in 227 CE, his grandson, Emperor Wu (r. 265–90 CE), kept the calendar and the *li* as they were. Renamed after the Grand Beginning period (265–74 CE), Yang Wei's *li* would continue in use into the fifth century at both Liu Song (420–79 CE) Jiankang and Tuoba Wei (386–535 CE) Luoyang, making it one of the longest-running *li* in Chinese history.

To what did it owe its political success? It was obviously not to its planetary models, which were scrapped in 317 CE, so that leaves calendars and eclipse prediction. At its core, Yang Wei followed the example of Han Yi in taking the Supernal Icon *li* and upping the Dipper parts – Han Yi went from $\frac{145}{589}$ (= 0.2462) to $\frac{1205}{4883}$ (= 0.2468), and Yang Wei to $\frac{455}{1843}$ (= 0.2469). It is difficult to say what this may have meant for the 'tightness' of a civil calendar prior to the 'fixing' of the mean lunation (Section 3.3.3 above), but his procedure text, preserved in the *Book of Song* and *Book of Jin*, offers much that is new in terms of eclipses.[21] Namely the Luminous Inception *li* is the first to feature a procedure for calculating 'first corner' (*qi jiao*) and the extent of 'depletion' (*kui*) – the first, that is, on record.[22]

4.2 The Oral Argument

4.2.1 *Transcript*

Having now set the stage and introduced the cast of characters, let us turn to the deliberation of 226 CE. Opening the debate is **Grand Clerk Xu Zhi**, who reiterates the rationale behind Han Yi's efforts to provide the Cao Wei court with an alternative to the Supernal Icon *li* as a replacement for the Quarter-remainder *li*:

[21] *Song shu*, 12.233–58; *JS* 18.536–52.
[22] On this procedure, see Liu Hongtao (2003, 204–22).

Liu Hong's procedure for lunar motion has been in use for more than forty years now, and [it] is repeatedly perceived to miss the mark (*shi*) by one *chen* and some odd [fraction] (*JS* 17.499).

This simple criticism provokes an outpouring of support for the late Liu Hong. The first to his defence is **Sun Qin**, who interjects with a history lesson:

[Grand] Clerk [Sima] Qian (*c*.145–*c*.86 BCE) constructed the Grand Inception.[23] Afterwards, Liu Xin (*c*.50 BCE–23 CE) considered it loose (*shu*) and thus made the Triple Concordance. In [87–8 CE], this was reformed to the Quarter-remainder. Using the heavenly *du* of the [sphere] sight to examine (*kao*) and match (*he*) foretoken and response (prediction and observation), [it was discovered that] the time experienced error and slip (*cha die*), overshooting solar eclipses by a half-day. By [172–8 CE], Liu Hong had reformed to the Supernal Icon,[24] the foretokens of the seven luminaries of heaven that it predicts (*buo^{B}) matching the order that between heaven and earth exists (*ziɑ^{B}) (*JS* 17.499).[25]

Sun Qin's point here is ambiguous. He would seem to say that we should respect Liu Hong's authority on matters of *li* by merit of demonstrated 'tightness', but it is the fact of progress and/or obsolescence that gave Liu the right to overturn the authorities of *his* day, and this holds just as well for Han Yi (if he can narrow the gap of 'one *chen* and some odd [fraction]'). Whatever his point, one notes that Sun feels the need to deliver it wrapped in rhyme and the mysterious language of prophecy literature, elevating the discourse on 'tightness' to a millenarian religious register.

Dong Ba responds to this in kind, turning the conversation to how the sage kings went about the business of *li*:

The sage men tracked the [sun] from gnomon shadows, verified (*xiao*) the [moon] from quarter and full moons, elucidated the [planets] from appearances and concealments, and settled right and wrong from the first and last days of the month. Quarter moons, full moons, concealments and appearances are the net threads (*ji*) of *li* numbers and are brilliant for inspection and verification (*jian yan*) (*JS* 17.499).

Here again we see sage gnosis presented in terms of the instruments and empirical methods of the *li* man, making 'the sagely method' an attainable norm. 'Let us look at the *numbers*', Dong would seem to say in response to Sun Qin, 'it is not the *man* but the *method* that counts'! In terms of rhetoric, one notes, an appeal to sagely authority seems to demand less by way of literary flourish to get one's point across.

23 This is not true; see Section 1.2.1 above.

24 On the confusion surrounding the Supernal Icon *li* and 'reform', see Section 1.2.2 above.

25 I would like to thank Jeff Tharsen for confirming the last two lines of Sun Qin's statement rhyme, pointing to Ting Pang-hsin (1975, 76), Luo & Zhou (2007, 142–3) and Coblin (1983, 100–5) as concerns *fu* 符 ('foretoken') and *xu* 敘 ('order'). The phonetic reconstructions here are per the Later Han Chinese of Schuessler (2007).

At this, **Xu Yue** enters the fray, redirecting Dong Ba's affirmation of the sagely method to support Sun Qin's argument from authority. He too begins with a history lesson:

Liu Hong absorbed [himself] in inner contemplation for more than twenty years with the *li*'s lag behind heaven, referencing and comparing the *li* procedures of Han experts in the Grand Inception, Triple Concordance and Quarter-remainder [*li*].

If one tests (*ke*) the quarter and full moons from the two sights' perimeter intervals, [one finds that] lunar motion terminates (repeats) once every nine years – what [we] call the 'nine roads'. In nine rules (*zhang*) – 171 years – the nine roads [makes a] small termination, and in nine nines – 81 rules – there are 567 parts and nine [small] terminations, advancing and retreating from four *du* and five parts before $\text{Ox}_{\text{.L09}}$ (the winter solstice).[26] Scholars duteously sought to match this with the Quarter-remainder by simply subtracting the 63 parts of one road such that the parts would not carry down. This is why it was loose and wide (*shu kuo*) – it was because the Dipper parts was too large. If one tests (*ke*) quarter and full moons via *du* measuring the position of the moon at dusk and dawn, then one will know whether the added hour is ahead or behind; it is not [however] advisable to use the two sights' perimeter intervals (*JS* 17.499).

Where Sun Qin grounds Liu Hong's credibility in a rhyming couplet about prophecy, Xu Yue digs into the specifics of the technical challenges that he faced in the 170s. The challenge, as discussed in Section 1.2.2 above, was that the 'nine-roads' model for lunar speed correction *worked*, but it did not work within the numerical superstructure – the *lü* – of the Quarter-remainder *li*, which was based on factors of two. To get it to work, one had to round off 'part' remainders here and there as a stop-gap, which invoked protestations of 'looseness'.[27]

Xu Yue identifies two factors behind the theoretical lock-in with which Liu Hong was faced: the reliance on *liang yi guo jian* ('the two sights' perimeter intervals' ?) to measure inequalities in lunar progress and the use of a 'Dipper parts' that was too large. The expression *liang yi guo jian* would seem to refer to perimeter readings taken from the equatorial and ecliptical 'sights' (rings) of an observational armillary sphere, namely the Luoyang observatory's 'Grand Clerk's Bronze Sight' of 103 CE.[28] We raised doubts about the observational effectiveness of sphere instruments in Chapter 2, and here Xu Yue would seem to confirm our suspicions: if you want reliable numbers, 'it is not advisable to use . . . sights' in place of rise–set timings against a simple horizon. As to the problem of Dipper parts (see Section 4.1.2 above), Xu Yue continues:

[26] Note that Xu Yue's description here of 'terminations' moving around the winter solstice comes word for word from Liu Xin's Triple Concordance *li* in *HS* 22.1007.

[27] See Ōhashi (1982, 12–14).

[28] The word *guo* 郭 (or *kuo* 廓) can mean 'rim' or 'outline', as in the outline of the moon or the raised perimeter of a round copper coin, and one notes that *HHS, zhi 2*, 3030, uses *guo* as a verb in the context of armillary sphere observation – 'that the Clerk's Office check quarter and full moon by *guo*-ing solar and lunar motion'.

Hong added twelve eras (*ji*) to the Grand Inception *li* origin, reduced ten from the fraction at the end of Dipper.$_{\mathrm{L08}}$, and started [his] origin from year *jichou*.$_{26}$.[29] In addition, [he] made procedures for the slow–fast crossing coincidence (speed correction) as well as yellow-road north polar distance (polar co-latitude) of lunar motion and for the [planets]. [He] reasoned out (*li*) their truths in a way that was pure and tight and that [I] believe can operate for a long time [to come].

Now, all of what Han Yi has constructed uses Hong's methods (*fa*). [He] slightly increases the fraction at the end of Dipper.$_{\mathrm{L08}}$, but there is barely any difference. As for Yi's additions and subtractions, their rendering was not without thought, but [his] ten procedures are newly established and are as yet imperfect, being sometimes less than completely effective (*xiao*) as regards solar eclipses.

Of the essentials of *li* testing (*xiao*), the [most] essential is the solar eclipse. In [172–8 CE], Hong was at the time a court gentleman and wished to reform the Quarter-remainder, [so he] first verified (*yan*) it above (i.e. in heaven and/or in writing to the emperor) via solar eclipse. The eclipse was on a clear day, the added hour was at *chen*.$_{\mathrm{B05}}$ (08:00–10:00), and the eclipse went from bottom to top, intruding two-thirds (over the disk of the sun). After the matter had been inspected, it was [found to be] like Hong said. [All] within the oceans recognised the truth (*zhen*) of it, and there was no one who did not hear of it. Ever since Liu Xin, there has been no match for Hong (*JS* 17.499–500).

At this point in the text comes a list of five eclipse observations from 221 to 223 CE to which are compared the instant of syzygy as calculated by the Yellow Inception *li*, the Supernal Icon *li*, and the Supernal Icon 'ebb-and-flow' (*xiaoxi*) method (Tables 4.2 and 4.3). Setting aside the question of what these numbers *mean*, let us focus on the score:

Out of a total of five tests (*ke*) of solar and lunar eclipses, the Supernal Icon had four (fars) [closes], and the Yellow inception had one close (*JS* 17.500).[30]

Han Yi is stupefied by these numbers, and it is in his response that the oral nature of this deliberation becomes apparent:

Yi challenged Xu Yue about the test results (*ke*): 'The Supernal Icon ebb-and-flow can only be subtracted, it cannot be added. There is no way to explain *adding* it. [This] is useless!'

Yue said, 'The original procedure itself has ebb-and-flow; [I] received the master's method (*shi fa*), and it is the ebb-and-flow that makes [it] astonishing. [I] am afraid that [this] cannot be changed (*gai*) and thus have arranged [here] the proper method of ebb-and-flow'.

Yi backed down (*JS* 17.500).

It was no secret that Han Yi's work was derivative – his *lü* are built on Liu Hong's, as we see in Section 4.3.1 – but the reform process would not have got

[29] For relevant Supernal Icon *li* numbers, see Section 3.3.1 above.

[30] The text here mistakenly reads 'far' for 'close' in the case of the Supernal Icon *li*, compare to the individual results in Tables 5.2 and 5.3.

Table 4.2 Book of Jin *solar eclipse data**

Item	Date	LAT	Assessment	Error	(cal)
SI(ef)		14:00	'close with heaven'	0	0
observed		14:00			
calculated	**5 Aug.**	**14:07**	**221 CE, magnitude 0.151**		
		†17:07			+18
SI		17:10	'1 $\frac{7}{12}$ *chen* behind heaven, close'	+19	+19
YI		19:10	'2½ *chen*, far'	+30	+31
QR	6 **Aug.**	14:48			+148
Item	Date	LAT	Assessment	Error	(cal)
SI		12:30	'2 $\frac{2}{12}$ *chen* ahead... far from heaven'	−26	−21
		†12:36			−21
SI(ef)		14:00	'1 $\frac{1}{12}$ *chen* ahead, ... far from heaven'	−13	−12
calculated	**30 Jan.**	**16:04**	**222 CE, magnitude 0.054**		
observed		16:10			
YI		17:50	'½ *chen* behind heaven, close'	+6	+11
QR		18:03			+12
Item	Date	LAT	Assessment	Error	(cal)
		†13:46			−7
SI		14:00	'1 *chen* ahead of heaven, far'	−12	−5
YI		14:10	'½ *chen* ahead of heaven, close'	−6	−4
calculated	**19 Jan.**	**14:58**	**223 CE, magnitude 0.915**		
observed		15:00			
SI(ef)		16:00	'close to hitting heaven'		+7
QR	20 Jan.	04:08			+79

* Above/below the reported time of syzygy ('observed'), I give the local apparent time (LAT) of maximum eclipse as calculated for Luoyang (34° 45′ N, 112° 28′ E, +144 metres) with the Besselian elements and values for ΔT provided by Fred Espenak and Jean Meeus' *Five Millennium Canon of Solar Eclipses* (http://eclipse.gsfc.nasa.gov/eclipse.html). As regards Supernal Icon (SI), Supernal Icon 'ebb-and-flow' (SI[ef]) and Yellow Inception (YI) predictions, note that reported ('Error') and calculated ('cal') errors are given in units of 1/12 *chen* (10 minutes) as per the original, but that I convert 'added hours' into the equivalent hour in the local apparent time (LAT) column, e.g. '17:10' for '*shen*.B09 and a half, strong' (see Appendix). The obelisks (†) indicate values calculated according to the supernal icon *li* procedure text. Contradictory numbers in the original are corrected as per Qu Anjing (1994).

this far were he and his office not confident of its superiority. After several years of development, testing and political consensus building, however, a private scholar appears from nowhere with game-changing numbers from the master himself. Four to one is not a narrow margin.

What is this 'ebb-and-flow' procedure that so undermined Han Yi's case for 'tightness'? We have no idea, and neither, for that matter, does Yixing

Table 4.3 Book of Jin *lunar eclipse data**

Item	Date	LAT	Assessment	Error	(cal)
observed		00:00	'and moon at [12:00]'		
calculated	**21 Aug.**	**00:09**	**221 CE, umbral mag. 0.920**		
SI(ef)		02:00	'1 *chen* behind, close'	+12	+12
SI		04:00	'2 *chen* behind heaven'	+24	+24
		†04:43			+28
QR		09:10			+55
YI		12:00	'6 *chen* behind heaven, far'	+72	+72
Item	Date	LAT	Assessment	Error	(cal)
SI		23:00	'2 *chen* ahead of heaven, close'	−24	−21
		†23:00			
SI(ef)		00:00	'1 *chen* ahead'	−12	−15
observed		02:00	'and moon at 14:00'		
calculated	**5 Jan.**	**02:31**	**223 CE, umbral mag. 0.291**		
YI		06:10	'2 1/12 *chen* behind heaven, far'	+25	+22
QR		09:45			+43

* Note that the observations give an 'added hour' for both sun and moon exactly opposite one another, that of the moon indicating position (see Section 2.1.4). Note that the 'added hour' *zi*.$_{B01}$ (00:00) for the eclipse of 21 August 221 has been emended from *ren*.$_{S09}$ (23:00) as per the reported errors of the predictions.

(683–727 CE), who reports that 'it was originally the ebb-and-flow that was [the most] astonishing', citing Xu Yue word for word, 'but the procedure was not passed down' (*XTS* 17B.622). Indeed, there is no mention of 'ebb-and-flow' in the procedure text for the Supernal Icon *li* as preserved in the *Book of Jin*, nor does Xu Yue or anyone else in the third century bother to explain what it is. Whatever it was, Xu Yue and Han Yi knew about it, but they knew different versions, and the fact of its apparent performance and Xu Yue's direct line to the master was enough to shut up the Assistant Grand Clerk.

In terms of rhetoric, Xu Yue has handed Han Yi a 'black box': a product of years of observation and calculation, the process behind which is effaced; an inscription in which complex questions about the overall quality of competing products is reduced to neat, commensurable quantities; a complete and self-contained data-set, which points to one inexorable conclusion; something, most importantly, so complex and well manufactured that it is impossible to deconstruct and contest without a considerable investment of time and resources, let alone *on the spot*, in the context of a debate (Latour 1987, 2–3). We too might be inclined to take his word for all this were it not for the way that he impugns himself along the way. As to the story Liu Hong's prediction in 172–8 CE, the first documented procedure for calculating the direction and magnitude of eclipse comes, as mentioned, in the Luminous Inception *li* of 237 CE.[31] It is possible that this, like 'ebb-and-flow', did not make it into the *received* Supernal Icon *li*, but this is not the only problem: Liu Hong was made gentleman-of-the-palace no earlier than 176 CE, and there was no eclipse between 176 and 178 CE that even remotely matches Xu Yue's description.[32] Whether or not 'there was no one who did not hear of it', the story smacks of legend.

Next, inserted into the text of the debate, is a list of the observed dates of fourteen planetary visibility phenomena to which are compared the predictions of the Yellow Inception and Supernal Icon *li* (Table 4.4). Again, what matters here is the score:

> Out of a total of fifteen appearances and hidings of four [planets], the Supernal Icon had seven closes and two hits, and the Yellow Inception had five closes and one hit (*JS* 17.501).

This too looks bad for Han Yi, who offers no explanation for these results. The next speaker is **Gentleman of the Interior Li En**, who turns the conversation back to the more immediate problem at hand:

> By comparison with the Grand Clerk's heavenly *du* (as calculated from the official Quarter-remainder *li*), on the other hand, [one notes that *there*] the full moons of 2-VII (21 August 221) and 3-XI (5 January 223) are not even on the same *day* as the [real] heavenly *du* – the added hour of lunar eclipse being six [double]-hours and a half (thirteen hours) behind heaven. This is not what one would call 'trailing by three *du*'; [we] determine [rather] a lag behind heaven in excess of a half-day (*JS* 17.502).

Whatever the fraction of a *chen* by which Liu Hong and Han Yi's *li* 'miss the mark', Li En reminds us, the official *li* was off by a unit of *days*, and thus *anything* would be an improvement in terms of reform. Li too is appealing to numbers, and so too are his numbers somewhat suspect upon further examination. First, whether or not prediction and observation are 'even on the same day' depends on one's definition of 'day', and, from procedure texts, we know that the date of lunar eclipse (and lunar eclipse alone) is counted from dawn. For the eclipse on 5 January 223 at 02:31, for example, the Quarter-remainder *li* is thus *more* 'off target' for predicting the eclipse several hours later, past dawn, than the Supernal Icon *li* several hours earlier, prior to

[31] On the history of eclipse prediction, see Steele (2000, 161–216) and Qu Anjing (2008, 390–531).

[32] The one solar eclipse that occurred at Luoyang during this period was that of 27 November 178, which did occurr at *chen*.$_{B05}$ (08:00–10:00), but which went from top to bottom, obscuring only 25.6 per cent of the solar disk (*Nasa Eclipse Website*). The two other eclipses recorded in the *Book of Later Han* for this period – 9 November 177 and 7 March 178 – are false reports (*HHS* 8.333–41).

Table 4.4 Book of Jin *planetary visibility data**

no	Phen.	Observed	SI pred.	err.	YI pred.	err.	Dif.
Jupiter							
1	FMR	222 Jun. 20	13 Jun.	−9	11 Jun.	−7	2
			† 11 Jun.	−7			
Mars							
–	–	–	–	–	–	–	–
Saturn							
2	FMR	221 Dec. 27	22 Dec.	−5	19 Dec.	−8	3
3	LES	222 Dec. 2	2 Dec.	0	28 Nov.	−4	4
4	FMR	223 Jan. 11	4 Jan.	−7	1 Jan.	−10	3
Venus							
5	LMR	222 Aug. 9	21 Jul.	−19	18 Jul.	−23	2
6	FES	222 Nov. 2	11 Oct.	−23	8 Oct.	−25	2
Mercury							
7	FMR	221 Dec. 18	14 Dec.	−4	13 Dec.	−5	1
8	LMR	222 Jan. 13	15 Jan.	+2	14 Jan.	+1	1
9	FES	222 Jun. 14	14 Jun.	0	13 Jun.	−1	1
10	LES	222 Jul. 9	16 Jul.	+7	15 Jul.	+6	1
11	FMR	222 Aug. 19	3 Aug.	−16	2 Aug.	−17	1
12	LMR	222 Aug. 31	4 Sep.	+4	3 Sep.	+3	1
13	LMR	223 Jan. 3	29 Dec.	−5	28 Dec.	−6	1
14	FES	223 Feb. 16	31 Jan.	−16	31 Jan.	−16	–
			† 1 Feb.	−15			

* The table notes four types of visibility phenomena: first morning rising (FMR; *chen xian*), last evening setting (LES; *fu*), and, for the inferior planets, last morning rising (LMR; *chen fu*) and first evening setting (FES; *xi xian*). I have marked the winner, by the text's accounting, in grey blocks, and added my own calculations for the Supernal Icon *li*, where it differs, marked by obelisk (†). The last column ('Dif.') gives the absolute difference (in days) between the two predictions.

midnight. (Were the definition of 'day' here the standard one, the tables would be turned). Second, as we see in Tables 4.2 and 4.3, the Quarter-remainder's lag was more in the order of 4½ *chen* (nine hours) at this time, which makes Li's 'half-day' an exaggeration. The juxtaposition of measuring units is clearly for rhetorical effect.

Out of nowhere, **Dong Ba** reinserts himself into the discussion to pontificate on first-month reform. He begins once again with a sage-time history lesson:

In the past, Fuxi first created (*zao*) the eight trigrams and invented (*zuo*) their three lines in order to symbolise (*xiang*) the twenty-four *qi*. The Yellow Emperor followed suit and first invented (*zuo*) the Adjusted *li*. Over eleven dynasties and five thousand successive years there was a total of seven *li*.

Zhuanxu made the first month of spring, month I; [we use] today his origin. At the time, on the first day of the first month, on the establishment of spring.$_{Q01}$, the five [planets] congregated at Heaven's Temple – that is, Hall.$_{L13}$ – as the ice first began to melt, the hibernating insects first came out, and the cocks first began to cry three times. It was the rise (*zuo*) of seasons in heaven, the rise (*zuo*) of prosperity on earth and the rise (*zuo*) of music/joy among man, and none of the birds, beasts or myriad creatures did not respond to one another in harmony. Thus is the sage man Zhuanxu the progenitor of *li*.

Tang invented (*zuo*) the Yin *li*, which no longer took [the coincidence of] the establishment of spring.$_{Q01}$ with the first day of the first month as its node, changing [instead] to an origin head [at the coincidence of] winter solstice.$_{Q22}$ with the first day of the eleventh month. Down to Zhou, Lu and Han, everyone followed this node, on which the corrections of the four seasons (regulated intercalation) have been [hitherto] based. The Xia won the support of heaven so as to succeed Yao and Shun by following Zhuanxu. *Dai the Elder's Record of Rites* says, 'The *li* of the Yu Xia established its first month on the first month of spring' (*Da Dai Liji*, 9.11b), that is what this refers to (*JS* 17.502–3).

Dong Ba is off topic. First-month reform has nothing to do with the 'tightness' of the eclipse and planetary models being discussed here, nor had the matter been up for debate since Emperor Wen's decree of 220 CE.

Whatever the point of this interjection, **Yang Wei** steps in to shut down Han Yi. Yang is a numbers man, like Xu Yue, but his closing statement is decidedly personal:

Looseness and tightness can be known in sixty days, one [need] not wait ten years. If you do not follow the/a method (*fa*), this is [like] abandoning the compass and T-square when checking (*jiao*) squareness and roundness; it is [like] forsaking the scale and balance when investigating (*kao*) lightness and heaviness; it is [like] abandoning the *chi* and *cun* (ruler) when testing (*ke*) length and shortness; and it is [like] turning away from distinctions and principles when discoursing upon right and wrong. If one does not first fix a 'root' method for checking *li*, carelessly entertaining 'branch' disputes that abandon method (*fa*), then this is what Mencius (fl. 320 BCE) refers to as 'being able to make a one-*cun* square of foundation reach a greater height than a tall building (by suspending it where it doesn't belong)' (*Mencius* VIB.1).

Now, from Han Yi's reliance on Liu Hong's procedures [we see that he] knows to esteem his procedures and treasure his methods, and yet [he] dismisses his discourse, forsakes his procedures, abandons his words and strays from his endeavour, the inevitable result of which will be to stop Hong's unique and marvellous model (*shi*) from being transmitted to future ages. If [the case be that he] is straying from it knowingly, then [he] has intentionally forsaken [the] master; if [it be that he] is basing himself upon it unknowingly, then [he] has ignorantly arrived at a confused understanding (*JS* 17.503).

Yang Wei has something to say about everybody. Yang dismisses the entire three-year process up to this point as a senseless waste of time. Who was to blame? – Who was *not* to blame? 'Looseness and tightness can be known in sixty days', he insists, and the only reason why something so bloody simple could drag on for this long is because of the idiots and liars trying to tell us that 'loose' is 'tight' and 'tight' is 'loose'. The worst, of course, is Han Yi, whose disrespect of Master Liu Hong's accomplishments threatens their very place in history.

What does Yang Wei divulge of his sixty-day miracle solution? One word: 'method' (*fa*). Turning to 237 CE, we get a glimpse of Yang's 'method' in action: use your position in the imperial secretariat to sneak your *li* onto the emperor's desk while dragons have him on a reform spree, thus bypassing all testing and policy review. Yang's criticism of Han Yi focuses on his 'straying from [Liu Hong's] endeavour' – the point of contention thus far being 'Dipper parts' and 'ebb-and-flow' – so what has Yang to offer some eleven years later? Yang Wei's Luminous Inception *li*, one notes, does exactly the same: it borrows 'Hong's unique and marvellous model', it increases the Dipper parts and it omits any mention of 'ebb-and-flow'. What it offers by means of novelty – procedures for calculating the direction and magnitude of eclipses – Xu Yue already mentions in relation to Liu Hong in 226 CE. Xu's account smacks of legend, but the fact that it anticipates *those exact procedures* might not be a coincidence. Either way, Yang is at least modest and deferential to Liu Hong's 'unique and marvellous model' in his own work, right? Yang Wei's preface recounts the history of *li* to his day, *omitting Liu Hong entirely from that history* and dismissing all his predecessors' work as 'first tight and later loose and, thus, inadvisable for use' (*JS* 18.536).

It is for this [reason] that Your humble servant has previously, in days off from attending to the canons [of state], calculated and tested the heavenly road, examining it in prior literature, and verifying it in eclipse syzygies; detailing and honing it, and establishing anew a tight *li* that be neither ahead nor behind, but on the mark in heaven, past and present . . . This Luminous Inception *li* established by Your humble servant, its methods and numbers are concise and essential, and its implementation and use are close and tight; work on it, and one will find it saves time; study it, and one will find it easy to understand. Now, even if one made [the legendary mathematicians Ji] Yan and Sang [Hongyang] perform mental calculations, [sage-time inventor of mathematics] Li Shou operate the counting rods, Chong and Li direct the gnomon, and Xi and He examine the shadow to investigate the road(s) of heaven, and calculate and verify the sun and moon, to plumb the extremes of fineness and subtlety, and to exhaust the limits of procedures and numbers (*shu shu*) – none of this would be nearly so marvellous as [the marvel that is] I, your humble servant. It is for this [reason] that the *li* numbers of the accreted ages are all loose and not tight, and why, since the time of the Yellow Emperor, they have been reformed without end (*JS* 18.536).[33]

[33] Cf. Guan Yuzhen (2015, 108).

4.2.2 What to Believe

We need not *believe* the speakers at this debate. Yang Wei, for one, makes it perfectly clear that we should expect liars (and hypocrites) in their midst. Nor would it be responsible to hold them to what they say, because the point here is not to construct a philosophy of science but to take apart a man – a man standing in the way of the greater good (and/or/as one's personal ambitions). Is Yang Wei inconsistent? Not in the slightest: here and in 227–33 CE he consistently sabotages Gaotang Long and his clique; and here and in 237 CE he consistently insists on a history of *li* (with or without Liu Hong) building to him, Yang Wei, as the absolute living authority. Is Xu Yue to be trusted? He can be trusted to defend his master's reputation (upon which his own depends): whether or not the story and the data he pulls from his pocket are 'real', as such, they have an undeniably real effect in turning the debate in his favour. What is *real* above all else on this stage is the *agon* suffocating any hope for reform.

Whether or not the participants in this (or any other) *li* debate are arguing what they *believe*, we can rest assured that they are arguing it *in a way they believe to be effective*, which reveals something of their epistemic culture. First of all, one notes that the majority of written and oral arguments here appeal to historical precedent. That precedent takes several forms: the actions of ancient sage kings, the actions of recent historical figures, and the desultory technical injunctions of classical literature. The past is supple. Actors reshape it to mirror their present, they pick and choose from it to support opposing sides, but they appeal to the past even in arguing that it has no relevance to the present. In a debate of 175 CE, for example, Cai Yong (133–92 CE) counters two opponents who 'talk only of diagram and prophecy [literature], and what [they] say does not convince' by appealing to modern precedence, in 104 BCE, that ancient precedence is flawed – 'This is a case where [an origin] was effective over its predecessors despite it not being an origin [that appears in] diagrams and prophecies'.[34]

There are other individual strategies at play in 226 CE. There is Sun Qin, Xu Yue and Yang Wei's argument from authority – 'Ever since Liu Xin, there has been no match for [Liu] Hong'. There is Xu Yue's argument from association – '[I] received the master's method, and it is the ebb-and-flow that makes [it] astonishing; [I] am afraid that [this] cannot be changed'. There is Sun Qin's appeal to rhyme and prophetic language – 'Liu Hong had reformed to the Supernal Icon, the foretokens of the seven luminaries of heaven that it predicts matching the order that between heaven and earth exists'. There is also Yang Wei's *ad hominem* – '[we see that he] knows to esteem his procedures and treasure his methods, and yet [he] dismisses his discourse, forsakes his procedures, abandons his words ... '.

[34] *HHS, zhi 2*, 3038–9; cf. Cullen (2007a, 259–60).

At the end of the day, however, the one thing that elicits a response from the assistant director (and consequently shuts him up) is numbers.

4.3 Keeping Score

4.3.1 Numbers That Speak

The *Book of Jin* debate transcript offers two sets of numbers: the one, a comparison of predicted and observed eclipse times, and the other, first and last planetary visibilities. The observations are dated 221–3 CE, which coincides with the time frame of Han Yi's own tests. That said, the way that Xu Yue and Han Yi interact over the 'ebb-and-flow' eclipse results that the latter 'lays out' (*lie*) suggests that the one, at least, represents an emended data set produced by Xu. The data suffer some textual corruption: the planetary scores tally 'fifteen appearances and hidings of four [planets]', when all we have is fourteen, and the eclipse tally gives the Supernal Icon *li* 'four fars' when it clearly means the opposite. Still, we have enough to make these numbers speak.

As to units, the eclipse and planetary data use the standard civil date format (e.g. 5 August 221 = 'Yellow Inception 2-VI-29, *wuchen*.$_{05}$') plus, in the case of eclipses, an 'added hour' (*jia shi*) in twenty-four 'corner-and-chronogram' notation. The 'hour' is divided into 'strong' and 'weak' 'less'–'half'–'more' fractions for a precision of ten minutes (see Appendix).[35] The fact that the moon, when eclipsed, is given its own 'added hour' opposite the sun's corroborates our analysis in Section 2.1.4 above that these 'hours' are spatial co-ordinates. As to how this 'hour' was observed, the unit itself points to the armillary sphere at which some 'desired that it be proven' (Chen Qun, Section 4.1.2 above), but it may well be converted from 100-notch water clock time. (There was calculation involved either way, because the instant of mid-eclipse must be interpolated from first and last contact). As to how this 'added hour' was *predicted*, Yang Wei's is the first extant procedure text to detail these co-ordinates, but this, like other of his procedures, was clearly in practice by 226 CE.[36]

As to the observations, comparison with modern calculations basically confirms what is reported. I say 'basically' because first and last planetary visibility is not something that you can predict with any reliability, depending as it does on factors like eyesight and atmospheric conditions at the horizon.[37]

[35] See Qu Anjing (1994).
[36] Yang Wei's 'Calculate the Added Hour' procedure is preserved in *Song shu*, 12.245–6 and *JS* 18.548–9.
[37] See Cullen (2011, 234–5).

What one *can* do is use software like Planetary, Lunar, and Stellar Visibility (v3.1) to falsify such data. By that count, the observations in Table 4.4 check out. The eclipses also check out, according to my calculations in Tables 4.2 and 4.3: four of five observation times fall within our subjects' ten-minute precision from the calculated moment of maximum eclipse, while one lunar eclipse (5 January 223 CE) is reported three units of precision (thirty-one minutes) early. All in all, these look like legitimate observations.[38]

As to the predictions, comparison with results calculated from the extant procedure text confirms what is reported for the Supernal Icon *li*. Using the 'numbers and procedures' there, as in the previous chapter, I found fourteen of the nineteen predictions to match my own calculations. Of the outliers, we might attribute the three-minute discrepancy on solar eclipse 1 to rounding and the fourteen-minute error on solar eclipse 3 to convoluted instructions for the last (twenty-eighth) day of the lunar 'slow–fast sequence', on which this date falls.[39] Once again, allowing room for rounding, confusion and human error, this data set checks out.

The predictions, conversely, allow us to reconstruct something of the other two contenders. 'Ebb-and-flow', one notes, appears only in the eclipse data. One gathers that it is some sort of correction – 'ebb-and-flow can only be subtracted, it cannot be added' (Han Yi, Section 4.2.1 above). One also gathers that it is applied to the mean sun: the term itself is a reference to the seasonal (i.e. solar) 'ebb-and-flow' of *yin* and *yang* in Han-era *Book of Changes* metaphysics;[40] furthermore, Li Chunfeng attributes 'creating the solar/daily motion slow–fast' and positing 'that the sun travels on the yellow road, experiencing advance and retreat in terms of red-road lodge *du*' to Liu Hong (Section 4.1.1 above). This points to two possibilities. One is that 'ebb-and-flow' is an 'advance–retreat' reduction to the equator (Section 2.1.4 above), which would explain Han Yi's insistence that it only goes in one direction. The other is that it represents a solar 'slow–fast' correction, which would indeed 'make [it] astonishing' (Xu Yue, Section 4.2.1 above) seeing that even Li Chunfeng credits *that* to Zhang Zixin (d. 577 CE).[41] What do the numbers say? My own analysis (Table 4.5) would seem to support Chen Jiujin's (2008, 112–15) conclusion that 'this correction should relate to the ebb-and-flow in the

[38] On Chinese eclipse observation, see Steele (2000, 161–215).

[39] On Liu Hong's 'slow–fast sequence' for lunar anomaly, see Liu Hongtao (2003, 129–32) and Cullen (2002, 27). See also the recalculations of Qu Anjing (1994, 165).

[40] See Nielsen (2003, 274–6).

[41] According to Li Chunfeng, 'Zhang Zixin . . . was the first to awaken [to the fact] that the crossed roads of sun and moon [both] have inside & outside (displacements in latitude) and slow and fast (pseudo equations of centre)' (*Sui shu*, 20.561). For examples of the later scholarly consensus on this point, see Jiang Xiaoyuan (1991, 382–3), Chen Meidong (2003, 298–303), Zhang Peiyu et al. (2008, 425–8) and Sivin (2009, 297–8).

Table 4.5 *Comparison of Supernal Icon 'ebb-and-flow' data with expected solar 'slow–fast' and 'advance–retreat'**

		Hour			Slow–fast		Advance–retreat	
Date	ecl.	SI	SI-ef	dif.	calc.	error	calc.	error
221 Aug. 5	☉	17:07	14:00	−187m	−187m	0m	−3m	184m
221 Aug. 21	☽	04:43	02:00	−163m	−211m	48m	−9m	154m
222 Jan. 30	☉	12:36	14:00	+84m	+193m	109m	−10m	94m
223 Jan. 4	☽	23:00	00:00	+60m	+106m	46m	−3m	63m
223 Jan. 19	☉	13:46	16:00	+134m	+172m	38m	−7m	141m

* 'Fast–slow' is the solar equation of centre as calculated according to Meeus (1998) and converted into a correction for the hour of syzygy (in minutes) as per the Supernal Icon usages. 'Advance–retreat' is the solar reduction to the equator calculated via interpolation between 0 a maximum of 3 *du* at 45° from the solstitial and equinoctial points and likewise converted into a correction for the time of syzygy (in minutes).

speed of solar motion'. The other, non-mathematical explanation is that Xu Yue has made up these numbers.

All the *Book of Jin* tells us about the Yellow Inception *li* is this: it is derivative, it 'used an era divisor of 4,883 and a Dipper parts of 1,205', and this it did in order not to 'slip ahead of heaven' (Section 4.1.2 above). Gautama Siddhārtha (fl. 729 CE) helps fill out some of the numbers, from which we learn that Han Yi used a solar year of 365 $\frac{1205}{4883}$ days, a synodic month of 29 $\frac{6409}{12079}$ days, and a high origin in year *renwu*.$_{19}$ at the coincidence of winter solstice.$_{Q22}$ and XI-1 at midnight in 31360 BCE.[42] The 'era' falls just short of an integer number of anomalistic months – 4,883 years = 64,728.95 anomalistic months, using Liu Hong's value (27 $\frac{3303}{5969}$) – from which we can likewise deduce that Han Yi was using a lunar 'slow–fast' of 27 $\frac{35817}{64729}$ days. Using these numbers, we arrive at the following observations. First, the combination of Han Yi's origin and lunisolar elements places the moment of mean syzygy for solar eclipse 1 (5 August 221 CE) 3.14 hours after that calculated according to the Supernal Icon *li*, which comports with reports that the latter was 'ahead of heaven' and 'missing the mark by one *chen* and some odd [fraction] (> 2 hours)'. Second, the difference between the syzygies predicted in Tables 4.2 and 4.3 and results recalculated from Han Yi's mean lunisolar elements is consistent with a Supernal Icon-style lunar 'slow–fast' – it is consistent, that is to say, in

[42] *Kaiyuan zhanjing*, 105.4a. As to the origin, 'from high origin [in year] *renwu*.$_{19}$ to today is 32,072 [years] outside the count (i.e. to be counted exclusively)', the 'today' (*jin*) here, judging from other of Gautama's counts in the same fascicle, indicating 713 CE.

Table 4.6 *Comparison of 'fast-slow' lunar speed corrections in 221–3* CE *eclipse data*[*]

Supernal Icon				Yellow Inception				
JD	elapse	t/m	s/f	JD	elapse	t/m	s/f	Dif.
1801 995.20	–	–	+9.90h	1801 995.29	–	–	+10.99h	+1.09h
1802 010.59	15.48d	+17.25h	−7.34h	1802 010.99	15.70d	+22.47h	−11.47h	−4.13h
1802 173.02	163.33d	−2.18h	−5.17h	1802 173.23	162.24d	−4.20h	−7.27h	−2.10h
1802 512.45	339.43d	−4.00h	−1.17h	1802 512.75	339.51d	−2.11h	−5.16h	−3.99h
1802 527.06	14.62d	−3.63h	+2.46h	1802 527.08	14.33d	−10.37h	+5.21h	+2.74h

[*] The 'JD' column gives the Julian day of the eclipse prediction. The 'elapse' column gives the time elapsed between successive phenomena in days. The 't/m' column gives the difference in hours between the previous column (time elapsed between 'true' or 'fixed' lunar phases) and that elapsed between mean lunar phases (according to each system's values for the mean synodic month). The 's/f' column gives the lunar speed correction as *calculated*, for the Supernal Icon *li*, and *as necessary to compensate for the previous column*, in the case of the Yellow Inception *li*. Note that the Yellow Inception 's/f' column is calculated from the 'high origin' (*shang yuan*) cited in *Kaiyuan zhanjing*, 105.4a. The 'Dif.' column gives the difference between the two 's/f' columns.

terms of positive/negative direction, if being somewhat skewed in terms of magnitude (see column 's/f' in Table 4.6). In other words, it looks like Han Yi may have also tinkered with Liu Hong's famous lunar model.

With a sense of what lies behind the data, let us consider the sort of 'scientific test' our *li* men thought to construct. Eclipses were the best sort of observational data with which the early imperial *li* man could hope to work. In the words of Ptolemy (*c*.90–*c*.168 CE), 'we should rely especially on those demonstrations which depend on observations which not only cover a long period, but are actually made at lunar eclipses. For these are the only observations which allow one to determine the lunar position precisely'.[43] Xu Yue insists that 'of the essentials of *li* testing (*xiao*), the [most] essential is the solar eclipse' (Section 4.2.1 above), but Xu, of course, did not know about parallax. You get a sense of how important was the solar eclipse from the lengths to which actors went to observe it. You only spot an eclipse of magnitude 0.151 (221 CE) or 0.054 (222 CE) if you're *really* looking for it, but there were *two* 'watchmen of the sun' (*hou ri*) whose job this was at the Luoyang observatory, and they had been reporting sunspots and eclipses of this magnitude for years (see Table 4.7).[44] This presents us with an interesting paradox. On the one hand, you have gentlemen 'watching' (*hou*) and 'testing' (*xiao*) eclipses as *li* data, the assumption being the predictable 'constancy' (*chang*)

[43] *Almagest* IV.1 H266; tr. Toomer (1998, 173).
[44] On sunspot observation, see Chen Zungui (2006, 768–86). On why we can assume that eclipse reports are issued from the capital, see Stephenson (1997, 230–2) and Steele (2000, 191–2).

Table 4.7 *Solar eclipses of magnitude < 0.2 observed in Eastern Han Luoyang (25–190 CE)**

No	Date	Mag.	Source (*HHS*)		Note
			ji	*zhi* 18	
1	40 Apr 30	0.175	1.66	3359	
2	46 Jul 22	0.171	1.74	3359	
3	54 Jul 23	0.009	–	–	
4	61 Oct 02	0.017	–	–	
5	90 Mar 20	0.101	4.170	3362	'Not seen by Clerk's Office but reported from Zhuo Commandery'
6	168 Jun 23	0.002	8.329	3369	
7	169 Dec 06	0.069	8.311	3369	'Heard about from Fufeng'
8	171 Apr 23	0.193	8.322	3369	

* Magnitude is calculated as per the previous tables.

of the phenomenon. On the other hand, we are in an age where this phenomenon is also subject to omen reading (*zhan*) and ritual prophylaxis. Indeed, Cao Wei officials independently memorialised the same solar eclipse of 5 August 221 as a 'disastrous anomaly' (*zai yi*) impeaching the Defender-in-Chief. The emperor, privy to both readings, ordered that '[everyone] go sincerely about their own job . . . and stop denouncing the Three Excellencies' (*SGZ* 2.78).

We are faced with a similar paradox when it comes to the selection of planetary phenomena. The 221–3 CE data eliminates Mars from consideration – 'fifteen appearances and hidings of four [planets]' – and this was probably a wise decision. They were bad at Mars. The second-century BCE manuscript *Wu xing zhan* concedes that its '[advancing and retreating] are without constancy (*wu heng*) and cannot be taken as a [standard]' (line 45), and Li Yexing (484–549 CE) complains some eight centuries later that 'the one [planet], [Mars], does not always correspond with the [predicted] *du*, as its hiding and appearance are inherently lacking any constancy (*wu chang*)' (*WS* 107B.2698).[45] Conversely, actors treat the 'hiding and appearance' of *the other planets* as a 'constant' sufficient for official ranking when this, as already noted, is not something that you can actually predict. One might deduce from this that, while they saw 'anomaly' in natural phenomena like comets, the ancients believed (just as naively) in the knowable mathematical regularity of others we no longer recognise as such.[46] The other possibility, of course, is that

[45] Note that Liu Lexian (2004, 44) fills out the deficiencies in the *Wu xing zhan* text here from parallel passages from contemporary sources like the *Huainanzi* (139 BCE).

[46] This is particularly true for Mercury, whose orbit is both very eccentric and very close to the sun, making it difficult to *see*, let alone model. On Zhang Zixin's first attempt to deal specially with the variabilities of this planet, see Jiang & Niu (2001, 223–39).

the fact that atmospheric conditions might send one's calculations off target introduced a shared element of chance that they may not have felt inimical to the goals of testing.

As to scope, one notes that the sample size adduced in 226 CE is rather limited. Ptolemy relied on a bare minimum of data to establish his geometric arguments out of a desire for elegance, but his, at least, draw equally from personal observation and ancient records.[47] The early imperial *li* man had a similar wealth of observational data at his disposal – 'clerk's records' (*shiji*), 'watch notes' (*hou zhu*), and the like – to which he was by no means averse. Generally speaking, our sources prioritise these records in the context of private practice and public argumentation, particularly the solar eclipse dates recorded in the *Spring and Autumn Annals* for 722–481 BCE.[48] When the Clerk's Office is handed a *li* for independent review, however, one notes that it is generally put to live trial, as the Three Excellencies decided to do in 223 CE.

Why live trial, specifically? Several advantages suggest themselves over those of ancient records in this regard. First, prior to the discovery and implementation of the 'year difference' (*sui cha*), there was a serious problem (to which no one personally admitted) of producing calculations that 'above matched with antiquity, and below responded to the present' (Li Chunfeng, Section 4.1.1 above). There was also the problem of observational standards. As the demand for accuracy increased from the day to the hour, as we see in 226 CE, (and from the hour to the minute, etc.) old records like *Spring and Autumn Annals* dates became 'loose'. More importantly, third-party live trial eliminated the possibility of cheating and it introduced an element of chance that gave successful predictions the status of public spectacle, and *that*, as Xu Yue shows us, is one way to guarantee that '[all] within the oceans recognised the truth (*zhen*) of it, and there was no one who did not hear of it'.[49]

At the end of the day, it is a score that 'proves' the winner 'tight', and the score here is 4 close to 1 close, and 7 close and 2 hits to 5 close and 1 hit, respectively, in favour of the Supernal Icon *li*. From the tables, we can discern the following rules for scoring. First, 'distance' is relative (i.e. it is the same to be 'further' by ten minutes as it is by ten hours). Second, an absolute match gets you a 'hit' or 'bull's eye' (*zhong*), as distinct from being 'closer' by an unspecified distance, but the two are tallied together. Third, there are no ties: the odd number of observational rounds guarantees a winner, and where there *is* a tie for solar eclipse 3 and visibility 14, for example, one *li* somehow walks

[47] See Graßhoff (1990, 198–216), Jones (2005) and Jones & Duke (2005).
[48] See, for example, Zhang Peiyu et al. (2008, 166–86) and Sivin (2009, 314–29).
[49] Here I would point to Brush (1989; 1994) for an example of how this continues to be true in modern science.

away with the better score (as if by coin toss). The 226 CE data may be the earliest 'test report' (*ke*) of its kind, but one finds similar language in earlier discussion of live trials. Jia Kui (30–101 CE), for example, speaks of an edict of 85 CE 'ordering two watches' to test opposing claims about the position of winter solstice.$_{Q22}$ over the course of five years, wherein 'the superior would be he who got the most counting rods (*suan*)' (*HHS, zhi 2*, 3028). The connection between 'hits', 'misses', tallies and success is even clearer in 62 CE:

An edict [was issued] ordering [Yang] Cen to [conduct] a general watch (*hou*) and rank (*ke*) [his system] against the official *li*. Of a total of five quarter and full moons from month VII to XI the official *li* was in each case off target (*shi*), and Cen was in each case on target (*zhong*). On *gengyin*.$_{27}$, an edict [was issued] ordering Cen to be appointed to the office for quarter and full and lunar eclipses; [it] furthermore ordered expectant appointees Zhang Sheng, Jing Fang, Bao Ye et al. to rank (*ke*) Cen['s system] against the quarter-remainder method. More than one year [later], [the number of] matters where [Zhang] Sheng et al. were on target (*zhong*) was greater (*duo*) than Cen by six. On 12-XI-*bingzi*.$_{13}$ (69 CE), an edict [was issued] ordering Sheng and Fang to replace Cen in predicting quarter and full, lunar eclipses, and added hours (*HHS, zhi 2*, 3025).

If the 226 CE results are any reliable indicator, the way that trials were scored may have been as adventitious as the choice of unpredictable phenomena. One notes first that the two ties in the 226 CE report are products of miscalculation, the tiebreaker going in each case to the wrong party. It balances out, in this case, but one error easily tips the balance in a sample of this size. After reversing the tiebreaker for a final score of 8 close and 2 hits to 4 close and 1 hit, Han Yi is the clear loser on planetary visibility (not that NASA could do any better). After reversing the other for a score of 3 close to 2 close, however, the verdict on eclipses is less clear-cut. If we were to count only solar eclipses (as Xu Yue might have us do), Han Yi would beat *both* versions of the Supernal Icon *li* by two to one. Instead, he loses by a narrow margin, and it is lunar eclipse 2 in particular that is to blame. There, Han Yi is by modern calculation a single unit of precision (ten minutes) 'further' than the Supernal Icon *li* (seven for 'ebb-and-flow'), but this is the eclipse whose recorded time is *three units of precision off*, and it is the *direction* by which the observer errs – and that alone – that 'proved' Han Yi 'loose'.

You begin to see why the historical loser here thought he had a hope – how his proposal made it through the secretariat, the Three Excellencies and years of testing, and why one more year may well have mattered. Han Yi was close to beating the Supernal Icon *li* on the 'essentials'; in the end, however, observational error and tiebreakers (let alone Xu Yue's surprise and suspect data) tipped the balance in a sample of only five. It was not as if this was all they had, mind you: the Clerk's Office had *centuries* of observational records at its disposal, but they preferred that everyone start from scratch. If the modern reader is at all perplexed as to why the Clerk's Office would choose *this* as their model of 'scientific testing', it is because what we are looking at, rather, is a *contest*, and contests run by different rules.

4.3.2 Masculinity, Missile Sports and Mathematical Precision

Man first reached heaven with an arrow; woman went there to play house, but men and women are different beasts. 'In the time of Yao', the legend goes, 'the ten suns rose at once, scorching the stalks and sow and killing the grasses and trees'. The people were left to starve, and the earth to be overrun with fiends. 'Yao', in his benevolence,

> sent Yi to punish Chisel Tooth in the meadows of Chouhua, to kill Nine Gullet on the banks of the waters Xiong, to snare Typhoon in the wilds of Greenhill, to shoot the ten suns above and kill the Yayu below, to split Long Snake at Cave's Court and to capture Mound Pig in Mulberry Grove – and the myriad people all rejoiced and installed Yao as the son of heaven.[50]

Yi's quiver empty, the earth strewn with ash and corpses and dying suns, the Queen Mother of the West granted him a drug of immortality for his feat. Before he could take it, however, '[his wife Chang']e stole it and fled to the moon', where she lives today, grinding elixirs with a toad and hare for company.[51] Whatever his own personal story, a 'gentleman of *li* calculation' was first and foremost a *man*, and you weren't a man in this culture if you didn't know how to shoot.[52]

'When a male child is born', says the *Record of Rites*, 'it is [given] a bow of mulberry and six arrows of fleabane by which to shoot heaven, earth and the four directions, [these] being where a man's business lies'.[53] One furthermore 'places a bow to the left of one's door [to announce] the birth of a male child . . . and, on the third day, one begins to carry the child on one's back, [going] shooting for a boy, but not for a girl'.[54] Shooting may have been third amongst the 'six arts' (*liu yi*) of the gentleman – 'ritual/politeness, music, shooting, driving, writing and numbers' – but it was the one that most marked him as a man.[55] So central were the bow and arrow to Chinese preoccupations with masculinity that it is there to which the gaze inevitably wanders in their hyper-masculinised accounts of the barbarian other. His *Book of Jin* biography reports that Xiongnu emperor Liu Yao (r. 318–29), for example, 'was $9^{ch}3^{c}$ (229 cm) tall . . . heroic and martial beyond anyone else, [he] could shoot and pierce

50 *Huainan honglie jijie*, 8.254–5; tr. modified from Major (2010, 275–6).
51 *Lingxian*, cited in *HHS, zhi* 10, 3216 (comm.). On the myth of Yi the Archer and Chang'e, see Birrell (1993, 138–45).
52 On early Chinese play and game culture, see Eggert (2002) and Finkel & Mackenzie (2004).
53 *Liji zhushu*, 28.534b. 54 *Liji zhushu*, 28.534a. 55 *Zhouli zhushu*, 10.160b.

through iron one *cun* (2.44 cm) thick'; his brother Cong (r. 310–18), moreover, 'studied swordsmanship at fifteen, he was an Ape-arm good shot, [his] crossbow [weighing] three-hundred *jin* (66 kg)'.[56] It is the trope of accuracy, specifically, that is the most common of the bunch – the Siberian Yilou, for example, the *Book of Later Han* describes as '[so] good at shooting that with [one] release [they can] penetrate a man's eye'.[57]

The gentleman was no stranger to the idea of accuracy, as that was the very essence of his art: 'the sport (*xi*) of shooting looks for looseness and tightness (*shu-mi*)', explains the historian Ru Chun (fl. 221–5 CE), 'and one takes alcohol to make the loser drink'.[58] One gets a sense of how 'looseness and tightness' was imagined when you look at popular exemplars. The term 'Ape-arm' (*yuan bi*), for example, is a reference to Yang Youji (fl. 597–558 BCE). Introducing him as 'a man of the six arts', *Mr Lü's Spring and Autumn Annals* (239 BCE) recounts how the Chu court was being visited by 'a divinity named Ape', whom 'the best marksmen of [Chu] were all unable to hit (*zhong*)', until Yang, that is, 'released, and Ape took the arrow and went down'.[59] The *Records of the Grand Clerk* (91 BCE) puts his prowess in more quantitative terms:

> The [Kingdom of] Chu had Yang Youji, who was a good shot. Shooting at a willow leaf from 100 *bu* (138.6 metres), [he] hit (*zhong*) it a hundred times out of a hundred releases, [leading] the several thousand observers (*guan zhe*) to the left and right to all say, '[What] a good shot!' (*SJ* 4.165).

Stories like this only get more fantastic with age. The *Book of Han* (111 CE) version, for example, adds that 'for big willow leaves he threw in another hundred hits (*zhong*)' (*HS* 51.2360). Similar lore arose around Yi, the author's favourite being that when challenged to shoot out the left eye of a sparrow, 'Yi . . . mistakenly hit (*zhong*) the right eye; [he] lowered his head from shame and never forgot it till the end of his days'.[60] The lesson is clear: 'A good marksman', in the words of the *Huainanzi*, 'does not shoot to miss (*shi*)'.[61]

Shooting had its place in war, and in hunting, spectacle and sport, but, like any mainstay of human experience, the gentleman also saw it as a mirror into the self. Nowhere is such reflection presented more elegantly than in the *Record of Rites*. In 'The Meaning of Shooting', the *Rites* tells us that the ancient kings saw in shooting the potential to 'observe the virtue' (*guan de*) of man, as only he who was 'inwardly correct of purpose was outwardly straight of body' and could also be true of target (*Liji zhushu*, 62.1014b).[62] Refining the art of contest with drinking, ceremony and music, they used shooting to channel human virtue to a higher end: the 'Big Shoot' (*da she*), where 'those who [deliver]

[56] *JS* 103.2683, 102.2657. [57] *HHS* 85.2812. [58] *SJ* 109.2872 (comm.).
[59] *Lüshi chunqiu*, 24.9b. [60] *Diwang shiji*, cited in *Taiping yulan*, 82.9a.
[61] *Huainan honglie jijie*, 16.529.
[62] For a translation of 'The Meaning of Shooting', see Legge (1967, vol. 2, 446–53).

the most hits (*zhong*) with their bodily comportment in conformity with the rites and their rhythm in conformity with the music shall be granted participation in sacrifice' (ibid., 62.1015b). For better or for worse, the honour (and political perks) associated with the royal sacrifices engendered a spirit of competition between officers and their lords alike, but Confucius reminds us that the real challenge is to best the self: father, son, lord and subject, 'each shooter shoots at the mark of his [ideal] self' (ibid., 62.1017b). At the end of the day, there is only heaven, you and the target, and it is here where the philosophy of sport and science coalesce:

> The shoot[er] seeks correctness in himself. Only after correcting himself does [he] release, and [if he] shoots and misses (*bu zhong*), then [he] is not angry with the winner but turns around and seeks [the cause of his failure] in himself (ibid., 62.1020a).[63]

In its abstraction, astronomy requires a good deal of metaphors to conceive ('belts', 'clouds', 'holes' and 'bangs'), and, for that, the early imperial gentleman often turned to sport. Presented with an instrument, pegs and armillary rings suggested themselves as 'sights' (*yi*) by which to 'target' (*zhun*) stars, and the water clock a game of 'pitch-pot' (*tou hu*), with an 'arrow' (*jian*) in a 'pot' (*hu*).[64] So too is his vocabulary of accuracy one that is familiar to any marksman: 'hitting' (*zhong*) and 'missing' (*shi*) the mark, landing 'close' (*jin*) and 'far off' (*yuan*) centre, and the 'looseness' (*shu*) and 'tightness' (*mi*) of one's grouping.[65] In archery, of course, whether or not you are 'good' (*shan*) comes down to numbers, and for that there were several procedures in place.

Thanks to archaeology, we now have a considerable body of evidence on the annual archery test to which military officers were subject in the Han. At Edsen-gol, for example, we find the following ordinance:

> Proficiency Ordinance no 45. Watch chiefs and gentleman clerks shall all be tested (*shi*) in archery. [They] shall shoot at targets at a remove from the butt using crossbows with the strength of operational bows. [They] shall release twelve arrows, six [registered] hits on the target being the standard [of proficiency]. If more than six, an award of fifteen days' service shall be granted in respect of each arrow (45.23).[66]

We find ample documentation at the same site confirming its administration in the form of individual reports, e.g.:

> Wang Wuhe, grandee of the eighth order, commanding officer of Nihu section, Jiaqu [company], Juyan (Edsen-gol), shot in accordance with the ordinance, in autumn of [57 BCE], releasing twelve arrows and hitting [the target with] six: qualified (312.9).[67]

63 Tr. modified from Legge (1967, vol. 2, 452).
64 On 'sights', see Section 2.1.4 above. On 'pitch-pot', see Bower & Mackenzie (2004).
65 On the vocabulary of accuracy, see Section 1.3.3 above.
66 Tr. modified from Loewe (1967, vol. 1, 118). 67 Tr. modified from ibid.

The reasoning behind Proficiency Ordinance no 45 is easy to understand: in archery, as in the astral sciences, you want government forces as accurate as they can be. Lawmakers knew better than to divine this from tortoise shells, to reason it from five-agent associations or to take men at their word – this was a matter of life or death, so it required a standardised test by which to quantify and empirically measure this value against a common metric. That metric, as in *li*, was an absolute number of 'hits'.[68]

That tells us something, but that was the military, and these were proficiency tests, so let us consider the rules of *contest*, properly speaking, a little closer to home. Whatever its relationship to Zhou kings, the 'Big Shoot' was still held in early imperial times (unless of course it was raining, as occurred on 2 October 492).[69] Administratively, it was the Ministry of Rites who was responsible for organising the 'Big Shoot', due to the ritual nature of the contest, the prize and the ministry's broader role in 'selecting and testing' (*xuan shi*) gentlemen for the civil service based on their 'ability' (*neng*).[70] As to how it worked, the *Book of Etiquette and Rites* describes this and the 'District Shoot' (*xiang she*) down to the very minutiae of where to put one's goblet, who goes to wash it and what stairs they should take.[71] There is, as Ru Chun suggests, *a lot* of drinking.

In brief, the competition consists of three pairs and three rounds, each round beginning and ending with drinking. Drinking and shooting can be dangerous, which is why there are special mats for 'guest/participants who are drunk' to lie down and a special rod to beat shooters who 'have an accident' (*Yili zhushu*, 7.221b; 5.149b). After drinking, the range is set up with three square butts (ten to eighteen *chi* (2.3–4.1 metres) at the centre) at 'fifty . . . seventy . . . and ninety' bow-lengths from the shooter (ibid., 5.151a; 7.187b). At each round, the pair alternate, shooting *once each* at the butts at fifty and seventy bow-lengths and *twice* at that at ninety, for a total of four arrows per person, and eight per pair. Round one is a warm-up, where any arrow that 'hits centre' is called, but 'results are not yet counted as results' (ibid., 7.203b). Round two is counted: 'the Great Clerk (i.e. Grand Clerk) counts the results' by 'filling' or 'solidifying' (*shi*) the tally rack by one 'counting rod' (*suan*) per registered hit (ibid., 7.206a). Round three, finally, is played to music, and '[if it's] not on drum (rhythm) [it] doesn't count' (ibid., 7.217b). The Grand Clerk having tallied the final score, the losers are made to drink by the winners, and the event concludes by unlimited 'cups-without-counting-rods' drinking, after which the winners have earned the right to participate in sacrifice (ibid., 7.221a).

[68] For sources on Proficiency Ordinance no 45, see ibid. (vol. 2, 50–60) and Li Junming (2009, 209–10). I thank Thies Staack for bringing these sources to my attention.
[69] See *WS* 7B.170. [70] *HHS zhi* 25, 3571 and Section 3.2 above.
[71] For a translation, see Steele (1917, vol. 1, 74–121, 150–88).

It makes sense that it would be the Grand Clerk who, 'in all matters of shooting, conducts ceremonial matters of proclaiming hits and lodging counting rods' (*Zhouli zhushu*, 26.403b) – he is, after all, the ministry's numbers man. It also makes sense that his office might apply their experience conducting one contest to the other – it is, after all, only a matter of tallying 'hits', 'the superior', in Jia Kui's words, 'being he who got the most counting rods'. The 'Big Shoot', of course, is a *game*. In more concrete terms, we might say that it embodies anthropologist Roger Caillois's (2001, 9–10) six criteria of 'play': (1) 'free' to join and leave; (2) 'separate', in its delineation in time and space; (3) 'uncertain', decided by initiative and/or chance; (4) 'unproductive' vis-à-vis real objects and wealth; (5) 'governed by rules'; and (6) 'make-believe', invoking a second reality parallel to one's own. The 'Big Shoot' is a *game*, for example, where Proficiency Ordinance no 45 is not, because it is voluntary and competitive, and adds elements of chance to augment the uncertainty of the results – the drinking and music for vertigo, and fewer arrows for chance (not to mention atmospheric conditions). By this count, the institution of the live trial, as examined in Section 4.3.1 above, meets every criteria of a game.

4.4 Conclusion

It is difficult to speak to the essence of 'Chinese science', because the phrase implies that there is such a thing as both. Writing of 400 BCE to 200 CE, Lloyd and Sivin (2002, 79) conclude, 'In China there was no tradition of public debate of the kind that was central in the Greek world. Philosophical and scientific argument tended to be written and indirect and was seldom confrontational'.[72] This is difficult to reconcile with the case of *li* as practised at Luoyang, at least, in 226 CE. *Yi* deliberation, for sure, was not as quotidian a spectacle as were Hippocratic debates, but, as an institution for deciding 'scientific' policy, it was decidedly 'public' and 'confrontational'. It is so confrontational, indeed, that it is hard to distil therefrom any one idea of 'Chinese science', because what we see are *individuals* at each other's throats. The reason, once you do a little digging, is not always 'scientific'. *Pride*, for example, was why the court wanted an alternative to the Supernal Icon *li*, and why Xu Yue and Yang Wei had to shoot it down. Yang Wei insists on authority and the limits of empirical knowledge, but he readily insists on the opposite, before and after, if it gets him what he wants.

[72] On this point, Lloyd & Sivin (2002) would seem to reflect the sort of generalisation about East Asian culture one sees, for example, in Becker (1986), but one notes that Lloyd (2008) and Sivin (2009, 19–59) individually allow for some of the sort of confrontation witnessed in this chapter as concerns *li*. On Chinese logic and argumentation more generally, see Harbsmeier (1998) and Olberding (2012).

If we are to draw a comparison between worlds, we might begin instead with rhetoric. In terms of rhetoric, the sort of critical thinking that we see here is not so different from that of the Hippocratic doctor described by Lloyd (1979, 97–8):

> Yet while the critical examination of other doctrines is sometimes well developed, this fertility in speculation is often not matched by a corresponding *self*-criticism. In a situation of competitive debate, however, this is readily understandable. The speaker's role was to advocate his own cause, to present his own thesis in as favourable a light as possible ... Given an interested but inexpert audience, technical detail, and even the careful marshalling of data, might well be quite inappropriate, and would, in any event, be likely to be less telling than the well-chosen plausible – or would-be demonstrative – argument.

When we take a step back from the individual, however, certain patterns do emerge. Early imperial actors voice unanimous support for the virtue of 'tightness' in the astral sciences – the problem, indeed, is that *everyone* claims this virtue, just like every military officer probably thinks himself qualified. When claims conflict, actors unanimously appeal in their defence to historical precedence and observationally derived numbers. These were not the only rhetorical strategies that saw use, but their frequency speaks to the epistemic culture in which they were thought convincing. The power of precedence is nothing new to the student of Chinese intellectual culture, as many of the sources like the standard histories and their monographs were compiled as canons thereof. Neither, for that matter, is the role of 'exemplars' (*shibiao*) like Liu Hong and his solution to 'slow–fast' lost on the modern philosopher of science:

> The resultant ability to see a variety of situations as like each other ... is, I think, the main thing a student acquires by doing exemplary problems, whether with a pencil and paper or in a well-designed laboratory. After he has completed a certain number ... he views the situations that confront him as a scientist in the same gestalt as other members of his specialists' group. For him they are no longer the same situations he had encountered when his training began. He has meanwhile assimilated a time-tested and group-licensed way of seeing (Kuhn 1996, 189).

What *is* strange, if anything, about the way that early imperial *li* men practised 'science' is how they used observational data. Observation played very different roles in Indian, Mesopotamian and Mediterranean astronomies, of course, and *guan* is not exactly *tērēsis* or *dṛś*, but there is a world of difference here, in third-century Luoyang, between public and private practice.[73] In private, one takes full advantage of the observational record, and in public one starts from scratch. The live trial, furthermore, introduced elements of chance otherwise absent in private testing. The Grand Clerk made the *li* numbers a *game* on the

[73] On the problem of observation in Mediterranean and Indian sources, see for examples Lehoux (2012, 106–32) and Plofker (2009, 113–20).

very model of the other for which he kept score – the 'Big Shoot'. Why? Perhaps because matters as serious as *li* policy and sacrificial attendance are best decided by play – because a *contest* is the one thing you cannot contest. The results of the World Cup finals are *final*; the matches do not play out differently by appealing to additional data, or by running further tests on the players, because the game is something separate from our reality, circumscribed within sacrosanct boundaries of time and space wherein different, arbitrary but consensual rules apply. Contests have a prize. That for archery was participation in imperial sacrifice, so what, by this logic, does the gentleman stand to win at *li*?

5 What the Ancients Had Yet to Learn

In 1109, the Ministry of Rites petitioned the throne to confer a round of posthumous titles on the sages, savants and worthies enshrined at King Wenxuan Temple. This was but the latest of many rounds, as King Wenxuan himself had begun as a lowly teacher with no title to his name – no title but Master Kong. Credited with redacting the *Documents, Changes, Odes* and *Annals*, people began to speak of Confucius (551–479 BCE) as a 'sage' and 'plainclothes king' within a matter of decades of his death. His was a royalty at once subaltern and transcendent: his birthright written not in blood but in the jade of a celestial proclamation borne by unicorn to his door; his reign exercised not in person but in writing, whence it would extend for an eternity beyond his mortal self. As fellow royals, and aspiring sages, the emperors of Han (206 BCE–220 CE) made pilgrimage to the local cult that had arisen around him in Qufu to sacrifice a 'big pen' (*tailao*) as they would to an ancestor. The more that subjects claimed him for their own, investing him with supernatural powers of prophecy, fertility and flight, the more honours and sacrifice the throne heaped upon his altar until moving that altar, in 241 CE, to the Imperial Academy. Promoted to First of All Sages in 628 CE, and King Wenxuan in 739 CE, Confucius rose through the pantheon of the Temple of *Wen* to sit as its godhead, his idol displacing that of the Duke of Zhou (r. 1042–1036 BCE).[1]

Such was the pantheon at King Wenxuan Temple, Kaifeng, as the Song (960–1279) had inherited it from the Later Zhou (951–60). The petition of 1109 is thus unremarkable except, perhaps, for the minor idols with which it deals:

> Those renowned from the days of yore for calculating numbers (*suan shu*) have their portraits painted on either corridor. It is requested that [they] be additionally bestowed noble titles of the five ranks and that their clothing be fixed (repainted) in accordance with their enfeoffment (*Song shi*, 105.2552).

In the list of sixty-six investitures that follows, we see names with which we are now familiar. Among the earls (*bo*), we find Xianyu Wangren (fl. 78–74 BCE),

[1] On the cult of Confucius, see Wilson (2002).

Geng Shouchang (fl. 52 BCE), Zhang Heng (78–139 CE), Zhou Xing (fl. 123 CE), Shan Yang (fl. 170–3 CE), He Chengtian (370–447 CE) and Zhang Zhouxuan (d. *c*.613 CE). Among the viscounts (*zi*), we find Deng Ping (fl. 104 BCE), Liu Hong (fl. 167–206 CE), Zu Chongzhi (429–500 CE), Liu Zhuo (544–610 CE), and Li Chunfeng (602–70 CE). Among the barons (*nan*), lastly, we find Luoxia Hong (fl. 104 BCE) and Jiang Ji (fl. 384 CE). Whatever his title, be it in this world, or the one beyond, each of these men had earned a place beside the sages and worthies of the past and a part of the victuals offered thereto. If the prize in archery was to assist in sacrifice, to return to the question posed in the previous chapter, that in *li* was to *receive it*.[2]

Bemoaning the continued outpouring of junk histories of science to our day, Agassi (2008, 128) lambastes 'inductivist histories' on the charge that 'their main function is ritual'. 'Inductivism', which he attributes to (Viscount) Sir Francis Bacon (1561–1626), 'divides thinkers into two categories variously characterised as right and wrong, scientific and superstitious, open-minded and dogmatic, observer of facts and speculator' (ibid., 125), the role of the Baconian historian being to mete out praise and blame of the individual's epistemic virtue. It is against the up-to-date textbook that one ultimately judges this virtue – the historian, in hindsight, being always on the right side (and the summit) of progress – the project of history being to provide textbook truths a human side and chronological order. The facts, of course, were *always* there, the mark of the hero being to 'observe and record facts as they are, as they appear to his eyes accidentally' (ibid., 125). This vision leaves very little to say: 'The formula they employ is: In year *x* scientist *y* made discovery *z*. Consequently, they have three kinds of problem: (a) chronological problems; (b) priority problems; and (c) authorship problems' (ibid., 131). At the end of the day, Agassi charges, 'The function of [such] histories is to stress that the field of study is important and that big marks must be given, at least as a token of gratitude, to some past scientists. The inductivist histories of science are, briefly, scientific ancestor-worship in pseudo-scholarly guise' (ibid., 129).

Agassi means this as an aspersion; in a culture where temple icons of mathematicians past were offered animal sacrifice, however, 'scientific ancestor worship' seems like an appropriate place to begin. In this chapter we will continue with the theme of 'tightness' (*mi*), shifting our focus from the individual to the aggregate. In Chapter 4 we watched a dozen men disputing its 'verification' (*yan*), and in Chapters 2 and 3 we saw how contemporary ideas of observation and time control only muddle the ordeal. Still, the question

[2] I thank Thomas H. C. Lee for explaining to me in a personal communication of 21 December 2015 that the enshrinement of these gentlemen at King Wenxuan Temple was likely part and parcel of Cai Jing's (1047–1126) educational reforms, which established an official 'school of mathematics' under the directorate of education between 1104 and 1110, for which see Lee (1985, 91–101).

remains: what happens with *hundreds of men* over the course of *a thousand years*? A priori, the answer probably depends upon the temple in the history and philosophy of science that you attend. In one – Sir Karl Popper's – they may mark stations of the cross, rising through tribulation to apotheosis, while in another – Thomas Kuhn's – we might find them arrayed in *maṇḍala*, turning on wheels of suffering and illusion, but these are not the only or the first temples of their kind. Barbarians have wrested control of the *ancestral hall*, Agassi complains, as the Jurchen did King Wenxuan Temple in 1127, but let us not forget the smell of flesh and incense that once permeated its walls as we consider how the ancients thought as 'moderns' (*jin*) about 'antiquity' (*gu*).

We begin in Section 5.1 with a careful reading of a memorial of 1065, the author of which – a *li* man in contemporary Bianjing – offers passing comments on the history of the field by way of justifying his work. Running through the pantheon of King Wenxuan Temple, he structures the past as a ladder of 'awakenings' mounting to the infinite possibilities of the present. After scrutinising his historiographic motives and *bona fides*, we turn in Section 5.2 to the figure of the historian. More specifically, we will take up the overlapping monographs of Li Chunfeng, Shen Yue (441–513 CE) and He Chengtian on the *tianwen* and *li* of 220–479 CE, attempting to tease out the ideological substance of the structural differences of their accounts of the same period. The idea of 'turns', 'accumulation' and 'improvement' is there more or less throughout, but where his predecessors hesitate to place human knowledge over the sages', I show, Li Chunfeng grinds them into dust. Faced time and again with the theme of presentist triumphalism, we then pose the question in Section 5.3 whether this idea – the idea of 'progress' – might not have roots that run broader and deeper through contemporary culture(s), namely the domain of orthodox religion. Whether or not 'progress' is a historical *fact*, I insist that it is a historical *belief*, and it is one every bit as real to its apostles as Saṃsāra.

5.1 The Analyst: Looking Forward and Looking Up

Rather than dig into the problematics of 'progress' in the history and philosophy of science, I would like to begin with a concrete example of how premodern *li* men looked back on the early imperial period. To this end, we shall focus on an excerpt from the 1065 memorial accompanying Grand Clerk Zhou Cong's Brilliant Heaven *li*. There is nothing extraordinary about its contents, belonging as it does to a centuries-long tradition of selling the throne on change – its claims, if anything, are more modest than Yang Wei's (fl. 226–37 CE), in Chapter 4. Its vision of history is perhaps archetypical of the genre, but our interest in this text is as a point of discussion to help navigate the pluralism we shall encounter in Section 5.2. If anything, my reason for focusing on this excerpt in particular, over, say, Guo Shoujing's

(1231–1316) near-identical statement in 1280,[3] is Zhou Cong's relative proximity to the figures discussed in this book – both in history and at temple.

Of the *li* of antiquity and today, there are inevitably procedures that surpass (*guo*) [those of their] predecessors, but only those that can become the model (*fa*) of a myriad generations are the [real] winners (*sheng*). Like Yixing (683–727 CE), who, in making [his] deliberation argument (*yi*) and summary examples for the Great Expansion *li*, collated and corrected successive generations [of data] in order to find the strengths and weaknesses of [historical] *li* models (*fa*), make summaries of [historical] *li* lineage experts, and obtain numbers of centre and balance, [so too shall we proceed]. Liu Zhuo awoke to [the realisation] that solar motion experiences a discrepancy of excess and deficit (*ying-suo*; i.e. inequality), and Li Chunfeng awoke to the model (*fa*) of fixed syzygy, combining *qi*, syzygy and intercalary remainder – this was all the same single procedure. Zhang Zixin awoke to [the realisation] that lunar motion [goes on] a crossed road, inside and outside [the ecliptic], and that the five [planets] have *qi*-entry addition and diminution (i.e. seasonal corrections for first and last visibility). He Chengtian of the [Liu] Song first awoke to [the idea] of fixing the sequence of *qi* via the measurement of shadows. Jiang Ji of the Jin first awoke to [the realisation that] the lodge opposite that of lunar eclipse is the *du* at which the sun is. Liu Hong of the Later Han created the Supernal Icon *li* having first awoken to [the realisation] that lunar motion experiences numbers of slow–fast. Zu Chongzhi of the [Liu] Song first awoke to [the realisation] of the year difference (i.e. precession of the equinoxes) . . . There is no latter-day *li* creator who does not honour and use these [models]. Those that are profoundly loose and erroneous [today] – namely the Supernal Origin *li* of Miao Shouxin (981 CE), the Adjusted Origin *li* of Ma Chongxu (938 CE) and the Five-Era *li* of Guo Shao (762 CE) – do not excel/emerge from (*chu*) this.

As such, all *li* creators must tally solar and lunar motion to have the numbers of dark and new moon and [must retrodictively] verify the solar eclipses of the *Spring and Autumn Annals* (722–481 BCE) to illustrate [their] strength and weakness. As to the sequence of *qi*, one takes verification from the [winter solstice records] of the traditions (of the *Spring and Autumn Annals*). As to the excess and deficit of solar motion, the slow and fast of lunar motion, the addition and diminution of the five [planets], the eclipse difference of the two luminaries, solar lodges and lunar displacement, meridian stars and gnomon shadows, establishing numbers and establishing models – all of this one roots in former accounts (*yu*). After that [comes] comparative verification: from all the way up to the 'the chronograms not gathering in Chamber.$_{L04}$'[4] in [King] Zhong Kang of Xia 5-IX down to today, for the [planets], chronograms, *qi* and syzygies and solar and lunar crossing and eclipse, etc., one makes [everything] within three thousand years align as if by level and plumb, and where one [is] ahead or behind, intimate (*qin*) or loose (*shu*), one makes numbers of centre and balance – [only] then can [one's procedures] spread to later generations.

One's comparative verification [should be] as per Yixing and Sun Sigong (eleventh century), taking many numbers – not going by few – for results that are intimate and tight. In comparing solar and lunar crossing and eclipse, [an error of] 0.12 notch (1m44s)

[3] See *Yuan shi*, 164.3848–9.

[4] Citing *Shangshu zhushu*, 7.102b. Later commentary to this passage in the *Documents* glosses 'the chronograms not gathering' as a solar eclipse.

or less is 'intimate'; 0.24 notch (3m27s) or less is 'close'; 0.35 notch (5m2s) or more is 'far'; and if the *li* notes (predictions) have eclipse, but heaven verifies there is no eclipse, or [vice versa], it is a 'miss'. In comparing [planetary] *du*, on the other hand, erring from heaven by two *du* or less is 'intimate'; three *du* or less is 'close'; and four *du* or more is 'far' ... The best (*zui*) is one who gets many of the numbers upon comparison with antiquity and is furthermore close today, establishing a model and establishing numbers upon the two to get at the natural order (*li*) and penetrate (*tong*) to [its] root (*Song shi*, 75. 1738–9).

Zhou Cong, like many of the *li* men encountered thus far, expresses an acute historical awareness of his predecessors. Many of his 'winners' are none other than those enshrined at King Wenxuan Temple – icons, one imagines, with an explanatory cartouche offering 'In year *x* scientist *y* made discovery *z*'. His formulae are no less boilerplate – that concerning Zhang Zixin's 'awakening' comes nearly word for word from Li Chunfeng – what stands out, however, is that some are *wrong*. Li Chunfeng did not *discover* the fixed lunation, he simply saw it into policy (see Section 3.3.3 above); nor was Zhang Zixin the first to note the obliquity of the lunar orbit (see Section 1.3.2 above). Zhou Cong is doing *li*, not history, of course, and textbook histories of science always leave something to be desired.

Zhou Cong's apparent misattributions make sense when viewed through the epistemological, moral and sociological assumptions he brings to bear. As to epistemology, Zhou Cong attributes procedures to spontaneous individual 'awakenings' (*wu*). There is nothing new here: Li Chunfeng describes Zhang Zixin as having 'first awoken [to the realisation] that the crossed roads of sun and moon see inside and outside and retardation and acceleration', for example, and He Chengtian describes himself as 'awakening ... to [the fact that] heaven's form is a perfect circle and that water completely [fills] its bottom [half]'.[5] The terminology would seem to imply a 'eureka moment', but Zhou Cong insists at once on a programme of 'comparative verification' (*jiao yan*) that is anything but spontaneous or individual. To this end, he insists upon established methods – the analysis of an all-inclusive 3,000-year dataset to find 'numbers of centre and balance' via minimum standard deviation – and lest there be any question of what counts as 'close', Zhou offers a strict definition of his terms. There is nothing inherently contradictory about this 'discovery–verification' epistemology, but its *dualism*, one notes, muddles the question of priority: what is the status of an unproven idea? And to whom, once proven, does it belong?

The fact of discovery implies a *discoverer* – a hero capable of 'awakening' from 'loose and erroneous' (*shu miu*) presuppositions to a truth dangling right before his eyes. The archetype here is the sage, *li* being 'possessed of six virtues

[5] *Sui shu*, 20.561; *Song shu*, 23.677; cf. Section 2.2 above.

of the sage';[6] and whatever instruments and methodologies the *li* man might project thereupon, the model of gnosis is one charged with moral implications. When faced with a discovery, it is the established 'winner' who often suggests himself as candidate. This can lead to the Matthew effect – 'the rich get richer, and the poor get poorer' – as then, as now, lead authors get all the credit, and household names get all the attention (Merton 1968). With the Grand Inception *li*, for example, the *Book of Han* acknowledges that *Li* worker Deng Ping worked with 'a total of twenty-plus people' (five of whom are named) but refers to it thereafter as 'Deng Ping's creation'.[7] In 226 CE, Sun Qin states that '[Grand] Clerk [Sima] Qian (*c.*145–*c.*86 BCE) constructed the Grand Inception [*li*]'.[8] Elsewhere, Luoxia Hong is sometimes cited as the sole author – because the collaborator behind 'sphere heaven', one assumes, must be behind the *li* – but the case for Sima Qian is rather special.[9] Sima Qian *dropped out*, if we remember, claiming that he 'couldn't do the maths', so the only reason to associate him with the project is by merit of his reputation.[10] The figure of Sima Qian poses a dilemma: as a Grand Clerk, and a *grand homme* in the history of science, he must have discovered *something*. If not the Grand Inception *li*, then what?

One thing that Zhou Cong's 'winners' have in common is enshrinement – the enshrinement of their images at King Wenxuan Temple and the enshrinement of their procedure texts in the standard histories.[11] In the history of science, we are often faced with the problem of obsolescence: the success of Ptolemy's (*c.*90–*c.*168 CE) *Almagest*, for example, has all but eradicated any written record of Hipparchus (*c.*190–*c.*120 BCE), and so too does Euclid's (fl. 300 BCE) *Elements* impose a veil on the early history of geometry. This is the case for *tianwen* omenology, as mentioned in Chapter 2, but *li* is a different beast. Zhou Cong and Yixing had *a lot* to work with. Judging from bibliographic monographs, 'winners' and 'losers' alike were well represented in imperial library holdings into the Song. When it came time to 'find the strengths and weaknesses of [historical] *li* models', however, it is only ever those we find in the *lü-li* monographs that enter into consideration.[12] History is for winners: as a rule, the monograph genre preserves only the *li* that entered into state

6 *HHS zhi* 3, 3057; cf. Section 1.1.2 above.
7 *HS* 21A.975–6; cf. Section 1.2.1 above.
8 *JS* 17.499; cf. Section 4.2.1 above.
9 Sources naming Luoxia Hong as the sole author of the Grand Inception *li* include, for example, Yan Minchu (fl. 594 CE), cited in *Sui shu*, 17.434; Xing Hepu (8th cent.), cited in *Jizuan yuanhai*, 3.37b; and *Song shi*, 76.1743.
10 *HS* 21A.975; cf. Section 1.2.1 above.
11 Note that Zhang Zixin is an exception: Zhang Zixin does not appear to have ever authored a *li*, but Li Chunfeng does 'enshrine' a list of his accomplishments in *Sui shu*, 20.561. Note also that Jiang Ji's procedure text, as preserved in *JS* 18.566–70, is incomplete.
12 Perhaps the best example of such work is the evaluation conducted by Guo Shoujing's team in the thirteenth century, by which they compared their Season-Granting *li* to data computed by no less than forty-three historical procedure texts; see Sivin (2009, 254–370).

policy, though Li Chunfeng, as we will see, is known to break the rules. Whether it falls to the court or to the historian to decide, the institutional award process further complicates the matter of priority: does an idea belong to its discoverer, or, if you will, the first to get it into peer-reviewed publication?

Having revealed some of the ambiguities in Zhou Cong's approach to the question of priority, let us consider his rather confusing treatment of Zu Chongzhi. At the end of the memorial we read that 'Zu Chongzhi of the [Liu] Song first awoke to [the realisation] of the year difference'. One might accuse him of having forgotten Yu Xi (fl. 307–46 CE), but Zhou Cong seems to know what he's doing. Earlier in the memorial we read the following:

The *Documents* raises the due-south stars (meridian stars) for rectifying the four directions, [this] being what the former kings used to elucidate the seasons and grant [them] to people, to serve heaven and nurture [the myriad living] things. However, the accounts of former scholars have similarities and differences between them. Yu Xi says: 'In the time of Yao, on the winter solstice.$_{Q22}$ – shortest day – the star was Mane.$_{L18}$; now, after 2,700-odd years, Eastern Wall.$_{L14}$ is centred, which [we] know was arrived at by gradual difference every year'. Moreover, He Chengtian says: ... 'By comparison with the centred star of today, the difference is twenty-seven or eight *du*, i.e. on winter solstice.$_{Q22}$ in the time of Yao the sun was at Maid.$_{L10}$ 10 *du*'. Thus, when Zu Chongzhi made the Grand Enlightenment *li*, he first established the annual difference, [his] *lü*-rate forty-five years and nine months per *du* of retreat (*Song shi*, 74.1689).

How does credit for 'establishment' (*li*) (and 'first awakening') 'thus' (*gu*) go to Zu Chongzhi? What, more specifically, is so special about his Grand Enlightenment *li*? The latter was not the last word on precession by Zhou Cong's day, as later *li* men like Liu Zhuo would substantially refine its *lü*. Where it stands out, rather, is that it was the vehicle by which the 'year difference' first entered into state policy. Ironically, Zu Chongzhi wasn't even there to see this happen, as *li* reform came at the Liang (502–57 CE) court in 510 CE – some forty-eight years after his own proposal was rejected, and some ten years after his death. 'Awakening', one might conclude, is bigger than life and death.[13]

Whether or not that is where he is going, the issue of (im)mortality is clearly one by which Zhou Cong is consumed. An 'awakening', as he describes it, is an event that ripples beyond a single human life, 'a myriad generations' (*wan shi*) into the future. 'Myriad' is the most mundane of Chinese hyperboles, but *li* men are one of the rare breeds who actually work with such numbers, so let's pretend that Zhou Cong is keeping count: at thirty years per generation, that's 300,000 years.[14] That's a *long* time, particularly when you look back at the

[13] On 'year-difference' precession, see Chen Meidong (1995, 261–70), Qu Anjing (2008, 167–95) and Sivin (2009, 99–101). Note that not all of Zhou Cong's apparent misattributions are due to the conflation of 'awakening' and 'publication'.

[14] Because 'thirty years makes one "generation" (*shi*)' (*Shuowen jiezi zhu*, 3A.7a).

dawn of human civilisation, which Yu Xi places a mere '2,700-odd years' before his day.[15]

'Awakening' is about more than posterity, however, as its effects ripple both forwards and backwards in time. Prior to the 'year difference', actors commonly speak of the impossibility of *tong* ('connecting' or 'communicating') with the past on high: in the words of Cai Yong (133–92 CE), 'today's procedures cannot *tong* up to antiquity any more than ancient procedures can *tong* down to today'.[16] The image recalls the legend of Chong and Li, ordered by Zhuanxu, in the time when man and god once freely mingled, to 'sever the *tong* of earth and heaven that there be no more descent and ascent [between]'.[17] But where ancient sages and mysterious 'differences' (*cha*) conspired to close a door on man and god, the *li* man would kick it down and off its frame. The answer to General Du Yu's (222–85 CE) historicist doubts offers us a rather dramatic example of what was brewing:

> In touching upon the events in the *Spring and Autumn Annals*, I once composed a *li* discourse as a culmination of talk about the *tong* principles (*li*) of *li*. The gist was this: heaven's motion never ebbs, and the sun, moon and [planets] each revolve [through] their lodges – all of these are moving things. The motion of things, however, is not one [and the same], [so] while [their] motion *du* can be in large part successfully delimited, as the days pile into months, and as the months pile into years, there is inevitably a difference (*cha*) of a hair's end in invoking the new against the old – this is a principle that is such in and of itself . . .
>
> After I made [this] *li* discourse, in [275–80 CE], the expert calculators Li Xiu and Bu Xian made procedures according to the discourse body, which were named the Supernal Degree *li* and memorialised to the court. The procedures . . . invoked the intention of 'changing constitutions every 300 years' [by] calculating with two different origins and multiplying by strong and weak (?) every seventy-odd years, the strong and weak difference being small but just enough to *tong* from afar [any] excess and deficit. At the time, the Masters of Writing, as well as the Clerk's Office, referenced and compared the Supernal Degree and Grand Beginning *li* (a.k.a. Luminous Inception *li*) against records and notes from antiquity and today, and the Supernal Degree *li* scored an outstanding win over the Grand Beginning *li*, winning over the official *li* above (in antiquity) on forty-five items. These procedures are both extant today, and when one combines [them] to investigate the ten *li* of antiquity and today in verifying the *Spring and Autumn Annals*, one [can] know the eminent looseness of the Triple Concordance (*c*.5 CE) (*JS* 18.563–4).

The canonical wisdom, given voice in the *Spring and Autumn Annals* weft *Bao qian tu*, was that the best of *li* was only good for 300 years.[18] Where Du Yu

[15] For more on contemporary archaeo-astronomy, see Cullen (2001).

[16] *HHS zhi* 2, 3038. For similar statements, see Jia Kui (30–101 CE), cited in *HHS zhi* 2, 3028, and Jiang Ji, cited in *JS* 18.567.

[17] *Shangshu zhushu*, 19.297b; cf. Birrell (1993, 91–5).

[18] Cited in *HHS zhi* 2, 3026; cf. Section 1.2.1 above.

accepts this as 'a principle that is such in and of itself' (*ziran zhi li*), however, our 'expert calculators' see but a technical challenge to be met with extra 'origins' (*yuan*) and new corrections. The results speak for themselves, and the general is happy to be proven wrong. A century later, *li* men like Jiang Ji had experienced such success with the two-origin approach as to toss all canonical wisdom aside: 'This model can be used for a long/eternal [number] of years (*yong zai*), so what of this "reforming the constitution of the Dipper *li* every 300 years"?' (*JS* 18.567). The doors were opening, the possibility of *tong* extending further and further beyond any limits once imagined; then, when the 'year difference' took hold, the *li* man's vision leapt at once from centuries and millennia to spans so large as to have no ready name. Working 3,000 years into the past, and a hundred times further into the future, the *li* man, some centuries later, had risen above all human time.

Zhou Cong and his forebears would seem to suggest that perfection was within reach. Their optimism is difficult to reconcile with the situation on the ground: the 398-year middle period between Han and Tang (618–907 CE) alone sees the authorship of at least forty-three new *li*, the average lifespan of the 'winners' among them shrinking to a matter of decades. Sivin (1986) collects examples of the sort of ambiguity and difference of opinion we should expect as concerns 'the limits of empirical knowledge' over fifteen centuries of *li*. It is difficult to distil a single narrative from such a sample, given that thinkers contradict both one another and themselves. There is the eternal problem of salesmanship, for one: 'We usually find doubts about the extension of knowledge expressed not with respect to a system presented for adoption but when the shortcomings of an established system have become apparent, and all the more when a competitor in the offing suggests that its tenure is limited' (ibid., 157). After cutting through the noise, Sivin claims, 'what emerges from the writings of fifteen hundred years is an abiding interest in the idea that the scale of the cosmos is too large, and the texture of nature is too fine, too subtle, too closely intermeshed ... for phenomena to be fully predictable' (ibid., 155). There is no denying that this is a common trope – one probably in the back of many people's *thoughts* – but it is not, I must insist, what Zhou Cong, Jiang Ji or Du Yu are *saying*.

This brings us to the question of 'progress' – perfection as historical process – which is so loaded a question in the history of science that we must begin with modern voices. No one has argued more vehemently against the possibility than Nathan Sivin. Sivin (ibid., 165) concludes that the idea of 'astronomical indeterminacy' played an enduring role in *li* that was 'opposed to the idea of progress', the aim of Chinese astronomy being to meet the practical needs of prediction on a provisional basis – 'this', he asserts, 'is the difference between astronomy as a collection of data and techniques, and astronomy as a science'. Where *li* men fall short of 'science', he cites, is in 'their abandoning the ideal of a rational astronomy governed by cosmology' (ibid., 153), the referent to which goes back to his earlier work:

> The Islamic and European reformers of Ptolemaic astronomy, of whom Copernicus considered himself one, were motivated by a desire to restore metaphysical rigor while maintaining or improving astronomical precision, in short to perfect the relation of cosmology and mathematical astronomy. This motivation, so decisive in the gestation of modern science, seems to have been in China an early casualty of a premature and unworkable relation between form and content (Sivin 1969, 69).

Walking the criteria of Ptolomaicity back, Sivin (2009) reframes the question in terms of anachronism. As to the *li* reform of 1280, Sivin acknowledges that 'ancient astronomers valued and sought to improve both accuracy and precision' (ibid., 552); that, in this regard, 'the astronomers achieved the aims with which the reform began' (ibid., 554); and that 'the section headings of the Evaluation [of 43 *li* over 13 centuries] amount to a list of the claims for improvement in technique that the evaluators tested' (ibid., 553). Still, we must ask what the actors themselves set out to accomplish, and in 1280, Sivin insists, they 'simply wished to improve on the faltering techniques of the Yuan system's predecessor' (ibid., 553). The reformers were not thinking in terms of *progress*, he argues, because these were terms that were not there: 'The doctrine that the long-term direction of history was upward, inevitably headed toward betterment, did not exist anywhere in the world before the European Enlightenment' (ibid., 131). To invoke the very word is moreover *dangerous*, because it implies the modern historian in the 'positivism' (or 'inductivism') behind the particularly insipid brand of nationalism with which we struggle in our field.[19]

Henderson (2006) entertains the possibility. Placing contemporaneous statements of 'astronomical triumphalism' side by side with the sort of 'astronomical indeterminacy' highlighted in Sivin (1986), Henderson argues that we are looking at a dialogue addressing a common paradox. The paradox is the *fact* of progress: 'astronomy was one of the few fields of knowledge in premodern China in which progress was so evident as to override the more deeply ingrained regressive and cyclical models of time' (Henderson 2006, 100). Indeed, the *fact* of progress is one that actors acknowledge in surprisingly modern (or 'positivist') language. Consider, for example, the preface to the *tianwen* monograph finished three years prior to Zhou Cong's memorial in 1060:

> As one proceeds to later ages, the methods (*fa*) become gradually tighter. [This is because] it is necessary to accumulate (*ji*) the knowledge of many people before it is possible to reach the ultimate in fineness and subtlety (*XTS* 31.805).[20]

[19] For an early voice against this strain of nationalism, see Jiang Xiaoyuan (1986).
[20] Tr. modified from Henderson (2006, 100).

It is surprising to hear this of *tianwen*, with its traditional emphasis on omenology, but, then again, *tianwen* too was changing.[21] Consciousness of change is all the more evident in *li*, where actors speak of things like a 'turn towards the fine and tight' and 'the pursuit of tightness through the accumulation of achievements' under the assumption that there were things 'that the ancients had yet to learn'.[22] One of those things was precession, Sivin's thirteenth-century reformers offering that 'it is only that that the difference is so supremely minute that [our] predecessors were not at first yet conscious [thereof]'.[23] The consciousness of progress, in turn, forced a sensitive issue: the obsolescence of the sages. It is *this* that is at the centre of the debate, Henderson (2006) argues, particularly as it developed in late imperial times. Rarely questioning the issue of micro-progress as experienced from the Qin (221–207 BCE) onwards, actors struggled to construct a macro-view of astronomical history in which the sages were both infallible and first. For some, the answer was to muddy the waters with 'indeterminacy'; for some, it was to cite the fall of man; but others found a happy compromise between infallibility and omniscience: they may not have *known* about precession, but 'the ancients did have the foresight to establish a general framework that made the later discovery of these anomalies and irregularities possible' (Henderson 2006, 101).

'Progress' is an anachronism, so let us speak of 'the pursuit of tightness through the accumulation of achievements' and ask where Zhou Cong stands. Some of his aims in the eleventh century are the same as those in the deliberation of 226 CE; one notes, however, that his definition of what counts as 'far' (*yuan*) in eclipse prediction – an error of ≥ 0.35 notch (5m2s) – is several orders of magnitude more precise than the (conservative) 'one *chen* and some odd [fraction]' (> 120 minute) error that Xu Zhi reports *of the winner* in Chapter 4 above.[24] That's all well and good, one might argue, but the history of Zhou Cong's day was marked by incessant policy reform, so any talk of a perpetual *li* would be wholly unrealistic. Indeed, Zhou's own Brilliant Heaven *li* of 1065 would be replaced only ten years later in 1075![25] Does he dream, then, only in the short term – only in the practicalities of besting his competition for a time? Does he know, deep down, like any marksman, that he can win the 'Big Shoot' once, but not forever? No. This, one notes, is the very criticism that Zhou Cong levels against recognised (but nevertheless 'loose and erroneous') reformers like Miao Shouxin, Ma Chongxu and Guo Shao: they 'do not excel/emerge from (*chu*)

[21] See Morgan (2016a); cf. Section 2.3 above.

[22] For 'turn towards the fine and tight' 轉爲精密, see *JS* 17.498 and *Song shu*, 12.231; cf. Section 4.1.1 above. For 'the pursuit of tightness through the accumulation of achievements' 累功以求密, see Dai Faxing (414–65 CE), cited in *Song shu*, 13.315. For 'what the ancients had yet to learn' 古人所未達/得, see *Sui shu*, 20.561; *XTS* 27A.593; *Song shi*, 81.1920; etc.

[23] *Yuan shi*, 52.1130; tr. modified from Henderson (2006, 101).

[24] *JS* 17.499.

[25] See *Song shi*, 13.257, 82.1929.

this'. No, Zhou is speaking of something at once bigger and smaller than that. When Zhou Cong and others speak about posterity, one notes, they do not speak in terms of '*li*' (let alone '*li* reform') but of the individual 'procedures' (*shu*) and 'models' (*fa*) thereof – the sort of 'Pythagorean theorem' one literally sees copied and pasted from one procedure text to the next. The Miao Shouxins of the world might content themselves with short-lived policy wins, but Zhou Cong, for one, has his eyes on the future, 'a myriad generations' from his day.

This is but one man's vision of the early imperial period – one offered in passing, no less, to get to a methodological point on data analysis. Zhou Cong is not doing history, he is arguing *through* it, and we should hardly expect a coherent philosophy of science to emerge therefrom. Indeed, there is much that is incoherent about his presentation: his idea of 'awakening', for example, would seem to confound observation, verification and publication, and the weight he gives publication in this mix belies his emphasis on 'models' transcending individual *li*. Were he writing in the twentieth century, moreover, we might take up his views on 'empiricism' and 'progress' to launch the last forty years of the history and philosophy of science at his 'positivist'/'inductivist' naivety, but there are things that the ancients had yet to learn, so let us forgive him of his sins. Zhou Cong is no historian, certainly not by our standards today, so the question then is how the *historian* dealt with history.

5.2 The Annalist: Looking Back and Looking Down

Most of what we know of the astral sciences in the early imperial period comes from the *tianwen* and *lü-li* monographs of the standard histories, a good half of which were written by two men: Li Chunfeng and Shen Yue. It is Li Chunfeng to whom we owe the respective monographs of the *Book of Jin* (265–420 CE) and the *Book of Sui* (581–618 CE). Written between 641 and 656 CE as part of the *History of the Five Dynasties* monograph project, Li's monographs go beyond the scope of the 'Five Dynasties' (502–618 CE) to pick up where the *Book of Later Han* leaves off. In between, we have Shen Yue's *Book of Song* (420–79 CE) monographs. Written *c*.492 CE at the Xiao Qi court (479–502 CE) at Jiankang, Shen Yue too starts from the 170s, relying to this end on the now lost writings of He Chengtian:

> In [424–53 CE], He Chengtian of Donghai received an edict to compile the monographs, in fifteen chapters, of a *Book of Song* so as to continue [Si]ma Biao's (*c*.237–306 CE) monographs of the [Later] Han (25–220 CE). The comprehensiveness and breadth of its evidence and citations [are why I] have gone to and followed them and also what places [He Chengtian] alongside Ban Gu (32–92 CE) and [Si]ma Qian (*c*.145–*c*.86 BCE) as a single expert lineage (of historiography). Where there are omissions and elisions, and were we come to events after Mr He, [I] patch things up as [I] go by supplying what [I myself] have gathered (*Song shu*, 11.205–6).

Read together, Li Chunfeng and Shen Yue's monographs provide us with a complete history of the astral sciences from the second to seventh centuries CE. We tend to turn to these sources as repositories of facts, but the monographs, we must remember, are *compositions*, and the intentions of their authors, however good, colour the facts they record.

These monographs were *written* as repositories of facts, so where are we to look for the individual's voice therein? The answer is boring repetition. The monographs are formulaic, each genre rooted in a tradition – an 'expert lineage' (*jia*) – going back primarily to the *Book of Han*: *tianwen*, for example, only ever covers the same three headings, and *li* always comes in the form of a chronicle. The *tianwen* monograph in particular is indeed so redundant that the historian Liu Zhiji (661–721 CE) would call for its abolition.[26] Here, much of Li Chunfeng's Sui monograph repeats what he wrote for the Jin; much of *that* is repeated from Shen Yue, and Shen, in turn, was only 'following' He Chengtian. In quantitative terms, 2,430 of the 3,185 graphs in Shen Yue's presentation of the instrument-cosmos (Chapter 2 above) consist of primary sources in citation; 1,931 of these appear expanded, abridged or word for word in Li Chunfeng's monograph, as do another 297 graphs of Shen's narrative and introduction (i.e. 'author *x*'s treatise *y* says . . . '). All that's left of Shen Yue is 956 graphs, 447 in his own (potentially borrowed) words. There is a lot of overlap, it's true, but there is something to learn about Li Chunfeng from what he does with his predecessors' work – what he changes, what he drops and how he acknowledges his source.

When you approach *tianwen* in terms of building blocks and generic rules you begin to see a world of difference between compilers. This is hardly surprising: Li Chunfeng and Shen Yue were different people. Li was accomplished in the astral and mathematical sciences, having constructed a 'sphere sight', overseen the creation of a mathematical canon, and authored a *tianwen* compendium, a winning *li*, and no less than four historical monographs on the subject; Shen, by contrast, acknowledges his reliance on previous writings and has no other relevant accomplishments to his name. Li was writing under a unified empire at relative peace; Shen, on the other hand, was writing in a period of disunion and domestic upheaval. Li was in the north, in Chang'an, while Shen was in the south. Li was a Daoist; Shen was a Buddhist. The list could go on, but perhaps the most important difference between the two was that, in 641–56 CE, one was alive, and the other was dead, meaning that one of them got the last word.[27]

[26] *Shitong*, 3.4a; cf. Damien Chaussende's forthcoming chapter, 'Treatises according to Tang historian Liu Zhiji', in Chaussende, Morgan & Chemla (forthcoming).

[27] On Li Chunfeng, see Guan Zengjian (2002) and Chen Meidong (2003, 350–7); on Shen Yue, see Lippiello (2001).

5.2.1 Instrument-Cosmos, 220–479 CE

Shen Yue begins his *tianwen* monograph with Cai Yong's memorial of 178 CE, cited in Section 2.2.1 above, which frames the history of 'heaven's form' (*tianti*) in terms of three imagined 'expert lineages' (*jia*): 'expansive night' (*xuanye*), 'umbrella heaven' (*gaitian*) and 'sphere heaven' (*huntian*).[28] Accordingly, expansive night 'has died out', umbrellism 'misses the mark' and spherism 'completely grasps the true circumstances'; still, spherism is threatened with extinction if Cai is not recalled to the capital to document the observatory sight of 103 CE (*Song shu*, 23.673). From there – from 178 CE – Shen Yue proceeds in seven blocks.

Block 1. Following Cai Yong's call to action are subsequent developments in spherism by Lu Ji (188–219 CE) and Wang Fan (228–66 CE). Lu is introduced as having 'first calculated the sphere heaven idea', which is followed by an excerpt of Wang's *The Sphere Heaven Effigy Explained* (*Huntian xiang shuo*) treating the dispositions, 'roads' and dimensions of heaven and earth.[29]

Block 2. We move to He Chengtian's *Discourse on the Sphere Effigy Form* (*Lun hun xiang ti*). Upon reading about 'observing' (*guan*) the instrument, Mr He 'awakens' to the fact that Wang Fan is correct, that the earth floats on an ocean filling half of heaven's interior, and that rivers are replenished by evaporation therefrom by the sun.[30]

Block 3. We move to Xu Yuan (395–475 CE) on the origins of the 'sphere sight' armillary sphere, presumably taken from his lost *Book of Song*. Xu cites Wang Fan's interpretation of the *Book of Documents* in support of a prehistoric jade armillary sphere that was lost only in the Qin. Acknowledging the diversity of opinions on the subject, Wang Fan goes with Zheng Xuan (127–200 CE), for which he offers a desperate apology:

> Zheng Xuan had talent in abundance, elegance, height, and reach and thinking of depth, calm fineness and subtlety – overtakingly, with [his] unique insight, did [he] correct [his predecessors'] explanations. The sage emerges once again, there is no changing his words (*Song shu*, 23.677).

Xu doesn't buy it. Citing Zhang Heng's biography to the effect that 'Zhang Heng created/invented (*zuo*) the sphere heaven sight' – probably referring to the treatise by that name – Xu concludes that 'one knows therefore that there was not yet such a sight prior to [Zhang] Heng'. Criticising Zheng Xuan for 'prejudiced belief without evidence not to be taken for granted', Xu returns to the common wisdom that the relevant phrase in the *Book of Documents* –

[28] On the retrospective drawing of *jia* 'expert lineages' or 'schools' in philosophy and medicine respectively, see Csikszentmihalyi & Nylan (2003) and Brown (2015).
[29] *Song shu*, 23.673. On Wang Fan's cosmology, see Kalinowski (1990).
[30] On He Chengtian's cosmology, see Cullen (1977, 332–6) and Chen Meidong (2007, 228–36); cf. Section 2.2.3 above.

'rotating mechanism and jade traverse' (*xuanji yuheng*) – refers to 'the seven stars of the Northern Dipper (UMa)' .[31]

Block 4. Shen Yue inserts himself into the text – '[Your] historian servant notes' – to redress Xu Yuan's claims. Citing Yang Xiong (53 BCE–18 CE) to the effect that Luoxia Hong, Xianyu Wangren and Geng Shouchang worked on 'sphere heaven' in the Western Han (206 BCE–9 CE), Shen insists that 'Western Han Chang'an already had the apparatus'. Turning to Zhang Heng, Shen suggests (seemingly unaware of the 103 CE sight) that Zhang made another because the earlier sight 'went missing amid the death and chaos' of the Wang Mang interregnum (9–23 CE). As to the fate of Zhang Heng's sphere, Shen Yue (incorrectly) states that it was lost with the fall of Luoyang in 311 CE and recovered by Jin forces from Chang'an in 417 CE.[32]

Block 5. We continue with the history of sphere instruments into the Liu Song, Shen Yue introducing Qian Lezhi's 'sphere sight' of 436 and 'small sphere heaven' of 440 CE.[33]

Block 6. Finished with the sphere, we turn to *Gnomon of Zhou* (*Zhou bi*) umbrella heaven. As concerns its antiquity, Shen Yue offers that it 'is said to have emerged from Duke Dan of Zhou's (r. 1042–1036 BCE) visit to the Yin-Shang, but this is due to false attribution; the book is called *Zhou bi, bi* meaning "gnomon," and *zhou* "the number of the circumference of heaven"'.[34] Shen cites an unknown description of umbrella heaven wherein 'the earth is tall in the centre and tapers at the four [edges]' in something of a departure from the *Gnomon of Zhou* and Wang Chong's (27–*c*.100 CE) planar model.[35] Shen follows the seventy-two-graph description with 265 graphs of criticism by Yang Xiong, Zheng Xuan and Liu Xiang (79–8 BCE) concerning, ironically, Han parallel-plane umbrellism. Umbrellism, Shen Yue concludes, is 'the creation of those fond of oddities (*yi*)'.[36]

Block 7. On the subject of 'oddities', Shen Yue provides one-line quotes from three post-178 CE theories. The first is Yu Xi's 'secure heaven' (*antian*), which posits that for heaven and earth to fit one into the other, they must be the same shape (be it round or square), and that heavenly bodies carry *themselves* around heaven. Second is Yu Song's (fl. *c*.265 CE) 'vault heaven' (*qiongtian*), which holds that heaven is like an overturned eggshell floating upon the world ocean and turning on an inclined axis (sparing heaven the indignity of passing below the earth, and the capital observer that of being anywhere but at its

[31] *Song shu*, 23.678. [32] *Song shu*, 23.678; cf. Section 2.2.2 above.
[33] Cf. Section 2.1.4 above.
[34] The *Gnomon of Zhou* indeed says much the same thing in *Zhoubi suanjing*, 1.38a; cf. Cullen (1996, 179).
[35] On umbrella heaven, see Qian Baocong (1983, 377–403), Qu Anjing (1993), Cullen (1996), Yuan & Qu (2008) and Wu & Li (2008).
[36] *Song shu*, 23.679.

centre). Third is Yao Xin's (fl. third century) 'floorboard heaven' (*xuantian*), which argues that the heaven–earth *hun* must rest on a second earth for support, and that the lodges cannot be as unevenly distributed vis-à-vis the pole as the sun's fluctuation in declination suggests. Shen Yue concludes with a simple judgement: 'These three explanations are all [nothing but] curious chatter and miss the mark by a great distance.'[37]

Let us consider how Shen Yue has structured his account. Spherism, as always, is a priori. Shen moves from the a priori in Blocks 1–5 to 'oddities' and 'curious chatter' in Blocks 6–7. The movement would appear to be chronological: 'curious chatter' (Block 7) is a product of the modern age, and it is the *Gnomon of Zhou* (Block 6) that opened the door thereto. Concerning the latter, Shen's philological argument for disassociating the *Gnomon* from the Zhou (1045–771 BCE) is tendentious – the text itself is built around the Duke of Zhou, and it elaborates upon its own title. The necessity of this argument, therefore, is likely the assumption that the ancients would have known better. As to the antiquity of sphere heaven, Shen Yue presents both sides of the argument, leaving the question open. Either way, what matters for Shen is that the instrument was there by 104 BCE, more than a half-century prior to any datable discussion of umbrellism. The necessity of *this* argument, in turn, would seem to be the assumption that good knowledge is ancient knowledge. Beginning as he does with Cai Yong's premature eulogy of the sphere, Shen Yue's vision of history here reads like one of loss and degeneration.[38]

Turning to **Li Chunfeng**, one notes that his *Book of Jin* and *Book of Sui* monographs divide the topic into 'heaven's form' (*tianti*) and separate chronologies for respective instruments. My analysis will thus focus on the former, citing the separate chronologies where relevant. The fact that his two monographs are accretive means that we can treat them as a single work, noting where one references or transitions to the other. For the sake of comparison, I divide Li Chunfeng's presentation according Shen Yue's block numbers, switching to letters where the former diverges from the latter.

Li Chunfeng's chronicle begins with Cai Yong's memorial of 178 CE. Where Shen Yue cites the memorial in full, however, Li excises Cai Yong's anxieties about the future. Again, the frame is that of 'three expert lineages', of which spherism is the a priori.

Block 6(a). We begin with the *Gnomon of Zhou*, to which Li Chunfeng attributes the highest possible antiquity: 'Its origins [began] with [Fu]xi's establishment of the *li du* of the circuits of heaven, and its transmission [began] with the Duke of Zhou receiving [it] from Yin Gao; men of Zhou

[37] *Song shu*, 23.680. On the third- and fourth-century theories, see Cullen (1977, 280–98) and Chen Meidong (2007, 158–60, 186–8, 219–28).

[38] Cullen (1977, 334–6).

made a treatise of it, which is why it is called the *Zhou bi*'.[39] This is followed by an expanded version of Shen Yue's description of non-planar umbrellism.

Block 6(b). We move to an extended citation from 'the *Gnomon of Zhou* expert lineage'. Accordingly, heaven is like a millstone upon which the luminaries, like ants, march in the direction opposite to its rotation, though 'the shape of heaven is like an inclined umbrella', tilted such that 'south is high, and north is down', the hour and place of the sun's 'emergence' (rising) depending on the seasonal prevalence of (dark) *yin* and (bright) *yang qi*.[40]

Block 6(c). In the *Book of Sui* version, we move to Yang Xiong's 'Eight Items on the Difficulties of Umbrella Heaven' (Nan gaitian ba shi) – a point-by-point rebuttal of umbrellism on logical and experiential grounds.[41] Inserting himself into the text, Li Chunfeng criticises that 'the later [attempts of] Huan Tan (*c*.43 BCE–28 CE), Zheng Xuan, Cai Yong and Lu Ji to each to examine and verify heaven's shape via exposition of the *Gnomon of Zhou* were largely contradictory'. Lastly, he dismisses Emperor Wu of Liang's (r. 502–49 CE) attempt to 'dream up his own heaven's form' as being 'completely the same as the text of the *Gnomon of Zhou*'.[42]

Block A. We then turn from umbrellism to expansive night. Despite its being 'lost', Li Chunfeng manages to summon a 186-graph description from 'Han Gentleman of the Palace Library Xi Meng's record of the transmission of its first master(s)'. Accordingly, heaven has no substance, its colour is an optical illusion, and celestial objects float in a void.[43]

Block 7. We now move to third- and fourth-century theories. Li's citations are more extensive here than Shen Yue's, but his conclusion is nearly word for word the same: 'from Yu Xi, Yu Song and Yao Xin [on], everything is whimsical and fantastical explanations, these are not people who discussed heaven by plumbing the numbers'.[44]

Block B. We move to Wang Chong's umbrellist criticism of sphere heaven in *Lunheng* as an example of how 'many scholars were dubious when faced with the subtle principles of sphere heaven'.[45] Li Chunfeng counters this with Ge Hong's (283–343 CE) point-by-point refutation, discussed in Section 5.3.1 below.

Block 4. Having discredited its adversaries, Li Chunfeng then turns to sphere heaven. The *Book of Jin* version begins by citing Zheng Xuan and the *Spring and Autumn Annals* weft *Wen yao gou* in support of the prehistoric armillary sphere, concluding:

[39] *JS* 11.278; *Sui shu*, 19.505; cf. Ho (1966, 49).
[40] *JS* 11.279; *Sui shu*, 19.506; cf. Ho (1966, 51–2). Note that later sources attribute this text to Ge Hong, e.g. *Yiwen leiju*, 97.28b–29a and *Taiping yulan*, 947.5b.
[41] See Cullen (1977, 150–9) and Chen Meidong (2007, 203–9).
[42] *Sui shu*, 19.507.
[43] *JS* 11.279; *Sui shu*, 19.507; cf. Ho (1966, 52).
[44] *JS* 11.280; *Sui shu*, 19.508; cf. Ho (1966, 53–4).
[45] *JS* 11.280; cf. Ho (1966, 54–5).

This, then, was the institution of the sight-effigy, its origins going back a long way. [This] was handed down through [each] continuous age, [but access to] the Clerk's Office was tightly forbidden, and scholars did not see [it], and thus did expansive [night] and umbrella [heaven] roil in competition (*JS* 11.284).

From there, the *Book of Jin* gives a fuller version of Han-era instruments, attributing Luoxia Hong, Xianyu Wangren and Geng Shouchang with a 'circle sight' (*yuan yi*) and (falsely) crediting Jia Kui (30–101 CE) with 'creating' (*zuo*) the ecliptic-fitted sight of 103 CE. The *Book of Sui* version, on the other hand, skips from Block B to Block 2; further down, under 'Sphere Sight', it then recycles Xu Yuan's argument *against* the prehistoric armillary sphere. Li Chunfeng identifies Zheng Xuan's teacher, Ma Rong (79–166 CE), as having instigated the hoax:

No brush-toting official (historian) ever makes the distinction. [Grand] Clerk [Sima] Qian and Ban Gu cast the matter into particular doubt (by explicitly identifying it with Northern Dipper), Ma [Rong] being the first to conceive of saying that [*xuanji*] *yuheng* was the sphere heaven sight . . . Today [I] note: Yu Xi says, 'Luoxia Hong turned the sphere heaven at the centre of the earth for Emperor Xiaowu of Han, fixing the seasonal nodes and creating the Grand Inception *li*' – [it] was probably his creation (*Sui shu*, 19.516).

Given that the *Book of Jin* monograph is referenced in the *Book of Sui*, it would appear that Li Chunfeng had finished the former first, and that his ideas on the matter had changed by the time he wrote the latter.

Block 2. The *Book of Sui* version moves from Ge Hong to He Chengtian's *Discourse on the Sphere Effigy Form*. The *Book of Jin* omits this passage, one imagines, because He Chengtian falls beyond its historical scope.

Block 1. On sphere heaven, we move to Wang Fan. The *Book of Jin* version gives the full text, matching word for word that in the *Book of Song*. The *Book of Sui* version, however, omits it, pointing the reader to the *Book of Jin*: 'From the outside, this (Block 2) is largely the same as Wang Fan. Wang Fan's *The Sphere Heaven [Effigy] Explained* is contained in the *History of Jin*'.[46]

This is where the *Book of Song* and *Book of Jin* stop. In the *Book of Sui*, however, Li Chunfeng ends with the history of a specific problem:

Old (*jiu*) explanations of sphere heaven relied on the sun, moon and stars; [they] did not inquire into [the seasons], [the sun's diurnal motion], nor whether the centre of the earth is the same distance from zenith and nadir, without [one] being further or closer [than the other] (*Sui shu*, 19.512).

This is followed by six excerpts. The first is an anecdote from the *Liezi*, wherein Confucius (fifth century BCE) intervenes in a squabble between two children, finding himself quickly out of his depth when the argument turns out to be

[46] *Sui shu*, 19.512.

whether the sun is closer at the horizon (where it appears the largest) or on the meridian (where it is hottest). The second is Huan Tan's recollection of a conversation with Yang Xiong (first century BCE), which adds to this that the stars appear more dispersed at the horizon than at the meridian, and that the heat of the noonday sun is likely the effect of accumulation. The third is Zhang Heng's *Lingxian* (second century CE), which reasons that the sun appears smaller at noon because of the relative brightness of its background, the way that fire appears smaller by day. The fourth is Shu Xi (third century CE), who reasons that the difference of objects' relative size at the horizon is due to their juxtaposition thereto, the way a bronze vessel appears larger in a smaller courtyard, or the way a current makes a stationary boat appear to be moving.

It is in the fifth excerpt that Jiang Ji (fourth century CE) takes us from reasoning from anecdotal experience to empirical falsification. For Jiang, 'the mess of explanations is due to the human eye' (*Sui shu*, 19.513). Using an armillary sphere – likely the Chang'an 'sphere sight' of 323 CE – he disproves the claim that the stars are any further apart at the horizon than they are at the meridian. Extending this to the sun, he reasons that its apparent diminution on the meridian is due to the disorienting effect of its brightness on the human eye – a brightness attenuated and reddened at the horizon by a layer of 'wandering *qi*' on the earth's surface. In the sixth excerpt, finally, Zu Geng (sixth century CE) caps this off with a forceful affirmation of the empirical approach:

Since antiquity, there have been many who have discoursed upon heaven, but the big names all disagree to the point of mutual destruction. I have humbly reviewed their differences, investigated the matter in the classics, looked up to observe (*guan*) the [stars] and pole, gazed out at the four corners (of the earth), perceived the rise and fall of sun and moon, inspected the appearance and hiding of the five [planets], checked it by means of sight and effigy, re-examined it by means of gnomon and [water clock], and [I can tell you that] the principles (*li*) of sphere heaven are credible and evidenced (*xin er youzheng*). [I] shall thus toss out the diverse explanations [of my predecessors] and stick to the sphere sight. From the [weft text] *Kao ling yao*, former scholars arrived at the figure of 178,500 *li* for the distance between heaven and earth,[47] [but] when verified by means of gnomon shadows it misses the mark in excess. Rather than reveal its procedure for attaining [this figure], [the *Kao ling yao*] emptily posits the number – probably just an expression of exaggerated nonsense – which is certainly not the directive of the sages. Most scholars hold stubbornly to this explanation without changing it – do [they] not know to seek out its principles, or that the reason for this number cannot be sought? Wang Fan's investigation, when compared to previous opinions, reduced [this figure] by no less than one-half. It was not something that [he] knew from observation (*kui*) but that [he] sought out [only] through reasoning (*li*), and thus [he] was truly unable to get at the substance of the matter from afar. Even so, might not [he] have been close and tight? (*Sui shu*, 19.514).

[47] Later sources tend to attribute this figure instead to the *Luoshu Zhen yao du*, e.g. *Fayuan zhulin* (*T.* no. 2122), 53:300a; *Kaiyuan zhanjing*, 3.2a and *Yuhai*, 1.43a. The figure appears also in the *Gnomon of Zhou* (*passim*).

In terms of structure, we can see that Li Chunfeng has appropriated the building blocks of Shen Yue's history to rearrange them. Instead of going from better to worse, Li moves sphere heaven – the a priori – to the end. This movement is likewise roughly chronological. Li insists upon the priority of umbrellism, tracing its Zhou transmission back to Fuxi. The sphere comes only later: in the *Book of Jin* he attributes it to Yao, and in the *Book of Sui* he walks it back to Luoxia Hong, in the second century BCE. Contrary to Shen Yue, Li Chunfeng presumes that there are things 'that the ancients had yet to learn' – that good knowledge must be modern knowledge. Such is clearly the lesson in his conclusion to the *Book of Sui* version: we are wiser now, we build better instruments, and by reasoning and experimentation alone, we can answer questions that once got the better of Confucius.

5.2.2 Li *Numbers, 220–479* CE

The *lü-li* monograph is a simpler matter to compare.[48] There is, for the purposes of historiography, only one 'school' of *li*, and only way to approach its history: absolute chronology. Introductory précis aside, where Shen Yue and Li Chunfeng's *lü-li* monographs overlap is as concerns the Cao Wei (220–65 CE) and Sima Jin (265–420 CE), Li having generously left the Liu Song (420–79 CE) alone. The focus of the period from 220 to 420 CE is, in both cases, the events covered in Chapter 4 above. The difference between our two historians is all the more striking here, if only for the simple reason that, where Li Chunfeng devotes 9,477 graphs to the lead-up to 237 CE, Shen Yue offers but 339:

In [178–84 CE], Watchman of the Wall at Gucheng Gate Liu Hong first awoke [to the realisation] that the Quarter-remainder [*li*] was loose and wide of heaven. [He] changed the era divisor to 589, and the Dipper parts to 145, and [he] created the Supernal Icon method. [He] furthermore formulated the slow–fast *li* to pace lunar motion, and only with this was there finally a turn towards the fine and subtle relative to the Grand Inception and Quarter-remainder.

In [220–6 CE], Assistant to the Grand Clerk Han Yi thought that the Supernal Icon *li* went too far in reducing the Dipper parts and that it would later slip ahead of heaven, so [he] constructed the Yellow Inception *li*, using an era factor of 4,883 and a Dipper parts of 1,205. Later, Prefect of the Masters of Writing Chen Qun (d. 237 CE) submitted a petition, which went: [Petition omitted, see Section 4.1.2 above]. The petition was approved.

In [226–39 CE], Master of Writing Yang Wei formulated the Luminous Inception *li*, which was used all the way into the Jin and [Liu] Song. As to the *li*-workers of antiquity, Deng Ping was able to repair the old and formulate the new, Liu Hong was first to

[48] Note that I exclude the tono-metrics (*lü*) fascicle of Shen Yue and Li Chunfeng's respective *lü-li* monographs for the reasons given in Section 1.1.1.

> diminish the Quarter-remainder (Dipper parts) and also to fix lunar motion slow–fast, and Yang Wei weighed both sides ... These three men were *li* experts of the Han and [Cao] Wei. Even so, Hong's slow–fast did not check out against the *Spring and Autumn Annals*, and Wei's five [planets] were largely contrary in later ages – the reason for this being that Hong's presence of mind was still loose, and that Wei restricted himself to everything emerging at a high origin of *renchen*.[29] (*Song shu*, 12.231–2).

If this sounds at all familiar it is because these lines appear nearly word for word in the *Book of Jin* chronicle translated in Section 4.1 above. Li is copying and pasting again. Indeed, 75 per cent (7,128 graphs) of his version is devoted to Supernal Icon procedure text – '[Liu] Hong's procedures were the exemplar of calculation for the subsequent age(s), and thus [do I] lay them out as follows' (*JS* 17.503). He is not making this up: we know from contemporary bibliographies that the Supernal Icon *li* was extant in multiple editions in his day.[49] Next in size is the debate transcript of 226 CE, translated in Section 4.2 above, coming in at 1,752 graphs (18 per cent). Where did he find this jewel? Are we to believe that the seventh-century historian had access to better information than his fifth-century counterpart? No, he had access to *the same information*: the He Chengtian monographs, which Shen Yue cites as his primary source for the third and fourth centuries.[50] These were extant in the seventh century, and Li Chunfeng has probably copied and pasted the transcript from there.[51] He certainly didn't make it up: the data from 226 CE correspond to real eclipses and real time-keeping conventions specific to the third and fourth centuries.[52] The question of the transcript, therefore, is not *how* Li Chunfeng has it but *why*.

The transcript is important for Li Chunfeng's history, one imagines, because it has Liu Hong emerge the victor. Li is explicit about his motivations: as 'the exemplar of calculation for the subsequent age', he wishes to give Liu the acclaim and attention he deserves. His editorial choices would seem to comport with this aim. In terms of copying and pasting, 75 per cent of his text is devoted to the *li*, and 18 per cent to evidence of its public triumph. Equally telling is what Li Chunfeng *adds*: all that is left after eliminating the aforementioned borrowings is 380 graphs (4 per cent), of which he devotes 183 to the praise of

[49] *Sui shu*, 34.1022; *JTS* 47.2037.

[50] On the question of *access*, the armillary sphere debacle of 418 CE does, however, speak volumes to the possibility of one having better access at a later date; see Morgan (2016c) and Section 2.2.2 above.

[51] For bibliographic records of Xu Yuan's original *Book of Song*, which included He Chengtian's monographs on *tianwen* and *li*, see *Sui shu*, 33.955; *JTS* 46.1989. Gautama Siddhārtha, for one, cites the *tianwen* monograph in 729 CE in *Kaiyuan zhanjing*, 1.28b–29a, the language reading more or less the same as *Song shu*, 23.678–9. Otherwise, one notes that Shu Xi (third century) Wang Yin (fourth century), He Fasheng (fifth century) and Zang Rongxu's (fifth century) respective histories of the Jin are on record as having likewise possessed 'monographs' (*zhi*) or 'records' (*ji*) from which these materials may have originated; see Ran Zhaode (2009) and Li Peidong (2009).

[52] See Qu Anjing (1994).

the Supernal Icon *li*, eighty-five to its transmission history and ninety to the Luminous Inception's failings. Qualitatively, quantitatively, internally and externally, everything points to the conclusion that Li Chunfeng is being forthright about his authorial intentions.

Why go to the effort of spelling this out? Because, in doing this, Li Chunfeng is actively violating the rules of the genre. This is not the only time he does this, as Sun Jiagan (1683–1753) points out:

> The Sovereign Pole *li* created by Liu Zhuo was never instituted in Sui times (581–618 CE), and when Fu Renjun (fl. 618 CE) created the *Wuxin*.$_{15}$ *li* in the early Tang, [he] adhered to Zhang Zhouxuan's successful model (*cheng fa*) as his sole ancestor, making no use whatsoever of what Zhuo had set down. However, this (the *Book of Sui*) monograph vacuously (*kong*) preserves his [work], devoting one separate fascicle [thereto]. [This] is like how the [*Book of*] *Jin* monograph appends the Triple-era *Jiazi*.$_{01}$-origin *li* created by Jiang Ji after the Supernal Icon and Yellow Inception *li*.[53]

Liu Hong has not earned the right to be the protagonist of his own fascicle of the *Book of Jin* (fascicle 17) any more than Jiang Ji (fascicle 18) or Liu Zhuo (*Book of Sui*, fascicle 18), because Liu Hong has contributed nothing by way of *policy* to any dynasty deserving a standard history. Shen Yue follows the rules, focusing on the 'successful model' of Yang Wei in fascicle 12 of the *Book of Song*, so what is Li Chunfeng thinking? As to the *Book of Sui*, Li's own reliance on Liu Zhuo's work to surpass Fu Renjun's *li* reform of 618 CE probably obliges him to bolster his legacy. There is something at stake with Liu Hong's legacy as well, it would seem, albeit less obvious than matters of 'expert lineage'. It could be a point of criticism against Yang Wei, rehashing the *politics* of Liu Hong's defeat, but maybe, as a fellow *li* man, Li Chunfeng is simply interested in the procedures.[54]

Returning to the question of 'progress', one notes that Li Chunfeng is appealing to the same themes of awakening and transcendence as Zhou Cong in Section 5.1 above. Liu Hong 'first awoke' (*shi wu*), he 'created' (*zuo*), and he 'constructed' (*chuangzhi*), the consequences of which extend beyond 'wish[es] to improve on the faltering techniques of the . . . predecessor' (Sivin 2009, 553) to lay bare hidden subtleties of the natural order and to act, in Liu Hong's absence, as the *primum movens* of 'the subsequent age' (*hou shi*). Liu Hong established a 'model' (*fa*) – an 'exemplar' (*shi biao*), no less – that 'turned' (*zhuan*) the course of history. If it were '*li* reform' that mattered, we might expect Yang Wei's political tenure to impress, but *that*, to Li Chunfeng, is small

[53] *Wuyingdian ben ershisan shi kaozheng*, 34.13a. I thank Li Liang for bringing this passage to my attention.

[54] On Li Chunfeng's indebtedness to and identification with Liu Zhuo as it colours his history writing, see Li Liang's chapter 'The compilation of the calendar part in the *Book of Sui* and its impacts', in Chaussende, Morgan & Chemla (forthcoming).

potatoes.[55] Yang Wei, whatever we make of his humility, is the first to honour Liu Hong's spirit: defending his presence-in-absence at the debate of 226 CE, Yang warns that 'forsaking his procedures . . . will stop Hong's unique and marvellous model from being transmitted to future ages'.[56] This brings us back to the debate and its inclusion in Li Chunfeng's history. The transcript is interesting, and it reveals invaluable information about early '*li* reform', but it is also the greatest possible testament to Liu Hong's transhistorical agency: it is his *spirit*, like Zu Chongzhi in Section 5.1 above, that wins an oral and mathematical contest against the living.

As contrarian as Li Chunfeng can be, our three historians all agree that Liu Hong is a 'turning point' in history. True, Shen Yue criticises his inability to *tong* the *Spring and Autumn Annals* as an example of how 'Hong's presence of mind was still loose', but he frames his chronicle with the following quote:

> He Chengtian says: 'As to the procedures of *li* numbers, if one's mind does not understand, then even if one has mastered others' previous knowledge, one will have no way to save it from their faults. This is why nothing has yet to be settled over the course of so many years. In heaven, the Quarter-remainder produces one day in excess every three hundred years. The accumulated ages did not awaken, vainly saying "the root to establishing *li* is the necessity of first establishing the origin" with spurious words of prophecy and weft [literature that they] then connected to order and chaos – this is a fault, and [it] is a sufficiently profound one at that. Liu Xin's Triple Concordance method was particularly loose and wide, adding yet another day every six thousand-plus years over the Quarter-remainder. Yang Xiong's mind was confused by [his] theory, adopting [it] for the [*Classic of*] *Supreme Mystery*; Ban Gu referred to it as "superlatively tight", writing [it] into the [*Book of*] *Han* monographs; Si[ma] Biao followed up with, "From first implementation in Grand Inception 1 (104 BCE), the Triple Concordance *li* was in effect for more than a hundred years." Not remembering that Liu Xin had yet to be born by the Grand Inception [period] (104–101 BCE), when these gentlemen speak of *li*, [they] speak foolishly out of near [total] ignorance' (*Song shu*, 12.231).

There are things that Liu Hong had yet to learn about precession, sure, but he managed to turn *this mess* 'towards the fine and subtle' (*Song shu*, 12.231) or, in Li Chunfeng' version, the 'fine and tight' (*JS* 17.498).

5.2.3 *The Mirror of History*

When the Jurchen sacked Bianjing (34° 48′ N) in 1127, eighteen years after the repainting of Liu Hong, He Chengtian and Li Chunfeng's sacrificial icons in King Wenxuan Temple, they stripped the Song capital of its riches, transporting the armillary sphere from Su Song's (1020–1101) famous clock tower of 1092 to their capital in modern Beijing (39° 55′ N). Stripped of its support structure,

[55] *JS* 17.498, 503; cf. Section 4.1.1 above. [56] *JS* 17.503; cf. Section 4.2.1 above.

moved 'more than a thousand *li*' to be there, its 'various apparatus having been long discarded and destroyed', the instrument became, like many of its predecessors, an expensive parody of its philosophical potential. Wobbly and misaligned, it sat useless on the observatory terrace until the storm of 1195, when the terrace collapsed, and 'the sphere sight lurched and fell off the terrace'. They managed to roll it back, but when the Jin (1115–1234) moved south in 1214, they ultimately 'melted the sphere sight down and cast it into [other] things'.[57] If history is a mirror, as the saying goes, Li Chunfeng's was forged in much the same way.

Li is good about his sources. He cites those included in Shen Yue's history, he cites others that Shen Yue misses, and *he*, for one, knows who wrote the monographs he is referencing:

> Liu Xin constructed the Triple Concordance to explain the *Zuo Tradition* [of the *Spring and Autumn Annals*], which, upon analysis, was not solid/true (*shi*); Ban Gu was fooled by it and selected [it] for [his] monograph . . . In [178–84 CE], Liu Hong and Cai Yong were ordered to co-operate on the compilation of [a] *lü-li* [monograph], which Sima Biao later used in following up Ban [Gu]'s history. Here [I] select words on past matters of *li* numbers from Emperor Wen of Wei's Yellow Inception [reign] (220–6 CE) on so as to continue from Sima Biao['s *Book of Later Han*] as follows (*JS* 17.498).[58]

Li Chunfeng is similarly careful in his other writings, but *nowhere*, in any of his monographs, does he acknowledge Xu Yuan or Shen Yue's *Book of Song* as the source from which he takes, word for word, the vast majority of his text.[59] The one place where he *does* acknowledge their histories is by point of direct criticism about the 'sphere sight' debacle treated in Section 2.2.2 above:

> Song Palace Aid to the Censor-in-chief He Chengtian and Superior Grand Master of the Palace Xu Yuan each wrote a history of the [Liu] Song wherein both took [the Kong Ting armillary sphere of 323 CE] as Zhang Heng's construction . . . Liang Master of Writing Shen Yue wrote a history of the [Liu] Song which also says as much – all of them miss the mark by a great distance (*Sui shu*, 19.518).

[57] *Jin shi*, 22.523–4. On Su Song's clock tower, see Needham, Wang & Derek (1986).

[58] Returning to the previous quotation by He Chengtian, one notes that Mr He misattributes Ban Gu's praise of Liu Xin. Ban uses 'superlatively tight' (*zui mi*) to describe *Deng Ping's Grand Inception li* in the context of the live trials of 104 BCE (*HS* 21A.976), praising Liu Xin for 'superlative detail' (*zui xiang*) as concerns his work on tono-metrics, which Ban acknowledges as having 'excised of its more spurious phrasing, taking [its] correct meaning and writing it into a piece' (ibid., 21A.955). As to He Chengtian's (therefore hypocritical) criticism of Sima Biao for having confused the Triple Concordance and Grand Inception *li*, Mr He treats Sima Biao as the monograph's author and fails to note that the offending passage is actually a quotation from Jia Kui in 92 CE (*HHS zhi* 3, 3025). All of this reflects rather poorly on his historiographic rigour. Shen Yue likewise fails to identify the authors of the *Book of Later Han tianwen* and *lü-li* treatises in *Song shu*, 11.203–6. On the authorship of these treatises, see Mansvelt Beck (1990).

[59] One finds a similar acknowledgement of Li Chunfeng's forerunners (correctly identifying the authors of the *Book of Later Han tianwen* monograph) in *JS* 11.278. On his citation policy in the omen compendium *Yisi zhan* of 645 CE, see Harper (2010).

They come up once again as concerns the 'rotating mechanism and jade traverse':

The *Spring and Autumn* [weft] *Wen yao gou* says: 'When Yao of Tang took the throne, the Xi and He [brothers] established the sphere sight'. However, some former scholars followed the star office books, [which state that] the Northern Dipper's second star is named 'Turner' (Xuan, β UMa), its third star is named 'Mechanism' (Ji, γ UMa), its fifth star is named 'Jade Traverse' (Yuheng, ε UMa), and thus that the talk about 'seven governors' [in the *Book of Documents* refers to] none other than the seven stars of the Northern Dipper. No brush-toting official (historian) ever makes the distinction, even though Clerk [Sima] Qian and Ban Gu cast [the matter] in doubt (ibid., 19.515–16).

What 'brush-toting officials' between Ban Gu and Li Chunfeng's day who wrote a history of astronomical instruments could he possibly be talking about?[60] *That* he leaves to the reader's imagination, silently reworking the *Book of Song* monographs into an argument against their authors' position on these very points. Ultimately, his strategy of vindictive plagiarism pays off, as later compendia and scholarship on the astral sciences are almost unanimous in citing Li Chunfeng over Shen Yue as concerns the events of 220–420 CE.[61]

Sinologists recognise that the standard histories are biased sources, pointing to the fact that 'the official histories, beginning with the *Shï-ki*, are written for officials by officials' (Franke 1950, 113) and, thus, reflect official interests.[62] They are *elitist*, focusing on the capital – the capital of the historical winners – and the nobles and gentlemen bureaucrats who constituted its rarefied upper class. They are *moralising*, sustaining the Confucian tradition of 'praise and blame' (*bao bian*) to cleave the past into good and bad examples. They are, moreover, *political*, arranging the goodies and baddies in time-worn patterns pointing to the moral and cosmic necessity of whatever regime is footing the bill. The *Book of Jin* is a case in point. From Taizong's (r. 627–49 CE) commission and personal participation in the project it is clear that his interest in the Sima Jin was as a mirror for reflecting on the sobering questions of his reign: how a fledgling dynasty was to survive succession struggles and the fracture of officialdom and how history was to judge their efforts.[63] Meeting the

[60] For Sima Qian and Ban Gu's – actually Ban Zhao's – identification of the *xuanji yuheng* with the Northern Dipper, see *SJ* 27.1291 and *HS* 26.1274. One notes that the *Book of Later Han* and *Book of Wei tianwen* monographs are silent on the issue, leaving Xu Yuan and Shen Yue's histories as the only plausible butt of Li Chunfeng's criticism.

[61] This argument is further developed in the author's forthcoming chapter, 'Heavenly patterns', in Chaussende, Morgan & Chemla (forthcoming). As to whether or not the label 'plagiarism' is appropriate to describe Li Chunfeng's appropriation of the Shen Yue *tianwen* monograph, I remind the reader that *Shen Yue* feels the need for transparency about what *he* is taking from He Chengtian's work.

[62] Cf. Balazs (1968, 47–67) and Twitchett (2002).

[63] Li Peidong (2009).

moral and political imperatives of historiography involved a certain flexibility with one's sources, and the *Book of Jin* was no exception: there, even the most defining battle of the period was rewritten.[64]

Given the context, it would be foolish to single out one monograph therein as a 'history of science', clear, objective, and chaste of ideology; the question, rather, is *where* the biases lie. It is not, for once, the broader issues of court politics that are at stake here. Consider the work required in Chapter 4 above: the *modern historian* must supply the political context of 220–37 CE, because the *medieval historian* has omitted details about ritual reform, political networks and personal tragedy as immaterial. Without this context, the story is essentially this: Liu Hong was great; Han Yi tried to beat him, but he failed the test of trial and debate; Yang Wei didn't beat him either, but he bypassed the review process to win state support; also, the throne changed the civil calendar, then they changed it back, all for no apparent reason. *Li*, to them, as they want it remembered, occurs in a vacuum where the only variable is the individual's epistemic virtue. Sombre and pristine, its history is a sacred space of luminaries, heavenly and human, separate in time and space from the 'world of dusts' (*chen shi*). History is a temple, and there are things that don't belong at temple.

5.3 The Ascetic: The Siddhis of Science and the Awakening of Man

How could the ancients have conceived of *progress*, you might ask, before there was such a thing as 'science' as separate from 'religion'? The answer is right there in the question: progress is a religious narrative. Indeed, the themes of sagehood, gnosis, virtue and transcendence explored thus far strike such an obvious chord with the history of religion as to leave one wondering where to begin. One could launch at this point into a résumé of East Asian religion, or 'religion' in general, citing anthropologists on archetypes of spiritual 'awakening' (*wu*/*jue*), but to avoid the dangers of essentialising, and the work of definition, let us focus instead on the individual. Of all the names that we have encountered up to this point in the astral sciences, there are two in particular that stand out as consequential figures in the history of religion: Ge Hong and Yixing. These men were many things in their lives, as underscored in Campany (2005) and Chen Jinhua's (2000–1) respective studies of their polymathy, so how they speak, as individuals, across different domains of knowledge, ought to give us an idea of how the ancient mind could reconcile what modernity teaches us are separate magisteria.

[64] See Rogers (1968).

5.3.1 *The Alchemist: Ge Hong (283–343 CE)*

Ge Hong was born into an aristocratic clan in Jurong County, halfway between the flourishing religious centre of Mt Mao and the beleaguered administrative centre moved to Jianye in 317 CE. Having lost his father at the age of twelve, Ge Hong describes his early life as one of poverty, isolation and independent reading. Zheng Yin, a master in classical learning and the occult arts, took him under his wing around the age of fifteen, sharing his library with the voracious autodidact and inducting him into the arcana of alchemy. Later pulled into state service upon the court's military retreat to the south, Ge Hong managed to balance the demands of high office with those of biospiritual cultivation, taking leave to collect ingredients for his elixirs, in seclusion, and requesting transfers, while in office, to areas where he might find more. His magnum opus, *The Master Who Embraces Simplicity* (*Baopuzi*), reveals the plurality of his intellectual world:

> In the 'Inner Chapters', [I] speak of such matters as the recipes and drugs of gods and immortals, the marvels and metamorphoses of ghosts and monsters, the extension of years via the nourishment of life, and the avoidance of disaster via the exorcism of evil – that belonging to the *dao* expert lineage (*dao jia*); in the 'Outer Chapters', [I] speak of success and failure among men and the goodliness or not of worldly matters – that belonging to the scholastic/Confucian expert lineage (*ru jia*).[65]

On the 'inner' end of things, we also have his *Traditions of Divine Transcendents* (*Shenxian zhuan*), a collection of success stories – hagiographies – of mortals who variously achieved longevity, immortality and superpowers via dietetics, sexual arts, talismans, ritual and alchemical supplements. Whatever their inherent novelty or impact at the time, as syntheses of the cult of immortality among southern intellectuals prior to the Celestial Masters' exodus there *c.*317 CE, and the Supreme Purity revelations of 364–70 CE, Ge Hong's writings are some of the most historically valuable in the Daoist canon.[66]

Ge Hong's principal interest in heaven was to fly there with feathered wings and a body lightened of death; maths, by contrast, were not his cup of tea: '[I] disliked star books as well as mathematical procedures . . . and the like, [so I] was unable to follow – it was because they were hard on a person while at once lacking in interest'.[67] Where Ge Hong does appear in the annals of the astral sciences (thanks to Li Chunfeng) is by point of criticism against Wang Chong. Wang Chong, in brief, argues that we must accept unintuitive (physical) explanations for umbrellism due to the (metaphysical) entailments of the sphere. Namely, spherism would have the sun pass under the earth and through

[65] *Baopuzi waipian jiaojian*, 50.698.

[66] On the *Master Who Embraces Simplicity*, see Ware (1966). On the *Traditions of Divine Transcendents*, Ge Hong's biography and immortality practices, see Campany (2002). For a broader look at the history of self-divinisation in early China, see Puett (2004).

[67] *Baopuzi waipian jiaojian*, 50.656.

the ocean, but *heaven : yang :: earth : yin* and *sun : fire :: ocean : water*, and since *yang > yin* and *water > fire*, this is (metaphysically) impossible. Ge Hong, as an outsider, offers what is probably the most forceful rebuttal of this position in Chinese history:

[Zhang Heng's] *Sphere Heaven Sight with Commentary* states: 'Heaven is like a chicken egg: earth is like the yolk, dwelling alone within heaven; and heaven is big while earth is small. Without and within heaven is water, heaven and earth each riding on *qi* for support and [one/both?] moving upon the water. The circuit of heaven is 365 *du* and ¼ *du*, which, moreover, is divided at [its] centre, half above, covering earth, and half below, encircling earth. Thus are the twenty-eight lodges half visible and half hidden [at any given time], heaven turning like the revolution of a wheel hub.'

There are many who discourse upon heaven, yet few who are refined in [matters of] *yin* and *yang*. Students of Zhang [Heng] and Lu [Ji] all pursue the *dao* of calculating the paths of the seven luminaries, *li*-ing and *xiang*-ing proven observations (*zheng hou*) of dusk and dawn with *du*, comparing with the *qi* of four and eight (indicator arrows?), examining with the fractions of the leak notch, divining the coming and going of gnomon shadows and seeking the shape [of heaven] in the truth of the matter (*shi qing*) – [for this] nothing is tighter (*mi*) than the sphere effigy. Zhang [Heng] created a bronze sphere heaven sight in a sealed-tight (*mi*) room, turned by leak water, ordering its attendant to shut the doors and sing it out. Its attendant announced to the observer(s) of heaven (*guan tian zhe*) of the Numinous Terrace, 'at [some] add[ed hour] (?) of the *xuanji*, such-and-such star is first visible, such-and-such star is already centred (culminated), and such-and-such star is currently setting' – all of which were like matching [the two halves of] a tally.[68] [The famous calligrapher] Cui Ziyu (77–142 CE) wrote the epitaph on his tombstone, which went: '[His] numbers and procedures (*shu shu*) excelled heaven and earth, [his] construction and invention matched Creation itself, [his] tall talent and extraordinary art matched tallies with the gods' – a fact (*gu*) proven (*you yan*) by [Heng's] sphere sight and seismoscope.

If heaven is indeed like the sphere, then the [risings] and [settings] of heaven must really (*di ran*) move through water. Thus does the *Book of the Yellow Emperor* state that 'beyond earth is heaven, and beyond heaven is water' – water being what floats heaven and carries earth. Moreover, the [*Book of*] *Changes* states that '[they] seasonally mount the six dragons' (Qian ䷀) – *yang* lines are called 'dragons', dragons are things that dwell in water, as a metaphor for heaven; heaven is a *yang* thing, and [it] furthermore exits and enters water, which is similar to dragons, thus is it compared with dragons. The sages looked up to observe and down to inspect, and [they] scrutinised that this is how it was. Thus in the Jin ䷢ hexagram is Kun ☷ (: *earth*) below and Li ☲ (: *fire*) above so as to prove (*zheng*) that the sun emerges from the earth; [thus], moreover, in the hexagram of Mingyi ䷣ is Li ☲ below and Kun ☷ above so as to prove that the sun enters into the earth. The Xu ䷄ hexagram has Qian ☰ (: *heaven*) below and Kan ☵ (: *water*) above, this also being an effigy/sign (*xiang*) of heaven entering water. Heaven is [the agent] metal, and metal and water are things that mutually generate, [so] what possible damage could it cause heaven to exit and enter the water to make one say that [it] cannot?[69]

[68] Cf. Li Chunfeng's alternative version in Section 2.1.4 above.

[69] *JS* 11.281–2; cf. *Sui shu*, 19.509–10; tr. modified from Ho (1966, 55–6).

Ge Hong keeps going, adding four points on experiential falsification from shadows, motions, apparent sizes and optics and another on prophecy literature and *yin-yang* correlations, running the full gamut of epistemic strategies before concluding that 'the principals of spherical heaven are thus credible and evidenced' (*xin er youzheng*).[70] The genius of this argument is its recognition that the opponent's rests on incommensurate grounds – that the only way to effectively counter an argument from metaphysics is via older and more sophisticated metaphysics. Citing the Yellow Emperor and the *Book of Changes*, Ge brings metaphysics in line with experience, demonstrating that, like dragons, the sun can be expected to pass beneath the earth.[71]

Is this parody? Evidence from his alchemical writings would seem to suggest that Ge Hong took *yin-yang* and five-agents reasoning rather seriously. Here he foregrounds a different type of argument, but it is one that the context demands of him, and it proves zero challenge to his metaphysical commitments or wit. If one were to ask him how he reconciled contradictions of 'science' and 'religion' in moving between these contexts, *that* might well stump him. To Ge Hong, they are more or less the same thing, which is to say that they are matters of learning. As to the astral sciences, he offers,

> To distinguish the signs of *yin* and *yang* by gazing in the distance at Net.$_{L19}$ (the Hyades) and to awaken (*jue*) the offset of the intercalary remainder from [the stirrings of] winter insects – what divinity (*shen*) is there in this? It is only study (*xue*).[72]

This rings as true for the individual as it does in the aggregate, as he argues of (secular) texts and technologies in the 'Outer Chapters':

> It is like how boats and carts substituted walking and fording, and how writing and ink replaced knots and ropes (*khipu*) – the one is a later invention, but better than the previous thing, the threads its achievements running in the thousands of myriads, this way and that, [such that] they cannot be undone. The people of [this] age all know that [things] are happier (*kuai*) than in the past, [so] how could it that be [modern] writings alone fall short of antiquity?[73]

So too is this how he approaches the subject of biospiritual cultivation in the 'Inner Chapters'. Ge Hong opens fascicle 12 with a challenge: 'If immortality (*xian*) is necessarily attainable, [then] the sages [would have] already practised it, [so] one knows from [the Duke of] Zhou and [Master] Kong's *not* having done it that there is no such *dao*'.[74] Ge Hong goes on from there to explain the

[70] *JS* 11.284; cf. *Sui shu*, 19.511; tr. modified from Ho (1966, 58).
[71] For Ge Hong's full rebuttal, see Ho (1966, 55–8); cf. Cullen (1977, 298–307) and Chen Meidong (2007, 214–19).
[72] *Baopuzi waipian jiaojian*, 3.129; tr. modified from Puett (2007, 105).
[73] *Baopuzi waipian jiaojian*, 30.78; tr. modified from Puett (2007, 108).
[74] *Baopuzi neipian jiaoshi*, 12.224.

difference between immortality and sagehood in terms of skill acquisition, because, here again, what we mistake for divinity is only study:

> The people of [this] age say that the sages fall from heaven, that they are divine and numinous beings, and that there is nothing they do not know, and nothing they cannot [do]. [So] profound is [their] obeisance and awe of their names [that they] do not dare to go back and assess them with regard to events. [They] say that what the sages did not do, no man will ever be able to do; that what the sages did not know, no man will ever know – isn't that a laugh?! [If], in the here-and-now, one compares (*jiao*) them against recent events, [I] think one can awaken (*wu*).[75]

Ge Hong returns to this theme throughout his writings, the thrust of which, Puett (2007, 114) argues, is 'that knowledge is human, that it is derived from sages, and that, since sages are limited, it is crucial that new sages be recognised such that the accumulation of human knowledge can continue'. This movement is not one that is limited to *sages* or *the individual* – for sages are but individuals who grew to greatness – nor, for that matter, is it one limited to any one domain of human knowledge. Ge Hong's message, rather, is one of total, collective salvation: 'for Ge Hong, humanity is slowly but surely transcending the natural world of the earthly and gaining ever more control over the heavenly realm as well' (ibid., 115).

It is probably safe to call this 'progress' – Ge Hong uses terms like 'the accretion of accomplishments and accumulation of hard work'[76] – and it is worth underscoring that this is *not* an idea that he develops in a 'scientific' context. The question of what constitutes 'science' and 'religion' is admittedly nebulous in this case: Ge Hong moves skilfully between '*dao*' and 'scholastic/ Confucian expert lineages', and elements of his practices resemble the 'scientific' diet fads of our day,[77] but such are the limitations of observers' categories. As to the question of 'progress' in the astral sciences, what matters here is *who* is speaking. Ge's relative disinterest and non-presence in the astral sciences would suggest that 'the accretion of accomplishments' was an idea whose roots ran deeper than astronomy.

5.3.2 *The* Ācārya*: Yixing (683–727* CE*)*

Yixing, for his part, was always fond of maths. Born into the Zhang clan of present day Hebei, into a line of scholars and statesmen with ties to the western frontier, Yixing (né Zhang Sui) was from an early age 'widely read in the classics and histories and particularly gifted in the studies of *li* and *xiang, yin* and *yang*, and the five agents'.[78] Drawn to a similar calling as Ge Hong, Yixing

75 *Baopuzi neipian jiaoshi*, 12.227; tr. modified from Puett (2007, 103).
76 積功累勤, see *Baopuzi neipian jiaoshi*, 13.240–1. 77 See Levinovitz (2015).
78 *JTS* 191.5112.

dabbled in *dao* before ultimately 'leaving the family' to study the precepts of Tiantai, meditation (*chan*) and discipline (*vinaya*) under the foremost masters of his day. Mixing with clergy, courtiers and the elite of Chang'an's elite, Yixing would rise to international prominence in the early Kaiyuan era (713–41 CE). Something new was coming out of the West in these years – something big enough to make the 'Big Vehicle' look small – the evangel borne to eastern lands by Śubhākarasiṃha (637–735 CE), Vajrabodhi (671–741 CE) and Amoghavajra (705–74 CE). What were these new and marvellous teachings? The Chinese would later call them Mi, which we translate as 'Esoteric Buddhism', but the *li* man might be tempted to read the word for what it is – 'the Religion of Tightness'. He would hardly be disappointed, upon reading a little further, as, although it becomes clear that *mi* is used in the sense of 'secret' (from 'tight' and 'tightly sealed'), the secrets therein are very much about the accuracy of rituals wrapped in number. That's beside the point, of course, because when Śubhākarasiṃha arrived in 716 CE there was nothing there to read.

Studying with Vajrabodhi, Yixing eventually came under the tutelage of Śubhākarasiṃha, with whom he collaborated to translate the *Mahāvairocanābhisaṃbodhi-vikurvitādhiṣṭhāna-vaipulyasūtra* (the *Mahāvairocana* or 'Great Sun' *sūtra*, for short) in 724 CE. The *sūtra* takes the form of a dialogue between Mahāvairocana and Vajrapāṇi, devoting its opening fascicle to the presentation of a coherent doctrinal framework for its ritual programme. Enlightenment – defined as 'knowing one's own mind as it truly is' – the text tells us, 'is rooted in compassion, caused by the *bodhi*-mind, and completed in skilful means'.[79] The latter are elaborated in the remaining six fascicles: the construction of the *garbhadhātu-maṇḍala* in order to summon, sacrifice to and ultimately embody the deities therein – manifestations of Vairocana's secrets of body, speech and mind – through the practice of ritual gestures (*mudrā*), recitations (*mantra*) and the sort of object-oriented meditation encountered in Chapter 2 above. This, in conjunction with Yixing's massive twenty-fascicle commentary, redacted from Śubhākarasiṃha's oral exegesis, would become the core of East Asian Vajrayāna for centuries to come. Read it how you like, but it was a *li* man who put a capital M on Mi.[80]

Yixing's position on 'scientific progress' is perhaps the easiest of any to gauge, for he composed a twelve-part 'deliberation' (*yi*), which, as Zhou Cong mentions, 'collated and corrected successive generations [of data] in order to

[79] *T* no. 848, 18: 1b–c.

[80] For more on Yixing, see Osabe (1963) and Chen Jinhua (2000–1). For the *Mahāvairocana sūtra*, see Hodge (2003) and Giebel (2005). For Yixing's commentary, see Müller (1976). For a less idiosyncratic exploration of the category 'Mi', see Payne (2006).

find the strengths and weaknesses of [historical] *li* models (*fa*)' (see Section 5.1 above). Drafting the deliberation in 727 CE, while finishing his commentary to the *Mahāvairocana sūtra*, Yixing, like so many others, would not live to see his findings published. It took Lord Specially Advanced Zhang Yue and *Li* Officer Chen Jingxuan to present the throne with their redaction of this and his Great Expansion *li* in 728 CE, which, in Yixing's absence, instigated one of the most celebrated *li* reforms in Chinese history. As to historical vision, it is to Meditation Master Yixing whom we owe the title of this chapter.[81]

In his 'Deliberation on the Syzygy' (Heshuo yi), Yixing recounts the history of lunar anomaly, bringing us back to the topic of Sections 1.2.2 and 3.3.3 above. 'When ancient people examined heavenly matters', he reminds the eighth-century reader, 'most of them did not know about fixing the syzygy'; nor *could* they, at the beginning, as 'the sages had yet to pose the question, [since] it was something at which [their] *li* calculations could not arrive' (*XTS* 27A.596). Once man 'awoke' to what the moon was doing, each generation then worked within the limits of their knowledge to capture its anomaly:

> To summarise the various *li* of the recent age (seventh century), levelling them at a *lü* of 1,000,000, their errors are, at the lowest, sometimes one part, and, at the highest, a miss of one part per ten numbers (?). In examination of the *Spring and Autumn Annals* [they] only erred a notch (14m24s), which would hardly amount to a difference of [a day] over a hundred and something years; shortly after implementation, [however], they soon proved loose and wide, because they had yet to learn (*wei zhi*) about seeking the step and distance mean syzygy from one another (i.e. the iterative correction of true solar and lunar position). Li Yexing, Zhen Luan et al. (sixth century) desired to seek heavenly verification, [but they] simply added and diminished the lunar fraction, shifting and changing [it] incessantly, breaching [one-day errors] back and forth – [this] again was because [they] had yet to learn (*wei zhi*) about the dusk and dawn limit on fixed syzygy. Yang Wei (third century) appropriated the Supernal Icon for the slow–fast and *yin–yang* sequences (i.e. lunar speed and latitude corrections), and though [he] knew that the added hour was behind heaven, and that the eclipse was not on new moon, [he] was as yet unable (*wei neng*) to do anything to change it (*XTS* 27A.596–7).

Step by step, they figured it out, narrowing the gap in theory and predictive accuracy, and the proof, Yixing shows us, is in the numbers: 'upon examination of the Clerk's Office's note records from [134 BCE] on, there being a total of thirty-seven instances of solar eclipse with added hour, [Li Chunfeng's] Unicorn Virtue *li* gets five, and [Yixing's] Kaiyuan *li* gets twenty-two' (*XTS* 27A.598).

That's 'science', but what is he thinking when he sets that down to work on scripture? In a way, it is the self-same issue that is at stake in the *Mahāvairocana sūtra* – 'truth' (*zhen*/*shi*), 'awakening' (*wu*/*jue*) and the progressive steps of its

[81] Specifically, Yixing uses the phrase 'this is what the ancients had yet to learn' in relation to the old approach to 'Dipper parts' and intercalation as cited in *XTS* 27A.593.

'accomplishment' (*gong*, or *siddhi*). This the Buddha Vairocana elaborates in a progressivist regime, taking the interlocutor step by step, day by day, from knowing to becoming God. The beauty of the *sūtra*'s words, in Yixing's Chinese, eclipses any wisdom that the author could hope to add:

Next, Lord of Mysteries, there is the practice of the Great Vehicle, whereby one generates the mind of the vehicle without any object [of cognition] and [understands] that *dharmas* have no self-nature. How? Just like those who practiced thus in former times, one observes (*guan cha*) the *ālaya* (substratum) of the aggregates and realizes that its own-nature is like an illusion, a mirage, a reflection, an echo, a whirling wheel of fire and an [imaginary] *gandharva* city. Lord of Mysteries, if one thus abandons no-self [in *dharmas*], the mind-lord being absolutely free, one awakens (*jue*) to the fact that one's own mind is originally unborn. Why? Because, Lord of Mysteries, the anterior and posterior limits of the mind cannot be apprehended. When one thus knows the nature of one's own mind, this represents the yogin's practice for transcending a second eon.

Next, Lord of Mysteries, bodhisattvas cultivating bodhisattva practices via the gateway of mantras accomplish all the immeasurable merit and knowledge accumulated during immeasurable and incalculable hundreds of thousands of *koṭis* of *nayutas* of eons and all the immeasurable wisdom and expedient means for fully cultivating all practices; they become a refuge for the worlds of gods and humans, they transcend the stages of all *śrāvakas* and *pratyekabuddhas*, and they are attended and revered by Śakra Devendra (King of Gods) and so on. So-called emptiness is dissociated from the sense organs and sense objects, has no [differentiating] characteristics and no [cognitive] objectivity, transcends all frivolous arguments (*prapañca*), and is boundless like empty space; all the *dharmas* (attributes) of a Buddha are successively born in dependence on it, and it is dissociated from the conditioned and unconditioned realms, dissociated from all activities, and dissociated from the eyes, ears, nose, tongue, body, and mind. [Then] is born the mind utterly without own-nature. Lord of Mysteries, such an initial mind the Buddha has declared to be the cause for becoming a Buddha; although liberated from karma and afflictions, it still has karma and mental afflictions at its base. The world will venerate [such a person] and should always make offerings to him.[82]

Maybe the world of *dharmas* and *devas* is separate from that of data, but it is one that spilled forth from the same pen, in the same years, and so too would it seem to bleed together. Looking back at Ge Hong, some four centuries earlier, he asks the following of cosmology: 'What benefit, really, are the likes of Wang [Chong] and Ge [Hong]'s trifling distinctions to the transformation of humanity?'[83]

5.4 Conclusion

Long before Kuhn's *Structure of Scientific Revolutions* shook our faith in progress, anthropologist Marcel Granet would assert in *La pensée chinoise* that 'revolutions' were how the Chinese had always thought of time:

[82] *T* no. 848, 18: 3b; tr. Giebel (2005, 15–16). [83] *JTS* 35.1307.

> La représentation d'un Espace complexe, clos et instable, s'accompagne-t-elle d'une représentation du Temps qui fait de la durée un ensemble de retours, une succession d'ères closes, cycliques, discontinues, complètes en soi, centrées, chacune, comme l'est l'Espace, autour d'une espèce de point temporaire d'émanation . . . Il ne faudrait point en conclure tout de suite que les Chinois ont édifié leur conception de la durée en se bornant à ne point distinguer le Temps (tout court) du temps qu'il fait ou du temps astronomique. Les saisons n'ont fourni que des emblèmes à la conception chinoise du Temps. Si elles ont été appelées à les fournir, c'est pour la raison que (l'Espace étant figuré comme clos) le Temps paraissait avoir une nature cyclique et que l'année, avec ses saisons, offrait l'image d'un cycle ainsi que des symboles propres à caractériser des cycles divers (Granet 1934, 96).

Needham (1965) would go on to show that there were, in fact, *plusieurs pensées* on the subject, but the capacity of 'Chinese thought' for linear time would pose a problem a decade later. Where once we thought to look for 'progress' in the ancient world, the idea fell out of fashion, and it fell, as Agassi (2008), Sivin (2009) and others rightly complain, into the most unfashionable of hands. Still, there is that one nagging question that will never go away: what do you *really* expect to happen if you throw *hundreds* of men for *thousands* of years at a high-stakes game of accuracy?

One thing that happens, as I have intended this chapter to show, is that the men in this game begin to *believe* they are improving. Another, one notes, is that they begin to perform comparative archaeo-astronomical data analysis in order to quantitatively evidence that belief. 'Progress' might not be the right word, but all of this talk about 'accumulating achievements' (*lei gong*) is hard to square with a circular idea of time. We see this with Zhou Cong and Yixing in this chapter, and Yang Wei in Chapter 4, in each and every reform they pitch to their respective courts. We see this in the way that Shen Yue, He Chengtian and Li Chunfeng structure the very histories upon which we rely to write our own. We see it, moreover, in the way that thinkers more and less involved with the astral sciences write on Buddhist and Daoist salvation. We see *so much* talk of 'accumulation' that we can make out different voices therein – micro versus macro, then versus now, and sage versus man – the contentious vacuum of the ideal, like those of 'observing', 'granting' and 'tightness', drawing one and all. 'Yes', one might ask, 'but is scientific progress *real*?' That is beside the point. If we, as modern historians, can humour Yixing's belief in 'religious' awakening, then we can humour his belief in 'scientific' awakening as well – not that such distinctions existed for him to make.[84]

The question that 'accumulation' poses for the history of the astral sciences is not whether our historical subjects are *wrong* (that, ironically, is a question

[84] On the question of scientific progress, Kitcher (1993) presents an interdisciplinary approach to the question of 'how science advances' using a metaphor of genetic diversity that I find wonderfully apropos a pluralist vision of scientific culture(s) in premodern China.

for 'positivism') but *what*, right or wrong, *they did about their beliefs*. The gentleman of *li* knew what was going on: he went to temple, he read the histories, he knew that icons are repainted and that new histories are written; several, like Liu Hong and He Chengtian, *wrote* those histories. One suspects that this knowledge, like any other element of *li* culture, may well have informed the gentleman's motivations – that fame and historical memory were part of a *reward structure*, as they are in any scientific culture. If this seems at all obvious, it is not, because we do not always allow for *individual* motivations in the case of China. Take, for instance, the individual's 'curiosity to explore the secrets of nature', as Jiang Xiaoyuan (1991, 152–3) explains:

> This type of situation is indeed extremely common in ancient Greek science and modern science, while in ancient China, however, we have yet to our day to discover any similar tradition, and, at the same time, we do not find any evidence to support this option in historical materials. Conversely, scholars early on have noticed the strong practical nature of all forms of knowledge in ancient China . . . From this we reveal the historical fact that *li was in service of astral omens*, which, as such, offers us a rational solution to the question of the purpose of *li*.[85]

[85] Emphasis is original. As to the claim that '*li* was in service of astral omens', Jiang Xiaoyuan (1992b, 173–4) clarifies that what he means is *forecasting*, though, 'at first glance, it seems that [omen literature] is all without exception in the form "*x* celestial phenomenon is a portent of *y* event" and, as such, that all that omenologists need do was be diligent at observation, waiting until after they had seen a certain celestial phenomenon before interpreting it according to omenological theory'. Classificatory studies such as Ho (1966), Jiang Xiaoyuan (1992b, 62–163; 2009) and Lu Yang (2007) confirm this 'first-glance' impression: the literature is heavily weighted toward phenomena that actors classify as 'anomalous' (*yi*), 'inconstant' (*bu chang*) and, thus, beyond the ability of contemporary *li* to predict, e.g. comets, vapours, 'guest stars', haloes and the colour, visibility, scintillation, altitude and 'encroachments' (*fan*) of otherwise predictable bodies. Having posited the historically 'exceptional' practice of omen forecasting as *li*'s one, true *raison d'être*, Jiang's (1991, 151–67; 1992b, 165–77) case rests on two examples, both of which are apocryphal. The first is an anecdote about 'Tang Grand Clerk Li Chunfeng' staking his life on a solar eclipse prediction to Taizong (r. 627–49 CE) while the latter 'observed at court and told Chunfeng "I release you to say farewell to your wife and children"' in *Taiping guangji*, 76.6a–b. Li, however, was only made Grand Clerk in 648 CE (*JTS* 79.2718), and the only solar eclipse reported and/or visible at court in 648–9 CE was that of Zhenguan 22-VII-*yiyou*.$_{22}$ (24 August 648 CE), when the emperor was away from court at Yuhua (*JTS* 3.61). Jiang's second example is that of Cui Hao (d. 450 CE) predicting the appearance and retrogradation of Mars in Well.$_{L22}$ in 415 CE as a forecast of Latter Qin emperor Yao Xing's (r. 394–416 CE) death in *WS* 35.808–9. Rather than the 'numbers and procedures' of *li*, one notes, Cui Hao uses hemerological five-agent associations to arrive at this forecast: 'One is invited to deduce it via the day-*chen* (sexagenary days): the "greatness" of *gengwu*.$_{07}$, the "court" of *xinwei*.$_{08}$, heaven has dark clouds, [Mars's] disappearance ought to have been within these two days,' etc. The reported positions of Mars that year are likewise physically impossible/disconfirmed by Alcyone Ephemeris v3.2. *Li* literature, for its part, is completely silent about its supposed purpose: in the ten *li* and *lü-li* monographs covering the 1,484-year period from the Han to the Song, by my counting, the word *zhan* in the sense of *tianwen* 'omen reading' occurs a mere sixteen times, ten in describing sage-time origins, two in relation to archaeo-astronomy, and four in juxtaposing *li* and *zhan* as distinct approaches limited, respectively, to 'constant' and 'inconstant' phenomena (compare this to 190 occurrences of *mi* 'tightness' in the same sample).

Perhaps this makes for better legend: *Greece : theoretical :: China : practical*, (because *theoretical > practical*), *ipso facto*, there must be a 'practical' reason for *li*; and since *China = nation*, that reason cannot be individual. This *must* make for better legend – the 'practical' other – because it is the same one we find in Assyriology and most every other 'non-Western' tradition, whether we justify it in terms of 'primitive thought' or less (overtly) value-laden claims of cultural/scientific relativism.[86] In the end, the power of this legend is such that it drowns out a voice like He Chengtian's:

> Emperor [Wen] of Song (r. 424–53 CE) was very fond/curious (*hao*) of *li*. Director of the Watches for the Heir Apparent He Chengtian privately wrote (*si zhuan*) a new method and submitted it in [443 CE]:
>
> 'Your servant's given nature is obstinate and indolent, and there is little that [I] truly understand, but ever since childhood have [I] been rather fond/curious (*hao*) of *li* numbers [too], and [I] have indulged [my] emotions and poured [my] attention [therein] until [my] head has grown white. Your servant's late uncle, the former director of the palace library, Xu Guang, had for a long time excelled in these matters. [He] had a Seven Luminaries of the Past *li*, and would always record its hits and misses, [which he did] from [between 365 and 371 CE] to [396 CE] for some forty-odd years. Your servant has continued to examine and compare every year now for another forty years, and thus [can I say that] its looseness and tightness, error and coincidence, can all be known.'[87]

Maybe there is no curiosity for curiosity's sake, but an eighty-year family observation programme requires some explanation other than 'the state'. Liu Hong too must have had *some* reason for 'absorbing [himself] in inner contemplation for more than twenty years' (*JS* 17.499), though it won him nothing in his life. Maybe it was selfless ideological devotion to the 'practicalities' of 'observing the signs and granting the seasons' – however impractical these had become – or maybe, as suggested in Chapter 4, it was the challenge, attention and thrill of competition that made him feel alive. Maybe he wanted to live *forever*, and maybe he realised whilst canonising his predecessors that *li* gave him that very chance. Whether or not it was his aim, it is there, in the temple of history, in the pantheon of progress, where Liu Hong's 'supernal icon' burnishes still.

[86] Against a near-identical version of this narrative as applies to cuneiform sources, see Rochberg (2004).

[87] *Song shu*, 12.260–1.

6 Conclusion

Mathematics and divinity are the Janus faces of the same human longing; one, and forever divided, theirs is an imbroglio that has outlived many of our gods. We are now perhaps most used to their opposition – to 'science versus religion' – and, in recent years, to thinking that this is our modern doing.[1] What indeed could be further from the *li* reforms of early imperial China than, say, those of the First Republic (1792–1804)? There, in revolutionary Paris, science and religion *were* on the table, and they were at the very heart of what Charles-Gilbert Romme's (1750–95) reform commission was trying to achieve.

Savants saw the Gregorian calendar, like the weights and measures of the period, as an arbitrary, parochial mess: twelve months of unequal length starting nowhere in particular, the months named after numbers that didn't apply (sept-, oct-, nov-, dec-), and the days after the feasts of Catholic saints. 'Universal' was the buzzword of the day, and such a thing could no more hope for universality in a divided Europe and a pluralist world than the county *pouces* and *pintes* of Lozère.[2] The first response to this (perceived) crisis came in 1788, with the publication of the anticlerical activist Sylvain Maréchal's (1750–1803) *Almanach des honnêtes gens*, which, in addition to straightening out the months (i.e. *gai zheng*), replaced the saints with scientists, discoverers and other secular figures. At the end of December (now month X), for example, one finds the following feasts:

24. Gama. 25. Jésus-Christ. Newton. 26. Helvétius. 27. Kepler. 28. Caton, cens. Bayle. 29. Wiclef. 30. Mar. de Brissac. 31. Swift, Boerhaave.[3]

The French Republican Calendar, settled in 1793, took this one step further, counting the years from the 'Era of Liberty', naming the months and days after natural phenomena (December now 'Snowous' (*nivôse*), and the 25th now the feast of Dog), and instituting the decimalisation of time (three *décades* to a month, ten hours to a day, and a hundred minutes to the hour). The reform

[1] For a history of the modern trinity of 'science', 'magic' and 'religion', see Tambiah (1990).

[2] On the origins of the metric system in the French revolutionary drive to replace regional weights and measures with a 'universal', 'scientific' standard, see Tavernor (2007, 62–87).

[3] *Almanach des honnêtes gens*, 11.

was the brainchild of Charles-François Dupuis (1742–1809), who, like Maréchal before him, was a vocal opponent of religion. In *Origine de tous les cultes, ou la Religion Universelle* (1795), most notably, Dupuis advances his famous thesis that Jesus is but the degenerate myth of a sun god appropriated piecemeal from more ancient cultures. Excoriating the idea of revelation – 'none but dunces will believe in revealed ideas and in ghosts'[4] – Dupuis insists that all religion is but the linguistic corruption of what was the true object of ancient philosophical speculation: Nature. Nature speaks, but she speaks in the language of numbers:

> All the Platonists admitted the unity of the Archetype or of the model, upon which God had created the World, also the unity of the Demiurgos, or the God artificer ('Dieu artiste') by a succession of the same philosophic principles, or in other words: by the very unity of the work, as may be seen in Proclus and in all the Platonists.
>
> Those, who like Pythagoras, employed the theory of numbers, in order to explain the theological verities, gave also to the Monad the title of Cause and principle. They expressed through the number One, or through unity the first Cause, and inferred the unity of God conformably to mathematical abstractions. The unity is reproduced everywhere in numbers: everything proceeds from unity. It was the same with the divine Monad. Subordinate to this unity were sundry Triads, which expressed faculties emanating therefrom and from secondary Intelligences.[5]

What was his inspiration for the new calendar? Dupuis explains: 'I took for my guide the ancient sages of the East, who made the year consist of twelve months, each of thirty days, plus five epagomenal ones. Like the Athenians and the Chinese I divided each month in three parts, called *Decades*. This was the calendar of ancient science.'[6] Apparently some things never change.

Time is not so easy a thing to redevise, nor, for that matter, is 'science' so easily divisible from 'religion'. Dupuis, Benes (2008) explains, was writing in the context of a broader 'linguistic turn' in eighteenth- and nineteenth-century European scholarship. Founded on the Protestant tradition of biblical exegesis, it was the German university that saw philology mature in the previous century into the basis of the human sciences, and it was philology that provided intellectuals seeking to construct a national language and/as national identity with a method. 'Seventeenth-century natural philosophers . . . searched for a universal system of representation, a mathematical language capable of depicting accurately the laws of a divinely authored universe', the aim of which, in Benes's (2008, 8) words, was to 'reverse Babel'. Casting their net

[4] *The Origin of All Religious Worship*, 216. [5] *The Origin of All Religious Worship*, 271.

[6] Cited in Buchwald & Josefowicz (2010, 52). For more on the French Republican Calendar, see Shaw (2011). On Dupuis, see Buchwald & Josefowicz (2010, 47–69). I thank Tuska Benes, on her visit of 11–12 February 2016, for helping me connect these threads.

far and wide, contemporary scholars – Protestant and Catholic, priest and atheist alike – sought to trace the family tree of human languages back to the root, back to the *Ursprache* before the communion of heaven and earth was severed, which led to such fantastical ideas as an ancient Germany at once the heritor of Hellas's rationalist tradition and descendants of Sanskrit-speaking poets. Dupuis may have favoured *mathematics* as the true original language of the poetry of Nature, but his was a quest every bit as defined by the shadow of Babel.

And what of Babel? There, where the eighteenth-century revolutionary longed to return, the temple scribes learned in their primordial language of lines and wedges that it was divinity, the giver of mathematics, and mathematics, the giver of divinity. In *Enlil and Sud*, for example, the god Enlil bestows Ninlil with the gift of literacy and numeracy on the occasion of his wedding, explaining his purpose in the following vow:

> The office of scribe, the tablets sparkling with stars, the stylus, the tablet board, Reckoning and accounts, adding and subtracting, the lapis lazuli measuring rope, the . . . The head of the peg, the 1-rod reed, the marking of the boundaries, and the . . . You have been perfected by them.[7]

And so too, to follow the orientalist thread, does the earliest of the *siddhāntas* – the *Paitāmaha-siddhānta* (fifth century CE?) – make such a gift and such a promise of mathematics:

> Bhṛgu, approaching the Bhagavān (Lord) who causes the creation, continuance, and destruction of the world, the reverend teacher of the moving and unmoving, said: Oh Bhagavān! The science of the stars is difficult to understand without calculation. Teach me, then, calculation.
>
> The Bhagavān said to him: Hear, my child, the knowledge of calculation. Time, which is Prajāpati and Viṣṇu, is an endless store; the knowledge of this by means of the motion of the planets is calculation . . .
>
> For the Vedas go forth for the sake of the sacrifices; the sacrifices are established as proceeding regularly in time. Therefore he who knows *jyotiṣa*, this science of time, knows all.[8]

The ancient mind knew not to distinguish 'God' from 'Nature', the modern might say, but she would be missing the true dilemma: if divinity is both the origin and outcome of mathematics, then where does that leave God and *man*? In the *Āryabhaṭīya* (499 CE), Āryabhaṭa (476–530 CE) opens with homage to Bhrāman before laying out one of the most innovative and consequential programmes in the history of *jyotiṣa*. Certain of his ideas were *so* innovative that they earned him centuries of criticism for spreading 'lies' and 'false

[7] *Enlil and Sud* (c. 1.2.2), lines 165–8; tr. cited from Robson (2007, 239).
[8] *Paitāmaha-siddhānta*, III.1, IX.8; tr. Pingree (1967–8, 476–7, 506).

knowledge' against the Bhrāmaṇic law of the *smṛtis* (Bongard-Levin 1977). The *Āryabhaṭīya*, however, was genius; Bhāskara's (*c*.600–*c*.680 CE) commentary thereto is thus forced to concede that Āryabhaṭa's could only be the work of revelation.[9] Perhaps there is a limit to inference and observation, but how is one to reconcile *contradictory* revelations? More importantly, how does the prerequisite of revelation work in a field whose acknowledged origins lie with the Yavanas – the Greeks?[10]

Such, ironically, was the same dilemma faced by the Yavanas, as much of their *mathēmata* was learned from the Egyptians and Mesopotamians before them. There, as concerns *technē*, we see sixth- and fifth-century BCE authors writing in the *prōtoi heuretai* genre identify three categories of 'first inventor': gods, historical figures and oriental peoples. The question, as always, was priority – 'who discovered what' – and the assumption, as always, was exclusivity – 'the new could either be learned from another, or found independently. Any thing that showed a superficial similarity with another, earlier one, could be declared a borrowing' (Zhmud 2006, 13). As to why authors like Hecataeus (*c*.550–*c*.476 BCE) and Herodotus (*c*.484–425 BCE) expressed such a 'sharpened interest in priority and consequently in the authorship of any achievements in every kind of creative activity', Zhmud (2006, 31–2) suggests that this was 'the product of forces that created Greek literature, art, philosophy, and science' in this period, namely 'the Greek agonistic spirit' and the culture of authorship that it inspired. In a culture that values discovery, in short, nice things imply discovery, and discovery implies a *discoverer*, the nicest things, as per the logic of sacrifice, going to the gods. Where we need an Enlil or a Bhagavān, Aeschylus (*c*.525–*c*.456 BCE) gives us Prometheus:

> I taught them to discern the rising of the stars
> and their settings, ere this ill distinguishable.
> Aye, and numbers, too, chiefest of sciences,
> I invented for them, and the combining of letters.[11]

But Prometheus was long ago, and the problem, as Xenophanes (*c*.570–*c*.475 BCE) explains, is time: 'The gods did not reveal to men all things in the beginning, but in the course of time, by searching, they find out better.'[12] Time changes the face of the human past and the place of the gods therein. By Aristotle's day in the fourth century BCE, Zhmud (2006, 26) shows, the *prōtoi heuretai* genre had experienced a 'gradual and incomplete replacement

[9] Shukla (1976, xvii–xxv).

[10] On *jyotiṣa* and the Greek connection more generally, see Pingree (1978) and Plofker (2009). On the topic of historical scepticism in Sanskrit sources, see Minkowski (2002). I thank Hirose Shō for helping me with the Sanskrit sources for this section.

[11] Aeschylus, *Prometheus Bound*, 457–60 (tr. Smyth).

[12] Xenophanes (21 B 18), tr. Guthrie (1962–81), vol. 1, 376.

of gods by semi-divine/heroic figures and next by people'. The obsession with discovery, moreover, had led to the 'secondary sacralization' of the first inventors – 'attempts to justify the activity of gods . . . by showing them as inventors', where once it was the god that justified the invention (ibid., 37–8). The quest for origins thus shifted from a dichotomy of 'god versus man' to one of 'indigenous versus borrowed', and many a historian, like Dupuis, turned from 'God' to 'the ancient sages of the East'. There were borrowings, for sure, but the *mythos of borrowing* was something else altogether, writers like Herodotus making 'persistent efforts to emphasize the non-Greek origin of many discoveries' while, 'with a few notable exceptions, the Greek tradition either passes over in silence things that were really taken over or attributes them to its own cultural heroes' (ibid., 40–1). It is likewise, Zhmud notes, that important thinkers like Thales (*c*.624–*c*.546 BCE), Pythagoras (*c*.570–*c*.495 BCE), Democritus (*c*.460–*c*.370 BCE) and Plato all accrued legends of having journeyed to the East.

Time changes everything when measured in terms of distance, and, in this sense, fourth-century writers speak of having come a long way. Isocrates (436–338 BCE), for one, expresses ambivalence towards first discoverers, placing the emphasis on how 'progress' (*epidosis, prokopē*) continues to shape our here and now:

> Progress is made, not only in *technai*, but in all other activities, not through the agency of those that are satisfied with things as they are, but through those who correct, and have the courage constantly to change anything that is not as it should be.[13]

Archytas (428–347 BCE), in discussing the utility of mathematics, goes so far as to claim that social progress indeed *depends* on *technē*:

> The invention of calculation put an end to discord and increased concord . . . A standard and a barrier to the unjust, it averts those who can calculate from injustice, persuading them that they would not be able to stay unexposed when they resort to calculation, and prevents those who cannot calculate from doing injustice by showing through calculation their deceit.[14]

The key to 'perfection', it turns out, lies not with Enlil, Prometheus or the Bhagavān; it lies with *man*. It is not some *prize* bestowed upon us out of pity or disdain; it is our *patrimony*, and it is our moral imperative. Perfection, Aristotle explains, is a multi-generational project only *begun* by the likes of Prometheus:

> In all discoveries, either the results of other people's work have been taken over and after having been first elaborated have been subsequently advanced step by step by those who took them over, or else they are original inventions which usually make progress which

[13] Isocrates, *Evagoras*, 7 (tr. van Hook). [14] Archytas (B 3), tr. Zhmud (2006, 71).

at first is small but of much greater utility than the later development which results from them. It is perhaps a true proverb which says that the beginning of anything is the most important; hence it is also the most difficult. For, as it is very powerful in its effects, so it is very small in size and therefore very difficult to see. When, however, the first beginning has been discovered, it is easier to add to it and develop the rest. This has happened, too, with rhetorical composition, and also with practically all the other arts. Those who discovered the beginnings of rhetoric carried them forward quite a little way, whereas the famous modern professors of the art, entering into the heritage, so to speak, of a long series of predecessors who had gradually advanced it, have brought it to its present perfection.[15]

And here something else has changed, because *mathēmata* are no longer among the *technai* and 'other arts' once used as a metaphor for what was happening in *epistēmē* – *now* they were counted therein, as pure, theoretical 'knowledge'. The transition occurred as Hippocrates' (*c*.460–*c*.370 BCE) systematisation of geometry vis-à-vis demonstration dovetailed with Plato's (*c*.429–*c*.347 BCE) quest for the axiomatic foundations of philosophy. Mathematics gave philosophy a methodology, and philosophy, in turn, gave mathematics a higher purpose.[16] The distinction between the theoretical (*epistēmē*) and the practical (*technē*) was officially redrawn with the establishment of the Lyceum *c*.335 BCE, where Aristotle placed *mathēmata* next to physics and theology and set students like Eudemus (*c*.370–*c*.300 BCE) upon an unprecedented historiographical project 'concerned exclusively', in the words of Zhmud's (2006, 16) study, 'with scientific discoveries, with the development of new theories and methods carried out within the framework of the professional community – *mathemata mathematicis scribuntur*'. Echoing Mencius (fl. 320 BCE) and Moritz Schlick in the Introduction, Ptolemy (*c*.90–*c*.168 CE) would later declare *mathēmata* the most exalted of the trinity:

From all this we concluded: that the first two divisions of theoretical philosophy should rather be called guesswork than knowledge, theology because of its completely invisible and ungraspable nature, physics because of the unstable and unclear nature of matter; hence there is no hope that philosophers will ever be agreed about them; and that only mathematics can provide sure and unshakeable knowledge to its devotees, provided one approaches it rigorously. For its kind of proof proceeds by indisputable methods, namely arithmetic and geometry.[17]

Still, the question is whether we own our own history, or whether we look back and to the East, because Ptolemy, like Hipparchus (*c*.190–*c*.120 BCE) before him, is building on 'Chaldean' records.[18]

[15] Aristotle, *On Sophistical Refutations*, 183b 17–32 (tr. Forster). [16] Lloyd (1979, 59–125).
[17] Ptolemy, *Almagest*, I.1 H7; tr. Toomer (1998, 36). For more on the idea of progress in ancient Greece, see Edelstein (1967).
[18] See Jones (1991).

There comes a time where, by learning, as Xenophanes suggests, man outgrows his gods. *Turning east*, one notes that Chinese philosophers wrote *prōtoi heuretai* of their own. Theirs featured different gods, but the formula – '*x* invented *y*' – was very much the same. Returning to the legend introduced in Section 1.1.2 above, whereby Fuxi 'observed the signs in heaven and ... observed the principles (*fa*) of earth' to 'invent' (*shi zuo*) the trigrams, the *Book of Changes* 'Appended Statements' continues:

When the [Fu]xi clan was gone, there arose the clan of the Divine Husbandman. [He] split wood to form the share, and bent wood into the handle, so as to teach all under heaven the advantages of ploughing and weeding. ... [He] made midday [the time of] market, to which he rendered [all] the people under heaven and assembled [all] the goods under heaven [so that they] made exchanges and retired, each obtaining what [they desired] ...

When the Divine Husbandman's clan was gone, there arose the clans of the Yellow Emperor, Yao and Shun. [They] bridged (*tong*) these switches (*bian*), making the people not tired, and [they] transformed (*hua*) them by [their] divinity (*shen*), making the people suited thereto ... [They] hollowed wood to form boats, and sharpened wood into the oars, [so as to teach] the advantages of boats and oars in fording where there is no bridge (*bu tong*) and in reaching distant parts to the advantage of [all] under heaven ... [They] domesticated oxen and rode horses, [teaching them] to pull heavy [loads] to distant parts to the advantage of [all] under heaven ... [They] strung wood to form bows, and sharpened wood into arrows, [teaching them] the advantages of bows and arrows in awing [all] under heaven.[19]

The list goes on, but where Confucius (551–479 BCE) taught man to respect his lot – 'to transmit but not create' (*shu er bu zuo*) – there is a world of difference in *Mr Lü's Spring and Autumn Annals* (239 BCE) between the gifts of divinity and what man had learned to do therewith:

Yang Youji (fl. 597–558 BCE) and Yin Ru (?) were both men of the six arts. The Jing (Chu) court was frequently [visited by] a divinity (*shen*) named Ape. The best marksmen of Jing were all unable to hit it, [so] the king of Jing asked Yang Youji to shoot it. Yang Youji lifted his bow, drew an arrow and went forth. Before shooting, [he] centred (*zhong*) him at the nock (i.e. in aiming), and with one release, Ape took the arrow and went down.[20]

Like Yin Ru, who learned charioteering in a dream where his master had taught him nothing, Mr Lü concludes that Yang Youji 'was what one might call an able learner, [or an] indomitable [spirit], which is why [he] was observed (*guan*) by later generations'.[21] Yang Youji slew a god to prove himself the better man, and

[19] *Zhouyi zhushu*, 8.167a–168a; tr. modified from Wilhelm (1967, 330–4).
[20] *Lüshi chunqiu*, 24.9b.
[21] *Lüshi chunqiu*, 24.10a. My translation of this and the previous citation is modified from Knoblock & Riegel (2000, 619).

such, as I have hoped to show, was the game that 'men of the six arts' learned to play with numbers.

The aim of this book, as much as it was promised to be comparative in the introduction, is to document how the sort of legend one might associate with 'the Greek miracle' or modernity arose likewise in first-millennium China. This is likely to be misconstrued, so allow me to reiterate what I mean and why it matters. The subject of this book is not 'empiricism', 'progress' or 'science' but the legend that these and similar ideas have inspired. The question as concerns *legend*, the history of religion should teach us, is not whether it is *true*, but how it is told, how it is acted upon, and how it is reimagined, renegotiated and repeated. To open our ears to the rather ubiquitous narrative of 'awakening' (*wu*/*jue*) and 'accumulation' (*lei*/*ji*) in the astral sciences, I believe, breathes motion into modern conceptions of a stagnant past, and it shifts some of the agency in Chinese history from the state back to the hands of the individual. Change implies conflict, and conflict implies a plurality of voices, as we might expect anywhere where there is a plurality of mouths. If the same legends arise from this din, in China as in Greece, then we might say that something similar was happening on the ground – if not in 'science', then at least on the level of its storytelling. And what if it did? The explanation that Zhmud (2006) offers for 'the Greek miracle' is 'the Greek agonistic spirit', so might we jettison this as a necessary condition, or perhaps claim it for the East? It is not my intention to answer such questions, but simply to show how easily the subject of legend allows us to arrive at them in the history of science, for legend saves us the effort of defining 'science', let alone some universal metric for measuring its progress in different contexts; legend asks not that you *believe* in progress any more than Prometheus to humour our subject's paeans thereto.

Things changed, and this change requires *some* explanation, if not that which our historical subjects provide, because nothing in these pages was preordained. The oldest and most 'venerated' (*shang*) of the classics, the *Book of Documents*, recounts the legend of how Sage King Yao presided over a golden age of man by establishing offices for the 'observation of signs' and 'granting of seasons' and by passing this, his patrimony, to the second and subsequent kings of man (Section 1.1.2 above). What little this or any other of the classics detail of 'signs' and 'seasons' comports with the sort of rudimentary practices we know to have been in place since well before the first millennium BCE: omen reading and the maintenance of a lunisolar calendar. This should have been enough. The very duality of the astral sciences indeed *ensures* that it is enough, because it is drawn at the line of 'constancy' (*chang*), and where the 'constancy' of *li* ends is where the 'anomalies' (*yi*) of *tianwen* begin. *This* is 'practical'. Are you unable to calculate a solar eclipse? Why not blame it on the king's policies to advance the ones you like? That, we are told, is why Confucius wrote the *Spring and Autumn Annals*, so shouldn't that be enough?

That was *never* enough. The very first human officers charged with these practicalities, the story goes, devolved into alcoholics. Here are the words of the Marquis of Yin on the eve of his campaign:

> The Xi and He [brothers] have now undermined their virtue, having deeply lost themselves in ale and abandoned office and quit camp, which has begun to disturb the net threads (*ji*) of heaven in the advanced abandonment of their charge. And so, on new moon's [day] of the last month of autumn, the chronograms did not gather in Chamber.$_{\text{L04}}$ (i.e. there was a solar eclipse); the blind musicians beat [their] drums, the [regional officers] galloped around, and the common people ran [in fear]. The Xi and He [brothers], impersonators to their office, heard nothing about it, being [now so] dim[-wittedly] astray in [matters of] heavenly signs, and [they] thus compel the punishment [appointed] by the former kings. The *Canon of Governance* says: 'He who is ahead of [the right] time shall be killed without mercy; he who is behind [the right] time shall be killed without mercy.'[22]

Their death may have come by sword or righteous arrow, but it was tedium, I would like to think, that ultimately did them in. Tedium is sometimes a greater force than wrathful gods and kings.

Some got drunk, others got organised. Moving into historical time, and the contents of the previous chapters, we see gentlemen of every station throw themselves at *li* to do it *better* – better than their competitors, better than their predecessors, and better, ultimately, than Yao, Shun or Confucius could have dreamed. Some of these men originated in the Clerk's Office, like the Xi and He of yore, but even those were mostly men who chose to be involved. *Better*, everyone agreed, meant 'tightness' (*mi*), the way you judge the better archer, and the state stepped in, as in archery, to provide them a social mechanism for its determination: competition and debate. Everyone was a winner: the gentleman of *li* got a public forum for his ideas and the opportunity to prove, once and for all, that *he* was the better man; the state, in turn, reaped the fruits of private learning brought willingly to its door. Sometimes, faced with tedium, the incentives of a simple game can prove stronger than 'killing without mercy', and *that* we can add to the list of things 'the ancients had yet to learn'.

The game played out, as in the Greece of Zhmud's (2006) study, with each generation claiming to have achieved, as much as possible, the perfection of their art, all the while chiding previous generations for the self-same presumption. Sivin (1986) is right to point out the contradiction – the asymmetries of salesmanship – but the same thing could be said of Ptolemy's claim that his geometry would show us God. Our subjects need not be perfectly consistent, nor need they deliver, for when such a claim is repeated so far and wide as to engender theological and historical debate, we are dealing with something larger than truth and human time. The legend of progress inspired men like

[22] *Shangshu zhushu*, 7.102b–103a.

Zhou Cong (eleventh century) to dream of 'models for a myriad generations' (Section 5.1 above), and it worried others like Yang Wei (fl. 226–37 CE) that one's methods might be 'stopped . . . from being transmitted to future ages' (Section 4.2.1 above). It upped the ante, rewarding the master of time a place *beyond it* in the temple and the annals of human accomplishment. It furthermore inspired Buddhist and Daoist writers, thinking larger than *li*, to dream of a better future through learning – one for which Isocrates, Archytas and Aristotle would surely be on board.

And whether or not we agree with their metrics for 'tightness', as I have done my best to critique, we see in China, as in Greece, a symptom far more telling of 'accumulation': the redrawing of intellectual boundaries. *Li* was once a simple matter of the calendar, but as generations of men worked to make it *better*, they made it into something *new*. Nowhere is the schizophrenia of 'granting the seasons' more evident than in the debate about 'fixing' (*ding*) the new moon via speed correction, discussed in Section 3.3: the practical thing to do, as concerned administration, copying and tradition, would be to leave the calendar moon as it was, but generations of *li* men lobbied to change everything for the sake of theoretical abstractions of 'tightness'. They, unlike Dupuis, actually won, and they fundamentally altered what it meant to 'grant the seasons'. Turning to *tianwen*, as discussed in Section 2.3, the self-same *li* men went about reshaping the field in their image. They brought in *li* to redraw the lines of nature, chastising the ancients' naivety about what in this world counts as 'anomalous'.[23] They introduced the tools and thought experiments of mathematics, demanding data over meaning, and they brought the work of 'observation' back indoors. More importantly, they transplanted their authorial culture into the wilds of anonymity and pseudepigrapha, which, over the course of time, would have *tianwen* talking progress too. 'Sphere heaven?!' – *none of this* is what the *Book of Documents* demands, which is where creative philologists like Ma Rong (79–166 CE) and Zheng Xuan (127–200 CE) enter to ensure that it *is*, and that it *always was* (Section 5.2.1 above).

If we must draw a single distinction between 'the Chinese' and 'the Greeks', let it be this: the Chinese did not look east for an explanation of their past. They looked west for some things, like paradise, scripture and sacred mountains, but not at all for maths. Why? One might point to the fact that there was no one such game-changing transmission into China prior to the eighth century as those experienced in India or Hellas. Faced with the negligible impact of those like Gautama Siddhārtha (fl. 729 CE) who *did* bring Meso-Greco-Indian *jyotiṣa* into court circles, however, we might just as well conclude that the legend of progress was by then too well established to open its doors to a parallel timeline. Whatever the reason, early imperial *li* men exercise only two of Zhmud's

[23] See also Morgan (2016a).

(2006) options for *prōtoi heuretai*: gods and historical figures. There is considerable debate about the weight that each is to receive, and the terms of the debate vary field-by-field (see Section 5.2 above), but the dilemma is very much the same as that presented by Isocrates and Aristotle: the importance of discovery versus that which you build upon it, or, in Chinese terms, 'origins' (*yuan*) versus 'course' (*lai*), 'ancient' (*gu*) versus 'modern' (*jin*), and 'old' (*jiu*) versus 'new' (*xin*). There is no single solution to the dilemma posed by these conflicting values, but we learn much about the mental universe of someone like Li Chunfeng (602–70 CE) as he struggles to find one.

Perhaps the more interesting question is why 'the Western mind' needs a third dimension to this dilemma: the 'Chaldeans', the 'ancient sages of the East'. Again, one might point to the *fact* of foreign transmission, which one sees, for example, in the papyri of Oxyrhynchus,[24] but there is something decidedly local about the excitement and hyperbole that it invites. Here we might well invoke orientalism: that 'almost from earliest times in Europe the Orient was something more than what was empirically known about it' (Said 1979, 55), because the Occident both constructs and is constructed upon the Orient of its desires. The Orient, for Dupuis, Herodotus and the like, serves a very specific function vis-à-vis the history of civilisation: origin and first discoverer. Since we are dealing in both cases with a sophisticated authorial culture which arrives at the Orient by asking 'from where does it come, who wrote it, when, under what circumstances, or beginning with what design?', let us call this its 'author function' – a construct that serves to organise a loose collection of written ideas by projecting upon it 'a "deep" motive, a "creative" power, or a "design," the milieu in which writing originates' (Foucault 1998, 213). Part of this projection certainly plays to Said's (1979, 56) critique, because, as 'first discoverer', 'to Asia are given the feelings of emptiness, loss, and disaster . . . and also, the lament that in some glorious past Asia fared better, was itself victorious over Europe'. On the other hand, to push Zhmud's (2006) study of the origins of this specific thread of historiography a little further, we might also say that the Orient represents a narrative surrogate for the divine.

Where in China the tension between 'ancient' (*gu*) and 'modern' (*jin*) was played out in two dimensions – between the (presumed) infallibility of the sages and the (presumed) advances of human knowledge – the Orient gave Greek and later European historiography a functional equivalent of the former with which to offload the mysterious workings of the divine into human hands and, thus, avert the problem of fallibility. In other words, where *Us : progress : modern* requires a correlate, *Orient : origin : ancient* works perfectly well without forcing a choice between man and God. If the Orient is to act as God, however, one needs an Orient that plays the part: one, pure, faceless,

[24] See Jones (1999).

unchanging, long ago, and far, far away. Such is the Orient we needed to come to terms with the divine, and such is the Orient, in many ways, that the twentieth-century historian has continued to fabricate by speaking of 'the Chinese', 'Chinese thought', 'Chinese science' (let alone its 'practical' or 'official nature') as if any of these were a single thing – *as if none of them*, ironically, *had a history*. We are making progress, if you will, towards rediscovering the messy, dynamic world of individual voices and contexts in ancient China obscured by monoliths like 'Chinese thought', but would that we were so modern as a Yang Youji or a Li Chunfeng to slay these untoward gods with our arrows and our ink.

'Man' versus 'god', 'ancient' versus 'modern', 'East' versus 'West' – we are used to thinking these distinctions as symptoms of the modern world. Latour (1993, 10) offers that, in contemporary usage, 'the adjective "modern" designates a new regime, an acceleration, a rupture, a revolution in time . . . defining, by contrast, an archaic and stable past . . . the word is always being thrown into the middle of a fight, in a quarrel where there are winners and losers'. Having devoted his work to 'hybrids' – to networks and collectives that weave like Ariadne's thread through nature, politics and discourse – he offers instead a definition of modernity as a constitution which acts in a dual process of mediation and purification to at once extend the interconnectivity of things and further fracture our intellectual world into discrete realms. Modernity, to Latour, is a problem – a problem to be reversed. By studying this constitution and the hybrids that it proliferates, he promises, we can realise that 'we have never been modern' and begin to study the seamless fabric of *our* world, in the here and now, with the eyes of the anthropologist. What lies on the other side of modernity, you ask? 'It is the Middle Kingdom, as vast as China and as little known' (Latour 1993, 48). To be clear, we are not speaking in 1993 about a Marxist China in the turmoil of Reform and Opening as an example of 'nonmodernity' – we are talking about *real* China, the Garden of Eden east. In the cosmopolitan and wagon-worn streets of eighth-century Chang'an, however, what you hear is simply more voices stuck between man and god, *gu* and *jin*, dreaming of simpler times in the Western Zhou (1045–771 BCE). And so it goes, 'turtles all the way down', to back before the Common Era. One might say that we have always been modern.

Appendix Co-ordinates

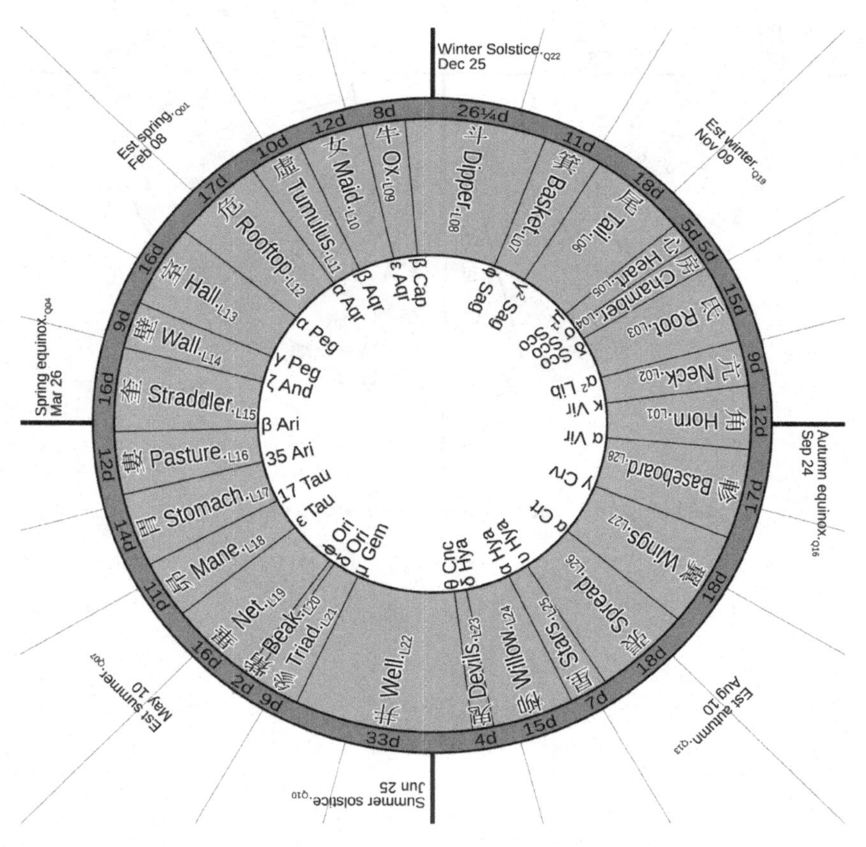

Appendix Figure 1 The twenty-eight equatorial lodges (epoch 5 CE).

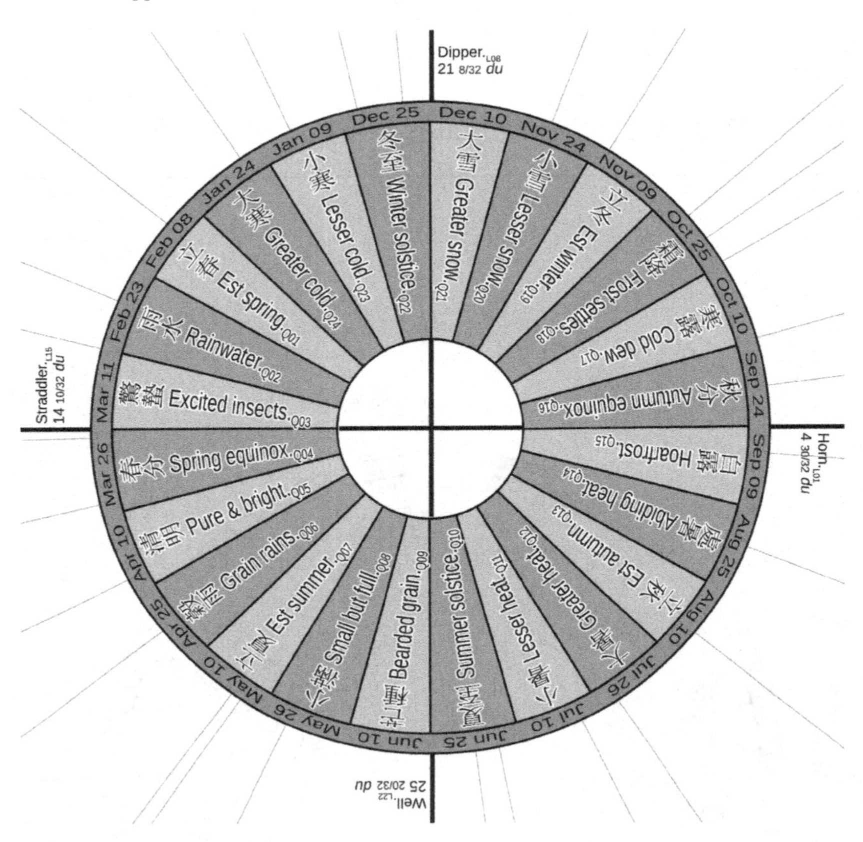

Appendix Figure 2 The twenty-four *qi* (epoch 5 CE).

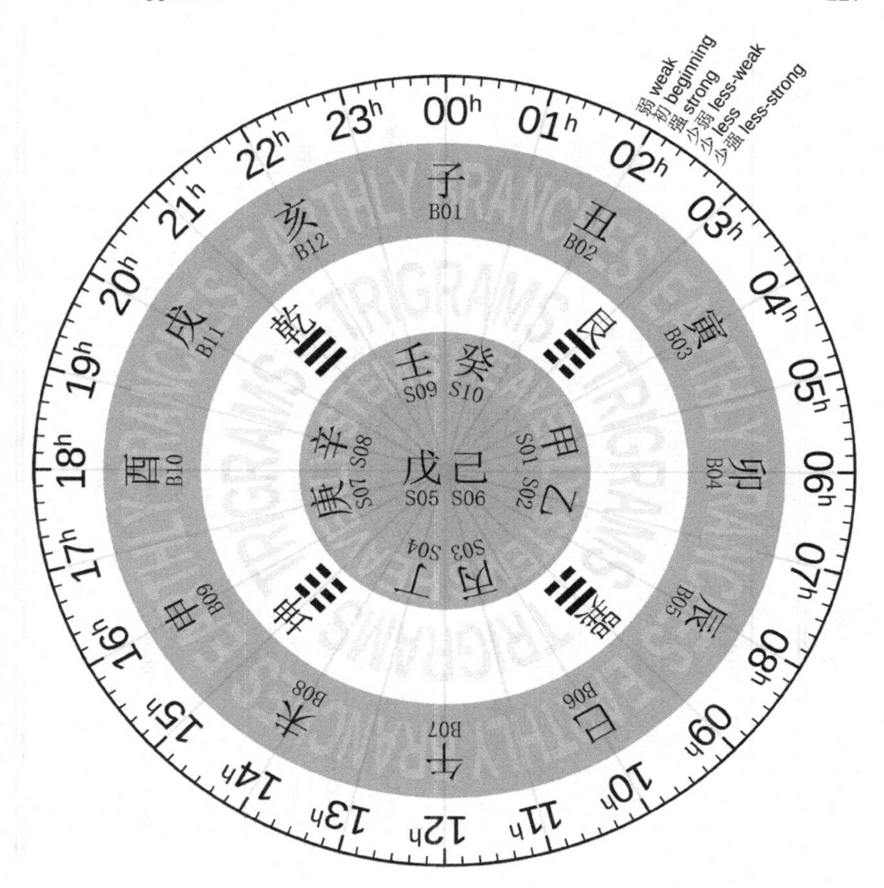

Appendix Figure 3 The twenty-four ‘added hours’.

Table Appendix 1. *Heavenly stems, earthly branches and the sexagenary cycle*

Stems	Branches	Binomes					
甲 *jia*.$_{S01}$	子 *zi*.$_{B01}$	甲子 *jiazi*.$_{01}$	甲戌 *jiaxu*.$_{11}$	甲申 *jiashen*.$_{21}$	甲午 *jiawu*.$_{31}$	甲辰 *jiachen*.$_{41}$	甲寅 *jiayin*.$_{51}$
乙 *yi*.$_{S02}$	丑 *chou*.$_{B02}$	乙丑 *yichou*.$_{02}$	乙亥 *yihai*.$_{12}$	乙酉 *yiyou*.$_{22}$	乙未 *yiwei*.$_{32}$	乙巳 *yisi*.$_{42}$	乙卯 *yimao*.$_{52}$
丙 *bing*.$_{S03}$	寅 *yin*.$_{B03}$	丙寅 *bingyin*.$_{03}$	丙子 *bingzi*.$_{13}$	丙戌 *bingxu*.$_{23}$	丙申 *bingshen*.$_{33}$	丙午 *bingwu*.$_{43}$	丙辰 *bingchen*.$_{53}$
丁 *ding*.$_{S04}$	卯 *mao*.$_{B04}$	丁卯 *dingmao*.$_{04}$	丁丑 *dingchou*.$_{14}$	丁亥 *dinghai*.$_{24}$	丁酉 *dingyou*.$_{34}$	丁未 *dingwei*.$_{44}$	丁巳 *dingsi*.$_{54}$
戊 *wu*.$_{S05}$	辰 *chen*.$_{B05}$	戊辰 *wuchen*.$_{05}$	戊寅 *wuyin*.$_{15}$	戊子 *wuzi*.$_{25}$	戊戌 *wuxu*.$_{35}$	戊申 *wushen*.$_{45}$	戊午 *wuwu*.$_{55}$
己 *ji*.$_{S06}$	巳 *si*.$_{B06}$	己巳 *jisi*.$_{06}$	己卯 *jimao*.$_{16}$	己丑 *jichou*.$_{26}$	己亥 *jihai*.$_{36}$	己酉 *jiyou*.$_{46}$	己未 *jiwei*.$_{56}$
庚 *geng*.$_{S07}$	午 *wu*.$_{B07}$	庚午 *gengwu*.$_{07}$	庚辰 *gengchen*.$_{17}$	庚寅 *gengyin*.$_{27}$	庚子 *gengzi*.$_{37}$	庚戌 *gengxu*.$_{47}$	庚申 *gengshen*.$_{57}$
辛 *xin*.$_{S08}$	未 *wei*.$_{B08}$	辛未 *xinwei*.$_{08}$	辛巳 *xinsi*.$_{18}$	辛卯 *xinmao*.$_{28}$	辛丑 *xinchou*.$_{38}$	辛亥 *xinhai*.$_{48}$	辛酉 *xinyou*.$_{58}$
壬 *ren*.$_{S09}$	申 *shen*.$_{B09}$	壬申 *renshen*.$_{09}$	壬午 *renwu*.$_{19}$	壬辰 *renchen*.$_{29}$	壬寅 *renyin*.$_{39}$	壬子 *renzi*.$_{49}$	壬戌 *renxu*.$_{59}$
癸 *gui*.$_{S10}$	酉 *you*.$_{B10}$	癸酉 *guiyou*.$_{10}$	癸未 *guiwei*.$_{20}$	癸巳 *guisi*.$_{30}$	癸卯 *guimao*.$_{40}$	癸丑 *guichou*.$_{50}$	癸亥 *guihai*.$_{60}$
	戌 *xu*.$_{B11}$						
	亥 *hai*.$_{B12}$						

Abbreviations

DT-TW	*Zhongguo kexue jishu dianji tonghui: tianwen juan*
Gongyang zhuan	*Chunqiu Gongyang zhuan zhushu*
Guliang zhuan	*Chunqiu Guliang zhuan zhushu*
HS	*Han shu*
HHS	*Hou Han shu*
JS	*Jin shu*
JTS	*Jiu Tang shu*
NQS	*Nan Qi shu*
SBCK	*Sibu congkan* 四部叢刊, Shanghai: Shangwu yinshuguan, 1919–36.
SGZ	*Sanguo zhi*
SJ	*Shiji*
SKQS	*Wenyuange Siku quanshu* 文淵閣四庫全書, 1782; reprint Taibei: Taiwan shangwu yinshuguan, 1983–6.
SSJ	*Shisanjing zhushu* 重刊宋本十三經注疏, edition 1815; reprint Taibei: Yiwen yinshuguan, 1965.
T	*Taishō shinshū daizō-kyō* 大正新脩大藏經, ed. Takakusu Junjirō 高楠順次郎 and Watanabe Kaikyoku 渡邊海旭. Tōkyō: Taishō issaikyō kankō-kai, 1924–34.
WS	*Wei shu*
XTS	*Xin Tang shu*
Zuo zhuan	*Chunqiu Zuo zhuan zhushu*

Bibliography

Pre-1850 Texts and Epigraphic Sources, by Title

Almanach des honnêtes gens, Sylvain Maréchal (1750–1803), 1788, reprint Nancy: Imprimerie de v[e] Hissette, 1836.

Baopuzi neipian jiaoshi 抱樸子內篇校釋, Ge Hong 葛洪 (283–343 CE), critical edition with annotations by Wang Ming 王明 (1911–92), edition Beijing: Zhonghua shuju, 1985.

Baopuzi waipian jiaojian 抱樸子外篇校箋, Ge Hong 葛洪 (283–343 CE), critical edition with annotations by Yang Mingzhao 楊明照 (1909–2003), edition Beijing: Zhonghua shuju, 1985 (2 vols.).

Chunqiu Gongyang zhuan zhushu 春秋公羊傳注疏, pre-Qin (< 221 BCE), commentary by He Xiu 何休 (129–82 CE), subcommentary by Xu Yan 徐彥 (fl. ninth century CE), edition SSJ.

Chunqiu Guliang zhuan zhushu 春秋穀梁傳注疏, pre-Qin (< 221 BCE), commentary by Fan Ning 范甯 (Jin), subcommentary by Yang Shiyun 楊士勛 (fl. 627/649 CE), edition SSJ.

Chunqiu Zuo zhuan zhushu 春秋左傳注疏, pre-Qin (< 221 BCE), commentary by Du Yu 杜預 (222–85 CE), subcommentary by Kong Yingda 孔穎達 (574–648 CE), edition SSJ.

Chuxue ji 初學記, compiled by Xu Jian 徐堅 (659–729 CE), edition SKQS.

Da Dai Liji 大戴禮記, compiled by Dai De 戴德 (fl. 43–33 BCE), reprint from SBCK.

Da piluzhena chengfo shenbian jiachi jing 大毘盧遮那成佛神變加持經 (*Mahāvairocanābhisaṃbodhi-vikurvitādhiṣṭhāna-vaipulyasūtra*), translated by Śubhākarasiṃha 善無畏 (637–735 CE) and Yixing 一行 (683–727 CE), 724 CE, edition *T* no 848.

Dunhuang tianwen lifa wenxian jijiao 敦煌天文曆法文獻輯校, ed. Deng Wenkuan 鄧文寬. Nanjing: Jiangsu guji chubanshe, 1996.

Dunhuang Xuanquan yueling zhaotiao 敦煌懸泉月令詔條, ed. Zhongguo wenwu yanjiusuo 中國文物研究所 and Gansu sheng wenwu kaogu yanjiusuo 甘肅省文物考古研究所. Beijing: Zhonghua shuju, 2001.

Erya zhushu 爾雅注疏, pre-Qin (< 221 BCE), commentary by Guo Pu 郭璞 (276–324 CE), subcommentary by Xing Bing 邢昺 (932–1010) et al., 999, edition SSJ.

Evagoras, Isocrates (436–338 BCE), with the translation of LaRue van Hook, edition Loeb Classical Library. London and Cambridge, MA: W. Heinemann and Harvard University Press, 1945.

Fayuan zhulin 法苑珠林, Daoshi 道世 (d. 683), 668, edition *T* no 2122.

Guanju Qin-Han mu jiandu 關沮秦漢墓簡牘, ed. Hubei sheng Jingzhou shi Zhouliang yuqiao yizhi bowuguan 湖北省荊州市周梁玉橋遺址博物館. Beijing: Zhonghua shuju, 2001.

Han guan liu zhong 漢官六種, compiled by 孫星衍 (1753–1818), critical edition with annotations by Zhou Tianyou 周天游 (b. 1944), edition Beijing: Zhonghua shuju, 1990.

Han shu 漢書, Ban Gu 班固 et al., 111 CE, commentary by Yan Shigu 顏師古 (581–645 CE), edition Beijing: Zhonghua shuju, 1962 (8 vols.).

Hou Han shu 後漢書, Fan Ye 范曄 (398–445 CE) et al., commentary by Li Xian 李賢 (654–684 CE), edition Beijing: Zhonghua shuju, 1965 (12 vols.).

Huainan honglie jijie 淮南鴻烈集解, Liu An 劉安 et al., 139 BCE, critical edition with annotations by Liu Wendian 劉文典 (1889–1958), edition Beijing: Zhonghua shuju, 1989 (2 vols.).

Huijiao Jiuzhang suanshu 匯校九章算術, anonymous, compiled probably in the Han (206 BCE–220 CE); critical edition with annotations by Guo Shuchun 郭書春, edition Shenyang: Liaoning jiaoyu chubanshe, 2004 (2 vols.).

Jin shi 金史, Tuotuo 脫脫 (1314–55) et al., 1345, edition Beijing: Zhonghua shuju, 1975 (8 vols.).

Jin shu 晉書, Fang Xuanling 房玄齡 et al., 648 CE, edition Beijing: Zhonghua shuju, 1974 (10 vols.).

Jizuan yuanhai 記纂淵海, compiled by Pan Zimu 潘自牧 (fl. 1195), edition SKQS.

Jiu jia jiu Jin shu jiben 九家舊晉書輯本, compiled by Tang Qiu 湯球 (1804–81); critical edition with annotations by Liu Xiaodong 劉曉東 et al., edition Ershiwu bieshi 二十五別史 Jinan: Qi Lu shushe, 2000.

Jiu Tang shu 舊唐書, Liu Xu 劉昫 (887–946 CE), 945 CE, edition Beijing: Zhonghua shuju, 1975 (16 vols.).

Kaiyuan zhanjing 開元占經, Gautama Siddhārtha 瞿曇悉達, 729 CE, edition SKQS.

Liji zhushu 禮記注疏, pre-Qin (< 221 BCE), commentary by Zheng Xuan 鄭玄 (127–200 CE), subcommentary by Kong Yingda 孔穎達 (574–648 CE), edition SSJ.

Lingtai miyuan 靈臺祕苑, compiled by Yu Jicai 庾季才 (d. 603 CE), critical edition with annotations by Ji Jiangjiang 紀江江, edition Zhongguo shida diwang cangshu 中國十大帝王藏書 5. Huhehaote: Nei Menggu renmin chubanshe, 2002.

Lunheng jiaoshi 論衡校釋, Wang Chong 王充 (27–*c.*100), 70/80, critical edition with annotations by Huang Hui 黃暉, edition Xinbian zhuzi jicheng 新編諸子集成, Beijing: Zhonghua shuju, 1990 (4 vols.).

Lunyu zhushu 論語注疏, pre-Qin (< 221 BCE), commentary by He Yan 何晏 (d. 249 CE), subcommentary by Xing Bing 邢昺 (932–1010) et al., edition SSJ.

Lüshi chunqiu 呂氏春秋, Lü Buwei 呂不韋 (291–235 BCE), 239 BCE, edition SBCK.

Mengxi bitan jiaozheng 夢溪筆談校證, Shen Gua 沈括 (1031–95), critical edition with annotations, Shanghai: Shanghai guji chubanshe, 1987.

Mohe zhi-guan 摩訶止觀, Zhiyi 智顗 (538–97 CE), compiled by Guanding 灌頂 (561–632 CE), 594 CE, edition *T* no 1911.

Nan Qi shu 南齊書, Xiao Zixian 蕭子顯 (489–537 CE), edition Beijing: Zhonghua shuju, 1972 (3 vols.).

Nan shi 南史, Li Yanshou 李延壽, 659 CE, edition Beijing: Zhonghua shuju, 1975 (6 vols.).

On Sophistical Refutations, Aristotle (384–322 BCE), with the translation of Edward Morgan Forster, edition Loeb Classical Library, London and Cambridge, MA: W. Heinemann and Harvard University Press, 1940.

The Origin of All Religious Worship. Translated from the French of Dupuis ... Containing also a Description of the Zodiac of Denderah, Charles-François Dupuis (1742–1809), 1795, anonymous translation New Orleans, 1872.

Prometheus Bound, Aeschylus (*c*.525–*c*.456 BCE), with the translation of Herbert Weir Smyth, edition Loeb Classical Library, London and Cambridge, MA: W. Heinemann and Harvard University Press, 1926.

Qin Han jinwen huibian 秦漢金文匯編, ed. Sun Weizu 孫慰祖, and Xu Gufu 徐谷富. Shanghai: Shanghai guji chubanshe, 1997.

Quan Tang wen 全唐文, compiled by Dong Gao 董誥 (1740–1818) et al., 1819, edition Beijing: Zhonghua shuju, 1983 (11 vols.).

Sanfu huangtu 三輔黄圖, anon., sixth century CE, edition SBCK.

Sanguo zhi 三國志, Chen Shou 陳壽 (233–97 CE), edition Beijing: Zhonghua shuju, 1959 (5 vols.).

Shangshu zhushu 尚書注疏, pre-Qin (< 221 BCE), commentary by Kong Anguo 孔安國, subcommentary by Kong Yingda 孔穎達 (574–648 CE), 653, edition SSJ.

Shiji 史記, Sima Qian 司馬遷 (*c*.146–*c*.86 BCE) et al., 109/191 BCE, commentary by Zhang Shoujie 張守節 (Tang), Sima Zhen 司馬貞 (eighth century CE), and Pei Yin 裴駰 (fifth century CE), edition Beijing: Zhonghua shuju, 1959 (10 vols.).

Shiliuguo chunqiu 十六國春秋, Cui Hong 崔鴻 (478–525 CE), 522 CE, edition SKQS.

Shitong 史通, Liu Zhiji 劉知幾 (661–721 CE), 710, edition SKQS.

Shushu jiyi 數術記遺, Xu Yue 徐岳 (fl. 226 CE), commentary by Zhen Luan 甄鸞 (fl. 535–70 CE), edition SKQS.

Shui jing zhu jishi ding'e 水經注集釋訂訛, Li Daoyuan 酈道元 (d. 527 CE), commentary/critical edition by Shen Bingxun 沈炳巽 (eighteenth century), edition SKQS.

Shuowen jiezi zhu 說文解字注, Xu Shen 許慎 (*c*.55–*c*.149 CE), 100, commentary by Duan Yucai 段玉裁 (1735–1815), 1815, edition Shanghai: Shanghai guji chubanshe, 1981.

Shuoyuan 說苑, Liu Xiang 劉向 (79–8 BCE), edition SKQS.

Song shi 宋史, Tuotuo 脱脱 (1314–55) et al., 1345, edition Beijing: Zhonghua shuju, 1977 (40 vols.).

Song shu 宋書, Shen Yue 沈約 (441–513 CE), 492/493 CE, edition Beijing: Zhonghua shuju, 1974 (8 vols.).

Sui shu 隋書, Wei Zheng 魏徵 (580–643 CE), Linghu Defen 令狐德棻 (582–666 CE) et al., 656, edition Beijing: Zhonghua shuju, 1973 (3 vols.).

Suizhou Kongjiapo Han mu jiandu 隨州孔家坡漢墓簡牘, ed. Hubei sheng wenwu kaogu yanjiusuo 湖北省文物考古研究所 and Suizhou shi kaogu dui 隨州市考古隊. Beijing: Wenwu chubanshe, 2006.

Taiping guangji 太平廣記, compiled by Li Fang 李昉 (925–96 CE) et al., 978, edition SKQS.

Taiping yulan 太平御覽, compiled by Li Fang 李昉 (925–96 CE) et al., 984, edition SBCK.

Tang liu dian 唐六典, compiled by Li Linfu 李林甫 (683–752 CE) et al., 738, edition SKQS.

Tang yulin jiaozheng 唐語林校證, compiled by Wang Dang 王讜 (fl. eleventh–twelfth centuries), critical edition with annotations by Zhou Yunchu 周勛初, edition Beijing: Zhonghua shuju, 1987 (2 vols.).

Tianyige cang Ming chaoben Tiansheng ling jiaozheng (fu Tang ling fuyuan yanjiu) 天一閣藏明鈔本天聖令校證 (附唐令復原研究), ed. Tianyige bowuguan 天一閣博物館 and Zhongguo shehui kexue yuan lishi yanjiusuo Tiansheng ling zhengli keti zu 中國社會科學院歷史研究所天聖令整理課題組. Beijing: Zhonghua shuju, 2006 (2 vols.).

Tongdian 通典, compiled by Du You 杜佑 (735–812 CE), 801, critical edition with annotations by Wang Wenjin 王文錦 et al., edition Beijing: Zhonghua shuju, 1988 (5 vols.).

Tonghu loujian zhidu 銅壺漏箭制度, Yan Yizhong 顏頤仲 (1187–1262), Qing manuscript of 1823, reprint in *DT-TW*, vol. 1, 943–57.

Wei shu 魏書, Wei Shou 魏收, 554 CE, edition Beijing: Zhonghua shuju, 1974 (8 vols.).

Weishu jicheng 緯書集成, compiled by Yasui Kōzan 安居香山 and Nakamura Shōhachi 中村璋八. Shijiazhuang: Hebei renmin chubanshe, 1994 (3 vols.).

Wenxuan zhu 文選注, compiled by Xiao Tong 蕭統 (501–31), commentary by Li Shan 李善 (d. 689), edition SKQS.

Wuyingdian ben ershisan shi kaozheng 武英殿本二十三史考證, Sun Jiagan 孫嘉淦 (1683–1753), edition Beijing: Beijing chubanshe, 1998 (2 vols.).

Xin Tang shu 新唐書, Ouyang Xiu 歐陽修 (1007–72) and Song Qi 宋祁 (998–1061), 1060, edition Beijing: Zhonghua shuju, 1975 (20 vols.).

Xin yixiang fayao 新儀象法要, Su Song 蘇頌 (1020–1101), 1094/1096, edition SKQS.

Xijian Tangdai tianwen shiliao san zhong 稀見唐代天文史料三種, ed. Gao Keli 高柯立. Beijing: Guojia tushuguan chubanshe, 2011 (3 vols.).

Yangzi Fa yan 揚子法言, Yang Xiong 揚雄 (53 BCE–18 CE), edition SBCK.

Yiwen leiju 藝文類聚, Ouyang Xun 歐陽詢 (557–641 CE), 624 CE, edition SKQS.

Yinqueshan Han mu zhujian (yi) 銀雀山漢墓竹簡（壹), ed. Yinqueshan Han mu zhujian zhengli xiaozu 銀雀山漢墓竹簡整理小組. Beijing: Wenwu chubanshe, 1985.

Yuhai 玉海, compiled by Wang Yinglin 王應麟 (1223–96), edition SKQS.

Yinwan Han mu jiandu 尹灣漢墓簡牘, ed. Lianyungang shi bowuguan 連雲港市博物館. Beijing: Zhonghua shuju, 1997.

Yisi zhan 乙巳占, compiled by Li Chunfeng李淳風 (602–70 CE), 645 CE, reprint from Shiwan juan lou congshu 十萬卷樓叢書 in *DT-TW*, vol. 4, 451–599.

Yuan shi 元史, Song Lian 宋濂 (1310–81), 1370, edition Beijing: Zhonghua shuju, 1976 (15 vols.).

Yuelu Qin jian 嶽麓秦簡, ed. Chen Songchang 陳松長. Shanghai: Shanghai cishu chubanshe, 2010–2015 (4 vols.).

Zhangjiashan Han mu zhujian (ersiqi hao mu) 張家山漢墓竹簡（二四七號墓), ed. Zhangjiashan ersiqi hao Han mu zhujian zhengli xiaozu 張家山二四七號漢墓整理小組. Beijing: Wenwu chubanshe, 2001.

Zhongguo gudai tianwen wenwu tuji 中國古代天文文物圖集*, ed. Zhongguo shehui kexueyuan kaogu yanjiusuo* 中國社會科學院考古研究所. Beijing: Wenwu chubanshe, 1980.

Zhongguo gudai tianxiang jilu zongji 中國古代天象記錄總集, ed. Beijing tianwentai 北京天文臺. Nanjing: Jiangsu kexue jishu chubanshe, 1988.

Zhongguo kexue jishu dianji tonghui: tianwen juan 中國科學技術典籍通彙·天文卷, ed. Bo Shuren 薄樹人. Zhengzhou: Hebei jiaoyu chubanshe, 1993 (8 vols.).

Zhong lun 中論, Xu Gan 徐幹 (170–218 CE), edition SBCK.

Zhoubi suanjing 周髀算經, anon., Han dynasty, commentary by Zhao Shuang 趙爽 (fl. 314 CE), recension by Zhen Luan 甄鸞 (fl. 535–70 CE), excursus by Li Chunfeng 李淳風 (602–70 CE), edition SBCK.

Zhouli zhushu 周禮注疏, pre-Qin (< 221 BCE), commentary by Zheng Xuan 鄭玄 (127–200 CE), subcommentary by Jia Gongyan 賈公彥 (seventh century CE); edition SJJ.

Zhouyi zhushu 周易注疏, pre-Qin (< 221 BCE), commentary by Wang Bi 王弼 (226–49 CE) and Han Kangbo 韓康伯 (fl. fourth century CE), subcommentary by Kong Yingda 孔穎達 (574–648 CE); edition SJJ.

Secondary Sources

Agassi, J., 2008. *Science and Its History.* Dordrecht: Springer.

Andersen, P., 1990. The practice of *bugang*, *Cahiers d'Extrême-Asie* 5, 15–53.

Arai Shinji 新井晋司, 1989. Chō Kō *Kontengi, Kontengi-chū* saikō 張衡『渾天儀』『渾天儀注』再考, in *Chugoku kodai kagakushi ron* 中國古代科學史論, ed. Yamada Keiji 山田慶兒. Kyōto: Kyōto daigaku jinbun kagaku kenkyūjo, 317–36.

Arrault, A., 2002. Les premiers calendriers chinois du IIe siècle avant notre ère au Xe siècle, in *Les calendriers: Leurs enjeux dans l'espace et dans le temps: colloque de Cerisy, du 1er au 8 juillet 2000*, ed. J. Le Goff, J. Lefort and P. Mane. Paris: Somogy, 169–91.

Arrault, A., 2003. Les calendriers de Dunhuang, in *Divination et société dans la Chine médiévale: Étude des manuscrits de Dunhuang de la Bibliothèque nationale de France et de la British Library*, ed. M. Kalinowski. Paris: Bibliothèque nationale de France, 85–123.

Balazs, E., 1968. *La bureaucratie céleste: Recherches sur l'économie et la société de la Chine traditionnelle.* Paris: Gallimard.

Barnes, B., D. Bloor & J. Henry, 1996. *Scientific Knowledge: A Sociological Analysis.* Chicago: The University of Chicago Press.

Becker, C.B., 1986. Reasons for the lack of argumentation and debate in the Far East, *International Journal of Intercultural Relations* 10, 75–92.

Beer, A., Ho Ping-Yü, Lu Gwei-Djen, J. Needham, E.G. Pulleyblank & G.I. Thompson, 1961. An 8th-century meridian line: I-Hsing's chain of gnomons and the pre-history of the metric system, *Vistas in Astronomy* 4, 3–28.

Benes, T., 2008. *In Babel's Shadow: Language, Philology, and the Nation in Nineteenth-Century Germany.* Detroit: Wayne State University Press.

Bielenstein, H., 1980. *The Bureaucracy of Han Times*. Cambridge: Cambridge University Press.

Birrell, A., 1993. *Chinese Mythology: An Introduction*. Baltimore: Johns Hopkins University Press.

Bloor, D., 1976. *Knowledge and Social Imagery.* London: Routledge & Kegan Paul.

Bo Shuren 薄樹人 (ed.), 1993. Han jian lipu 漢簡曆譜, in *Zhongguo kexue jishu dianji tonghui: tianwen juan* 中國科學技術典籍通彙·天文卷. Zhengzhou: Hebei jiaoyu chubanshe, 221–61.

Bokenkamp, S.R., 1994. Time after time: Taoist apocalyptic history and the founding of the T'ang Dynasty, *Asia Major*, 3rd ser. 7(1), 59–88.

Bongard-Levin, G.M., 1977. Āryabhaṭa and Lokāyatas, *Annals of the Bhandarkar Oriental Research Institute* 58–9, 187–93.

Bower, V. & C. Mackenzie, 2004. Pitchpot: the scholar's arrow-throwing game, in *Asian Games: The Art of Contest*, ed. I.L. Finkel & C. Mackenzie. New York: Asia Society, 275–81.

Brown, D., 2000. The cuneiform conception of celestial space and time, *Cambridge Archaeological Journal* 10(1), 103–22.

Brown, M., 2006. Neither 'primitives' nor 'others,' but somehow not quite like 'us': the fortunes of psychic unity and essentialism in Chinese studies, *Journal of the Economic and Social History of the Orient* 49(2), 219–52.

Brown, M., 2015. *The Art of Medicine in Early China: The Ancient and Medieval Origins of a Modern Archive*. Cambridge: Cambridge University Press.

Brown, M. & U. Bergeton, 2008. 'Seeing' like a sage: three takes on identity and perception in early China, *Journal of Chinese Philosophy* 35(4), 641–62.

Brush, S.G., 1989. Prediction and theory evaluation: the case of light bending, *Science* 246(4934), 1124–9.

Brush, S.G., 1994. Dynamics of theory change: the role of predictions, *PSA: Proceedings of the Biennial Meeting of the Philosophy of Science Association*, ed. D.L. Hull, M. Forbes and R.M. Burian. East Lansing, MI: Philosophy of Science Association 1994, 133–45.

Buchwald, J.Z. & D.G. Josefowicz, 2010. *The Zodiac of Paris: How an Improbable Controversy over an Ancient Egyptian Artifact Provoked a Modern Debate between Religion and Science*. Princeton: Princeton University Press.

Cai Wanjin 蔡萬進, 2006. Yinwan shi hao mudu Shi Junxiong dai Shi Zixia quanwen chutan 尹灣十號木牘師君兄貸師子夏券文初探, in *Jianbo yanjiu* 簡帛研究, ed. Bu Xianqun 卜憲群 & Yang Zenhong 楊振紅. Guilin: Guangxi shifan daxue chubanshe, 242–7.

Caillois, R., 2001. *Man, Play, and Games*. Urbana: University of Illinois Press.

Campany, R.F., 2002. *To Live as Long as Heaven and Earth: A Translation and Study of Ge Hong's Traditions of Divine Transcendents*. Berkeley: University of California Press.

Campany, R.F., 2005. Two religious thinkers of the early Eastern Jin: Gan Bao and Ge Hong in multiple contexts, *Asia Major* 3rd ser. 18(1), 175–224.

Chao Fulin 晁福林, 2008. Jinwen 'mieli' yu Xizhou mianli zhidu 金文「蔑曆」與西周勉勵制度, *Lishi yanjiu* 歷史研究 2008.1, 33–42.

Chapman, A., 1995. *Dividing the Circle: The Development of Critical Angular Measurement in Astronomy 1500–1850*, 2nd ed. Chichester: John Wiley & Sons.

Chaussende, D., D.P. Morgan & K. Chemla, eds., forthcoming, *Monographs in Tang Official History: Perspectives from the Technical Treatises of the Book of Sui*. Dordrecht: Springer.

Chemla, K., 1997. Qu'est-ce qu'un problème dans la tradition mathématique de la Chine ancienne ? Quelques indices glanés dans les commentaires rédigés entre le IIIe et le VIIe siècles au classique Han *Les neuf chapitres sur les procédures mathématiques, Extrême-Orient, Extrême-Occident* 19, 91–126.

Chemla, K., 2009. On mathematical problems as historically determined artifacts: reflections inspired by sources from ancient China, *Historia Mathematica* 36(3), 213–46.

Chemla, K., 2010. Mathematics, nature and cosmological inquiry in traditional China, in *Concepts of Nature: A Chinese–European Cross-cultural Perspective*, ed. G. Dux, H.U. Vogel & M. Elvin. Leiden: Brill, 255–84.

Chemla, K., 2013. Shedding some light on a possible origin of a concept of fractions in China: division as a link between the newly discovered manuscripts and 'The Gnomon of the Zhou [Dynasty]', *Sudhoffs Archiv* 97(2), 174–98.

Chemla, K., 2014. Observing mathematical practices as a key to mining our sources and conducting conceptual history: division in ancient China as a case study, in *Science after the Practice Turn in the Philosophy, History, and Social Studies of Science*, ed. L. Soler, S. Zwart, M. Lynch & V. Israel-Jost. New York: Routledge, 238–68.

Chemla, K. & Guo Shuchun, 2004. *Les neuf chapitres: le classique mathématique de la Chine ancienne et ses commentaires*. Paris: Dunod.

Chen Hao 陳昊, 2007. Tulufan Taizangta xinchu Tangdai liri yanjiu 吐魯番臺藏塔新出唐代曆日研究, *Dunhuang Tulufan yanjiu* 敦煌吐魯番研究 10, 207–20.

Chen Jinhua 陳金華, 2000–1. The birth of a polymath: the genealogical background of the Tang monk–scientist Yixing (673–727), *T'ang Studies* 18–19, 1–39.

Chen Jiujin 陳久金, 1982. Jiudaoshu jie 九道術解, *Ziran kexue shi yanjiu* 自然科學史研究 1(2), 131–5.

Chen Jiujin 陳久金, 1994. Beidouxing doubing zhixiang kao 北斗星斗柄指向考, *Ziran kexue shi yanjiu* 自然科學史研究 13(3), 209–14.

Chen Jiujin 陳久金 (ed.), 2008. *Zhongguo gudai tianwenxuejia* 中國古代天文學家 (Zhongguo tianwenxueshi daxi 中國天文學史大系). Beijing: Zhongguo kexue jishu chubanshe.

Chen Meidong 陳美東, 1986. Liu Hong de shengping, tianwenxue chengjiu he sixiang 劉洪的生平、天文學成就和思想, *Ziran kexue shi yanjiu* 自然科學史研究 5(2), 129–42.

Chen Meidong 陳美東, 1995. *Gu li xin tan* 古曆新探. Shenyang: Liaoning jiaoyu chubanshe.

Chen Meidong 陳美東, 2003. *Zhongguo kexue jishu shi: tianwenxue juan* 中國科學技術史：天文學卷. Beijing: Kexue chubanshe.

Chen Meidong 陳美東, 2006. Zhongguo gudai de loujian zhidu 中國古代的漏箭制度, *Guangxi minzu xueyuan xuebao (ziran kexue ban)* 廣西民族學院學報（自然科學版) 12(4), 6–10, 23.

Chen Meidong 陳美東, 2007. *Zhongguo gudai tianwenxue sixiang* 中國古代天文學思想 (Zhongguo tianwenxueshi daxi 中國天文學史大系). Beijing: Zhongguo kexue jishu chubanshe.

Chen Xiaozhong 陳曉中 & Zhang Shuli 張淑莉, 2008. *Zhongguo gudai tianwen jigou yu tianwen jiaoyu* 中國古代天文機構與天文教育 (Zhongguo tianwenxueshi daxi 中國天文學史大系). Beijing: Zhongguo kexue jishu chubanshe.

Chen Zungui 陳遵嬀, 2006. *Zhongguo tianwenxue shi* 中國天文學史, 2nd ed. Shanghai: Shanghai renmin chubanshe.

Christianson, J.R., 2000. *On Tycho's Island: Tycho Brahe and His Assistants, 1570–1601*. Cambridge: Cambridge University Press.

Coblin, W.S., 1983. *A Handbook of Eastern Han Sound Glosses*. Hong Kong: Chinese University Press.

Csikszentmihalyi, M. & M. Nylan, 2003. Constructing lineages and inventing traditions through exemplary figures in early China, *T'oung Pao*, 2nd ser. 89(1/3), 59–99.

Cullen, C., 1976. A Chinese Eratosthenes of the flat Earth: a study of a fragment of cosmology in *Huai Nan tzu* 淮南子, *Bulletin of the School of Oriental and African Studies, University of London* 39(1), 106–27.
Cullen, C., 1977. Cosmographical discussions in China from early times up the T'ang dynasty, Ph.D. diss., University of London.
Cullen, C., 1980–1. Some further points on the *shih*, *Early China* 6, 31–46.
Cullen, C., 1993. Motivations for scientific change in ancient China: Emperor Wu and the Grand Inception astronomical reforms of 104 BC, *Journal for the History of Astronomy* 24(3), 185–203.
Cullen, C., 1996. *Astronomy and Mathematics in Ancient China: The* Zhou Bi Suan Jing. Cambridge: Cambridge University Press.
Cullen, C., 2000. Seeing the appearances: ecliptic and equator in the Eastern Han, *Ziran kexue shi yanjiu* 自然科學史研究 19(4), 352–82.
Cullen, C., 2001. The birthday of the old man of Jiang County and other puzzles: work in progress on Liu Xin's *Canon of the Ages,* Asia Major, 3rd ser. 14(2), 27–60.
Cullen, C., 2002. The first complete Chinese theory of the moon: the innovations of Liu Hong *c.* A.D. 200, *Journal for the History of Astronomy* 33, 21–39.
Cullen, C., 2005. Translating ancient Chinese astronomical systems with EXCEL: how not to stew the strawberries?, *Journal for the History of Astronomy* 36(3), 336–8.
Cullen, C., 2007a. Actors, networks, and 'disturbing spectacles' in institutional science: 2nd century Chinese debates on astronomy, *Antiqvorvm philosophia* 1, 237–67.
Cullen, C., 2007b. Huo Rong's observation programme of AD 102 and the Han li solar table, *Journal for the History of Astronomy* 38(1), 75–98.
Cullen, C., 2009. People and numbers in early imperial China: locating 'mathematics' and 'mathematicians' in Chinese space, in *Oxford Handbook of the History of Mathematics*, ed. E. Robson & J.A. Stedall. Oxford: Oxford University Press, 591–618.
Cullen, C., 2011. Understanding the planets in ancient China: prediction and divination in the *Wu xing zhan*, *Early Science and Medicine* 16, 218–51.
Cullen, C., 2017. *The Foundations of Celestial Reckoning: Three Ancient Chinese Astronomical Systems*. New York: Routledge.
Cullen, C. & A.S.L. Farrer, 1983. On the term *hsüan chi* and the three-lobed jade discs, *Bulletin of the School of Oriental and African Studies* 46(1), 53–76.
Daston, L. & P. Galison, 2007. *Objectivity*. New York: Zone Books.
Deane, T.E., 1989. The Chinese Imperial Astronomical Bureau: form and function of the Ming dynasty Qintianjian from 1365 to 1627, Ph.D. diss., University of Washington.
Deng Wenkuan 鄧文寬, 2002. *Dunhuang Tulufan tianwen lifa yanjiu* 敦煌吐魯番天文曆法研究. Lanzhou: Gansu jiaoyu chubanshe.
Donner, N.A. & D.B. Stevenson, 1993. *The Great Calming and Contemplation: A Study and Annotated Translation of the First Chapter of Chih-I's Mo-Ho Chih-Kuan*. Honolulu: University of Hawai'i Press.
Drège, J.P., 1991. *Les bibliothèques en Chine au temps des manuscrits: Jusqu'au Xe siècle* (Publications de l'École française d'Extrême-Orient 161). Paris: École française d'Extrême-Orient.
Dunhuang-xian wenhuaguan 敦煌縣文化館, 1984. Dunhuang Suyoutu Handai fengsui yizhi chutu de mujian 敦煌酥油土漢代烽燧遺址出土的木簡, in *Han jian yanjiu wenji* 漢簡研究文集, ed. Gansu-sheng wenwu gongzuodui 甘肅省文物工作隊 & Gansu-sheng bowuguan 甘肅省博物館. Lanzhou: Gansu renmin chubanshe, 1–14.

Eberhard, W. & R. Mueller, 1936. Contributions to the astronomy of the Han period III: astronomy of the Later Han period, *Harvard Journal of Asiatic Studies* 1(2), 194–241.

Edelstein, L., 1967. *The Idea of Progress in Classical Antiquity*. Baltimore: John Hopkins University Press.

Eggert, M., 2002. Transcendent, transgressive, expressive: games and playing in premodern Chinese culture, in *The Chinese at Play: Festivals, Games and Leisure*, ed. A. Hansson, B.S. McDougall & F. Weightman. London: Kegan Paul, 137–50.

Finkel, I.L. & C. Mackenzie (eds.), 2004. *Asian Games: The Art of Contest*. New York: Asia Society.

Forte, A., 1988. *Mingtang and Buddhist Utopias in the History of the Astronomical Clock: The Tower, Statue and Armillary Sphere Constructed by Empress Wu* (Serie orientale Roma v. 59). Rome and Paris: Istituto italiano per il Medio ed Estremo Oriente and École française d'Extrême-Orient.

Foucault, M., 1998. *Aesthetics, Method, and Epistemology* (Essential works of Foucault 2). New York: The New Press.

Franke, H., 1950. Some remarks on the interpretation of Chinese dynastic histories, *Oriens* 3(1), 113–22.

Frazer, J.G., 1890. *The Golden Bough: A Study in Magic and Religion*. London: Macmillan and Co.

Frazer, J.G., 1911. *The Golden Bough: A Study in Magic and Religion*, 3rd ed. London: Macmillan and Co.

Gassmann, R.H., 2002. *Antikchinesisches Kalenderwesen: die Rekonstruktion der chunqiu-zeitlichen Kalender des Fürstentums Lu und der Zhou-Könige*. Bern: Peter Lang.

Giebel, R.W., 2005. *The Vairocanābhisaṃbodhi Sutra*. Berkeley: Numata Center for Buddhist Translation and Research.

Giele, E., 2006. *Imperial Decision-Making and Communication in Early China: A Study of Cai Yong's* Duduan (Opera sinologica 20). Wiesbaden: Harrassowitz.

Goodman, H.L., 1998. *Ts'ao P'i Transcendent: The Political Culture of Dynasty-Founding in China at the End of the Han*. Seattle: Scripta Serica.

Goodman, H.L., 2005. Chinese polymaths, 100–300 AD: the Tung-kuan, Taoist dissent, and technical skills, *Asia Major* 3rd ser., 18(1), 101–74.

Goodman, H.L., 2010. *Xun Xu and the Politics of Precision in Third-Century* AD *China*. Leiden: Brill.

Granet, M., 1934. *La pensée chinoise*. Paris: La Renaissance du livre.

Graßhoff, G., 1990. *The History of Ptolemy's Star Catalogue* (Studies in the history of mathematics and physical sciences 14). New York: Springer-Verlag.

Greene, E.M., 2012. Meditation, repentance, and visionary experience in early medieval Chinese Buddhism, Ph.D. diss., University of California, Berkeley.

Guan Yuzhen 關瑜楨, 2015. Eclipse theory in the *Jing chu li*: part I. The adoption of lunar velocity, *Archive for History of Exact Sciences* 69(1), 103–23.

Guan Zengjian 關增建, 1989. Chuantong 365¼ fendu bushi jiaodu 傳統365¼分度不是角度, *Ziran bianzhengfa tongxun* 自然辯證法通訊 11(5), 33, 77–9.

Guan Zengjian 關增建, 2002. Li Chunfeng ji qi *Yisi zhan* de kexue gongxian 李淳風及其『乙巳占』的科學貢獻, *Zhengzhou daxue xuebao (zhexue shehui kexue ban)* 鄭州大學學報（哲學社會科學版) 35(1), 121–4, 131.

Guthrie, W.K.C., 1962–81. *A History of Greek Philosophy*. Cambridge: Cambridge University Press (6 vols.).

Han Wei 韓巍, 2015. Beida cang Qin jian *Lu Jiuci wen shu yu Chen Qi* chudu 北大藏秦簡『魯久次問數于陳起』初讀, *Beijing daxue xuebao (zhexue shehui kexue ban)* 北京大學學報（哲學社會科學版) 52(2), 29–36.

Harbsmeier, C., 1998. *Science and Civilisation in China, vol. 7, pt. 1: Language and Logic*. Cambridge: Cambridge University Press.

Harper, D., 1998. *Early Chinese Medical Literature: The Mawangdui Medical Manuscripts*. London: Kegan Paul International.

Harper, D., 1999. Warring States natural philosophy and occult thought, in *The Cambridge History of Ancient China: From the Origins of Civilization to 221* B.C., ed. M. Loewe & E.L. Shaughnessy. Cambridge: Cambridge University Press, 813–84.

Harper, D., 2007. Communication by design: two silk manuscripts of diagrams (*tu*) from Mawangdui tomb three, in *Graphics and Text in the Production of Technical Knowledge in China: The Warp and the Weft*, ed. F. Bray, V. Dorofeeva-Lichtmann & G. Métailie. Leiden: Brill, 169–89.

Harper, D., 2010. The textual form of knowledge: occult miscellanies in ancient and medieval Chinese manuscripts, fourth century B.C. to tenth century A.D., in *Looking at It from Asia: The Processes That Shaped the Sources of History of Science*, ed. F. Bretelle-Establet. Dordrecht: Springer, 37–80.

Hart, R., 1999. Beyond science and civilization: a post-Needham critique, *East Asian Science, Technology, and Medicine* 16, 88–114.

Hasebe Eiichi 長谷部英一, 1991. Gi Shin Nanbokuchō no rekiron 魏晉南北朝の曆論, *Chūgoku tetsugaku kenkyū* 中國哲學研究 3, 1–43.

Hasebe Eiichi 長谷部英一, 1993. Zuidai no rekiron 隋代の曆論, *Chūgoku tetsugaku kenkyū* 中國哲學研究 6, 1–21.

Hashimoto Keizō 橋本敬造, 1979. Seidō no shisō to dentō Chūgoku no tenmongaku 精度の思想と傳統中國天文學, *Kansai daigaku shakai gakubu kiyō* 關西大學社會學部紀要 11(1), 93–114.

He Maohuo 何茂活, 2015. Jianshui Jinguan chutu *Han Jushe yuan nian lipu* zhuihe yu kaoshi 肩水金關出土『漢居攝元年曆譜』綴合與考試, *Kaogu yu wenwu* 考古與文物 2015.2, 61–8.

Henderson, J.B., 1984. *The Development and Decline of Chinese Cosmology*. New York: Columbia University Press.

Henderson, J.B., 2006. Premodern Chinese notions of astronomical history and calendrical time, in *Notions of Time in Chinese Historical Thinking*, ed. C. Huang & J.B. Henderson. Hong Kong: Chinese University Press, 97–113.

Hirase Takao 平勢隆郎, 1996. *Chūgoku kodai kinen no kenkyū: tenmon to koyomi no kentō kara* 中國古代紀年の研究：天文と曆の檢討から. Tōkyō: Kyūko shoin.

Ho Peng Yoke 何丙郁, 1966. *The Astronomical Chapters of the Chin Shu*. Paris: Mouton.

Ho Peng Yoke 何丙郁, 2003. *Chinese Mathematical Astrology: Reaching out to the Stars* (Needham Research Institute series). London: RoutledgeCurzon.

Hodge, S., 2003. *The Mahā-Vairocana-Abhisaṃbodhi Tantra*. London: RoutledgeCurzon.

Hsing I-t'ien 邢義田, 1998. Yueling yu Xihan zhengzhi: cong Yinwan jibu zhong de 'yi chunling chenghu' shuoqi 月令與西漢政治——從尹灣集簿中的「以春令成戶」說起, *Xin shixue* 新史學 9(1), 1–54.

Hua Tongxu 華同旭, 1991. *Zhongguo louke* 中國漏刻. Hefei: Anhui kexue jishu chubanshe.

Huang Yi-long 黃一農, 1992a. Jixing yu gudu kao 極星與古度考, *Tsing Hua Journal of Chinese Studies* 22(2), 93–117.

Huang Yi-long 黃一農, 1992b. Zhongguo shi libiao shuorun dingzheng juyu 中國史曆表朔閏訂正舉隅, *Hanxue yanjiu* 漢學研究 10(2), 279–306.

Huang Yi-long 黃一農, 1999. Cong Yinwan Han mu jiandu kan Zhongguo shehui de zeri chuantong 從尹灣漢簡牘看中國社會的擇日傳統, *Guoli zhongyang yanjiuyuan lishi yuyan yanjiusuo jikan* 國立中央研究院歷史語言研究所集刊 70(3), 589–625.

Huang Yi-long 黃一農, 2001. Qin wangzheng shiqi lifa xinkao 秦王政時期曆法新考, *Huaxue* 華學 5, 143–9.

Hucker, C.O., 1985. *A Dictionary of Official Titles in Imperial China*. Stanford: Stanford University Press.

Hulsewé, A.F.P., 1985. *Remnants of Ch'in Law: An Annotated Translation of the Ch'in Legal and Administrative Rules of the 3rd Century B.C., Discovered in Yün-Meng Prefecture, Hu-Pei Province, in 1975*. Leiden: E.J. Brill.

Jiang Xiaoyuan 江曉原, 1986. Aiguo zhuyi jiaoyu bu ying chengwei kejishi yanjiu de mudi 愛國主義教育不應成爲科技史研究的目的, *Daziran tansuo* 大自然探索 5(4), 143–8.

Jiang Xiaoyuan 江曉原, 1991. *Tianxue zhen yuan* 天學真原. Shenyang: Liaoning jiaoyu chubanshe.

Jiang Xiaoyuan 江曉原, 1992a. Lishu qiyuan kao 曆書起源考, *Zhongguo wenhua* 中國文化 1992.6, 150–9.

Jiang Xiaoyuan 江曉原, 1992b. *Xingzhanxue yu chuantong wenhua* 星占學與傳統文化. Shanghai: Shanghai guji chubanshe.

Jiang Xiaoyuan 江曉原, 2009. *Zhongguo xingzhanxue leixing fenxi* 中國星占學類型分析. Shanghai: Shanghai shudian chubanshe.

Jiang Xiaoyuan 江曉原 & Niu Weixing 鈕衛星, 2001. *Ouzhou tianwenxue dongjian fawei* 歐洲天文學東漸發微. Shanghai: Shanghai shudian chubanshe.

Johns, A., 1998. *The Nature of the Book: Print and Knowledge in the Making*. Chicago: The University of Chicago Press.

Jones, A., 1991. The adaptation of Babylonian methods in Greek numerical astronomy, *Isis* 82(3), 440–53.

Jones, A., 1999. *Astronomical Papyri from Oxyrhynchus*. Philadelphia: American Philosophical Society.

Jones, A., 2005. In order that we should not ourselves appear to be adjusting our estimates ... to make them fit some predetermined amount, in *Wrong for the Right Reasons*, ed. J.Z. Buchwald & A. Franklin. Dordrecht: Springer, 17–39.

Jones, A. & D. Duke, 2005. Ptolemy's planetary mean motions revisited, *Centaurus* 47 (3), 226–35.

Kalinowski, M., 1990. Le calcul du rayon céleste dans la cosmographie chinoise, *Revue d'histoire des sciences* 43(1), 3–34.

Kalinowski, M., 1996. Astrologie calendaire et calcul de position dans la Chine ancienne: Les mutations de l'hémérologie sexagésimale entre le IVe et le IIe siècles avant notre ère, *Extrême-orient, Extrême-occident* 18, 71–113.

Kalinowski, M. (ed.), 2003. *Divination et société dans la Chine médiévale: Étude des manuscrits de Dunhuang de la Bibliothèque nationale de France et de la British Library*. Paris: Bibliothèque nationale de France.

Kalinowski, M., 2004. Technical traditions in ancient China and *shushu* culture in Chinese religion, in *Religion and Chinese Society*, ed. J. Lagerwey. Hong Kong and Paris: The Chinese University Press and École française d'Extrême-Orient, 223–48.

Karlgren, B., 1957. *Grammata Serica Recensa*. Stockholm: Museum of Far Eastern Antiquities.

Kawahara Hideki 川原秀城, 1991. The world-view of the *Santong-li*, *Historia Scientiarum* 42, 67–73.

Kawahara Hideki 川原秀城, 1996. *Chūgoku no kagaku shisō: ryō Kan tengaku kō* 中國の科學思想：兩漢天學考 (Chūgoku gakugei sōsho 1). Tōkyō: Sōbunsha.

Keightley, D.N., 2000. *The Ancestral Landscape: Time, Space, and Community in Late Shang China, ca. 1200–1045* B.C. (China research monograph 53). Berkeley: Institute for East Asian Studies, University of California, Berkeley.

Kern, M., 2002. Methodological reflections on the analysis of textual variants and the modes of manuscript production in early China, *Journal of East Asian Archaeology* 4(1–4), 143–81.

Kern, M., 2005. The Odes in excavated manuscripts, in *Text and Ritual in Early China*, ed. M. Kern. Seattle: University of Washington Press, 149–93.

Kim Yung Sik 金永植, 2004. The 'why not' question of Chinese science: the Scientific Revolution and traditional Chinese science, *East Asian Science, Technology, and Medicine* 22, 96–112.

Kitcher, P., 1993. *The Advancement of Science: Science without Legend, Objectivity without Illusions*. New York: Oxford University Press.

Knoblock, J. & J.K. Riegel, 2000. *The Annals of Lü Buwei: A Complete Translation and Study*. Stanford: Stanford University Press.

Kuhn, T.S., 1996. *The Structure of Scientific Revolutions*, 3rd ed. Chicago: The University of Chicago Press.

Lai Swee Fo 賴瑞和, 2003. Tangdai de Hanlin daizhao he Sitiantai 唐代的翰林待詔和司天臺, *Tang yanjiu* 唐研究 9, 315–42.

Lao Gan 勞乾, 1970. Shangsi kao 上巳考, *Minzuxue yanjiusuo jikan* 民族學研究所集刊 29(1), 243–62.

Latour, B., 1987. *Science in Action: How to Follow Scientists and Engineers through Society*. Cambridge, MA: Harvard University Press.

Latour, B., 1993. *We Have Never Been Modern*, tr. C. Porter. Cambridge, MA: Harvard University Press.

Latour, B., 1999. *Pandora's Hope: Essays on the Reality of Science Studies*. Cambridge, MA: Harvard University Press.

Latour, B. & S. Woolgar, 1986. *Laboratory Life: The Social Construction of Scientific Facts*, 2nd ed. Princeton: Princeton University Press.

Lau, D.C., 1970. *Mencius*. Harmondsworth: Penguin.

Lau, D.C., 2000. *Confucius: The Analects*. Hong Kong: The Chinese University Press.

Lee, T.H.C., 1985. *Government Education and Examinations in Sung China*. Hong Kong: Chinese University Press.

Lee, T.H.C., 2000. *Education in Traditional China: A History* (Handbuch der Orientalistik, Vierte Abteilung, China 13). Leiden: Brill.

Legge, J., 1967. *Li Chi: Book of Rites. An Encyclopedia of Ancient Ceremonial Usages, Religious Creeds, and Social Institutions*, 2nd ed. New Hyde Park, NY: University Books (first published 1885).

Lehoux, D., 2012. *What Did the Romans Know? An Inquiry into Science and Worldmaking*. Chicago: The University of Chicago Press.

Levinovitz, A., 2015. *The Gluten Lie: And Other Myths about What You Eat*. New York: Regan Arts.

Lewis, M.E., 2005. *The Construction of Space in Early China*. New York: State University of New York.

Li Feng 李峰 & D.P. Branner (eds.), 2011. *Writing & Literacy in Early China Studies from the Columbia Early China Seminar*. Seattle: University of Washington Press.

Li Guowei 李國偉, 1991. Zhongguo gudai dui jiaodu de renshi 中國古代對角度的認識, in *Shuxue shi yanjiu wenji* 數學史研究文集, ed. Li Di 李迪. Huhehaote: Nei Menggu daxue chubanshe; Taiwan Jiuzhang chubanshe, 6–14.

Li Hai 李海, 1994. Beiwei tie hunyi kao 北魏鐵渾儀考, *Yanbei shifan xueyuan xuebao* 雁北示範學院學報 10(2), 79–83.

Li Hongfu 李洪甫, 1982. Jiangsu Lianyungang shi Huaguoshan chutu de Handai jiandu 江蘇連雲港市花果山出土的漢代簡牘, *Kaogu* 考古 1982.5, 476–80.

Li Jiancheng 李鑒澄, 1989. Guiyi: Woguo xiancun zui gulao de tianwen yiqi 晷儀: 我國現存最古老的天文儀器, in *Zhongguo gudai tianwen wenwu lunji* 中國古代天文文物論集, ed. Zhongguo shehui kexue yuan kaogu yanjiusuo 中國社會科學院考古研究所. Beijing: Wenwu chubanshe, 145–53.

Li Junming 李均明, 2009. *Qin Han jiandu wenshu fenlei jijie* 秦漢簡牘文書分類輯解. Beijing: Wenwu chubanshe.

Li Ling 李零, 2006. *Zhongguo fangshu zheng kao* 中國方術正考. Beijing: Zhonghua shuju.

Li Ling 李零, 2008. Shiri, rishu he yeshu: san zhong jianbo wenxian de qubie he dingming 視日、日書和葉書: 三種簡帛文獻的區別和定名, *Wenwu* 文物 2008.12, 73–80.

Li Peidong 李培棟, 2009. *Jin shu* yanjiu 『晉書』研究, in *Jin shu,* Ba shu, Er shi *yanjiu* 晉書、『八書』、『二史』研究, ed. 周文玖, 1982, rpt. Beijing: Zhongguo da baikequanshu chubanshe, 44–76.

Li Tianhong 李天虹, 2012. Qin Han shifen jishi zhi zonglun 秦漢時分紀時制總論, *Kaogu xuebao* 考古學報 2012.3(186), 289–314.

Li Xueqin 李學勤, 2003. Meixian Yangjiacun xin chu qingtongqi yanjiu 眉縣楊家村新出青銅器研究, *Wenwu* 文物 2003.6, 66–73.

Li Zhichao 李志超, 2014. *Tianren gu yi: Zhongguo kexueshi lungang* 天人古義：中國科學史論綱, 3rd ed. Zhengzhou: Daxiang chubanshe.

Li Zhonglin 李忠林, 2010. Zhoujiatai Qin jian lipu xinian yu Qin shiqi lifa 周家臺秦簡曆譜係年與秦時期曆法, *Lishi yanjiu* 歷史研究 2010.6, 36–53.

Lien, Y.-C.E., 2011. Zhang Heng, Eastern Han polymath: his life and works, Ph.D. diss., University of Washington.

Lien, Y.E., 2012. Zhang Heng's *Huntian yi zhu* revisited, *T'oung Pao*, 2nd ser. 98(1–3), 31–64.

Lippiello, T., 2001. *Auspicious Omens and Miracles in Ancient China: Han, Three Kingdoms and Six Dynasties (Monumenta Serica monograph series 39)*. Sankt Augustin: Monumenta Serica Institute.

Liu Changdong 劉長東, 2012. Luoxia Hong de zushu zhi yuan ji huntian shuo, huntian yi suo qiyuan de zushu 落下閎的族屬之源暨渾天說、渾天儀所起源的族屬,

Sichuan daxue xuebao (zhexue shehui kexue ban) 四川大學學報（哲學社會科學版) 2012.5, 30–45.

Liu Ciyuan 劉次沅, 1987. Zhongguo gudai tianxiang jilu zhong de chi cun zhang danwei hanyi chutan 中國古代天象記錄中的尺寸丈單位含義初探, *Tianwen xuebao* 天文學報 28(4), 105–12.

Liu Ciyuan 劉次沅, 2015. *Zhu shi tianxiang jilu kaozheng* 諸史天象記錄考證 (Ershisi shi jiaoding yanjiu congkan 二十四史校訂研究叢刊). Beijing: Zhonghua shuju.

Liu Guosheng 劉國勝, 2009. Guanyu Zhoujiatai Qin jian 69–130 hao de jianxu bianpai wenti 關於周家臺秦簡69–130號的簡序編排問題, *Jianbo* 簡帛 4, 27–35.

Liu Hongtao 劉洪濤, 2003. *Gudai lifa jisuanfa* 古代曆法計算法. Tianjin: Nankai daxue chubanshe.

Liu Jinyi 劉金沂, 1984. Cong 'yuan' dao 'hun': Han chu ershiba xiu yuanpan de qishi 從「圓」到「渾」—漢初二十八宿圓盤的啟示, in *Zhongguo tianwenxue shi wenji* 中國天文學史文集, ed. Zhongguo tianwenxue shi wenji bianjizu 中國天文學史文集編輯組. Beijing: Kexue chubanshe, 205–13.

Liu Lexian 劉樂賢, 2002. *Jianbo shushu wenxian tanlun* 簡帛數術文獻探論. Wuhan: Hubei jiaoyu chubanshe.

Liu Lexian 劉樂賢, 2004. *Mawangdui tianwen shu kaoshi* 馬王堆天文書考釋. Guangzhou: Zhongshan daxue chubanshe.

Liu Lexian 劉樂賢, 2007. Weishu zhong de tianwen ziliao: *Hetu Dilanxi* wei li 緯書中的天文資料——『河圖帝覽嬉』爲例, *Zhongguo shi yanjiu* 中國史研究 2007.2, 71–82.

Lloyd, G.E.R., 1979. *Magic, Reason, and Experience: Studies in the Origin and Development of Greek Science*. Cambridge: Cambridge University Press.

Lloyd, G.E.R., 1996. *Adversaries and Authorities: Investigations into Ancient Greek and Chinese Science*. Cambridge: Cambridge University Press.

Lloyd, G.E.R., 2006. *Ancient Worlds, Modern Reflections: Philosophical Perspectives on Greek and Chinese Science and Culture*. Oxford: Oxford University Press.

Lloyd, G.E.R., 2007. *Cognitive Variations: Reflections on the Unity and Diversity of the Human Mind*. Oxford: Oxford University Press.

Lloyd, G.E.R., 2008. The Varying agenda of the study of the heavens: Mesopotamia, Greece, China, *Asia Major*, 3rd ser., 21(1), 69–88.

Lloyd, G.E.R. & N. Sivin, 2002. *The Way and the Word: Science and Medicine in Early China and Greece*. New Haven: Yale University Press.

Loewe, M., 1959. Some notes on Han-time documents from Chüyen, *T'oung Pao*, 2nd ser. 47(3–5), 294–322.

Loewe, M., 1965. The wooden and bamboo strips found at Mo-chü-tzu (Kansu), *Journal of the Royal Asiatic Society of Great Britain and Ireland* (1–2), 13–26.

Loewe, M., 1967. *Records of Han Administration*. Cambridge: Cambridge University Press (2 vols.).

Loewe, M. (ed.), 1993. *Early Chinese Texts: A Bibliographical Guide*. Berkeley: Institute of East Asian Studies, University of California.

Lu Yang 盧央, 2007. *Zhongguo gudai xingzhanxue* 中國古代星占學. Beijing: Zhongguo kexue jishu chubanshe.

Lü Zifang 呂子方, 1983. *Zhongguo kexue jishu shilun wenji* 中國科學技術史論文集. Chengdu: Sichuan renmin chubanshe (2 vols.).

Lü Zongli 呂宗力, 2003. *Power of the Words: Chen Prophecy in Chinese Politics,* AD *265–618*. Oxford: Peter Lang.

Luo Changpei 羅常培 & Zhou Zumo 周祖謨, 2007. *Han, Wei, Jin, Nanbeichao yunbu yanbian yanjiu* 漢魏晉南北朝韻部演變研究. Beijing: Zhonghua shuju.

Luo Jianjin 羅見今, 1999. Dunhuang Han jian zhong lipu niandai zhi zai yanjiu 敦煌漢簡中曆譜年代之再研究, *Dunhuang yanjiu* 敦煌研究 61(3), 91–100.

Luo Jianjin 羅見今 & Guan Shouyi 關守義, 2000. Dunhuang, Juyan Hanjian zhong yu shuorunbiao bu he zhu jian kaoshi 敦煌、居延漢簡中與朔閏表不合諸簡考釋, *Wen shi* 文史 (1), 57–72.

Ma Yi 馬怡, 2006. Handai de jishiqi ji xiangguan wenti 漢代的計時器及相關問題, *Zhongguo shi yanjiu* 中國史研究 2006.3, 17–36.

Maeyama Yasukatsu 前山保勝, 1975. On the astronomical data of ancient China (*ca.* −100 ~ +200): a numerical analysis (part I), *Archives internationales d'histoire des sciences* XXV/XCVii, 247–76.

Maeyama Yasukatsu 前山保勝, 1976. On the astronomical data of ancient China (*ca.* −100 ~ +200): a numerical analysis (part II), *Archives internationales d'histoire des sciences* XXVi/XCVii, 27–58.

Maeyama Yasukatsu 前山保勝, 1977. The oldest star catalogue of China, Shi Shen's *Hsing ching*, in *Prismata: Naturwissenschaftsgeschichtliche Studien: Festschrift Für Willy Hartner*, ed. W. Hartner, W.G. Saltzer & Maeyama Yasukatsu 前山保勝. Wiesbaden: F. Steiner, 211–45.

Major, J.S., 1993. *Heaven and Earth in Early Han Thought: Chapters Three, Four and Five of the Huainanzi*. Albany: State University of New York Press.

Major, J.S. (ed.), 2010. *The Huainanzi: A Guide to the Theory and Practice of Government in Early Han China*. New York: Columbia University Press.

Makeham, J., 2002. *Balanced Discourses*. New Haven: Yale University Press.

Mansvelt Beck, B.J., 1990. *The Treatises of Later Han: Their Author, Sources, Contents, and Place in Chinese Historiography*. Leiden: E.J. Brill.

Martzloff, J.-C., 2009. *Le calendrier chinois: Structure et calculs, 104 av. JC–1644: Indétermination céleste et réforme permanente: La construction chinoise officielle du temps quotidien discret à partir d'un temps mathématique caché, linéaire et continu* (Sciences, techniques et civilisations du Moyen Âge à l'aube des Lumières 11). Paris: Champion.

Maspero, H., 1939. Les instruments astronomiques des Chinois au temps des Han, *Mélanges chinois et bouddhiques* 6, 183–370.

Maspero, H., 1951. Le *ming-t'ang* et la crise religieuse chinoise avant les Han, *Mélanges chinois et bouddhiques* 9, 1–71.

Meeus, J., 1998. *Astronomical Algorithms*. Richmond, VA: Willmann-Bell.

Merton, R.K., 1968. The Matthew effect in science, *Science* 159(3810), 56–63.

Meyer, D., 2012. *Philosophy on Bamboo: Text and the Production of Meaning in Early China*. Leiden: Brill.

Minkowski, C., 2002. Astronomers and their reasons: working paper on Jyotiḥśāstra, *Journal of Indian philosophy* 30(5), 495–514.

Mo Zihan 墨子涵 (Daniel P. Morgan), 2011. Cong Zhoujiatai *Rishu* yu Mawangdui *Wu xing zhan* tan rishu yu Qin-Han tianwenxue de huxiang yingxiang 從周家臺『日書』與馬王堆『五星占』談日書與秦漢天文學的互相影響, *Jianbo* 簡帛 6, 113–37.

Mo Zihan 墨子涵 (Daniel P. Morgan) & Lin Lina 林力娜 (Karine Chemla), 2016. Ye you lunzhe xiede: Zhangjiashan Han jian *Suan shu shu* xieshou yu bianxu chutan 也有輪着寫的：張家山漢簡『筭數書』寫手與編序初探, *Jianbo* 簡帛 12, 235–51.

Morgan, D.P., 2015. What good's a text? Textuality, orality, and mathematical astronomy in early imperial China, *Archives internationales d'histoire des sciences* 65(2), 549–72.

Morgan, D.P., 2016a. Mercury and the case for plural planetary traditions in early imperial China, in *The Circulation of Astronomical Knowledge in the Ancient World*, ed. J.M. Steele. Leiden: Brill, 416–50.

Morgan, D.P., 2016b. The planetary visibility tables in the second-century BC manuscript *Wu xing zhan* 五星占, *East Asian Science, Technology, and Medicine* 43, 17–60.

Morgan, D.P., 2016c. Sphere confusion: towards a textual reconstruction of instruments and observational practice in first-millennium CE China, *Centaurus* 58(1–2), 87–103.

Müller, W.K., 1976. Shingon-Mysticism Śubhākarasiṁha and I-Hsing's Commentary to the Mahavāirocana-Sūtra, Chapter One an Annotated Translation, Ph.D. diss., University of California, Los Angeles.

Nagata Hidemasa 永田英正, 1972. Kandai no shūgi ni tsuite 漢代の集議について, *Tōhō gakuhō* 東方學報 43, 97–136.

Nakayama Shigeru 中山茂, 1964. *Senseijutsu: sono kagaku shijō no ichi* 占星術：その科學史上の位置. Tōkyō: Kinokuniya shoten.

Nakayama Shigeru 中山茂, 1965. Characteristics of Chinese calendrical science, *Japanese Studies in the History of Science* (4), 124–31.

Nakayama Shigeru 中山茂, 1966. Characteristics of Chinese astrology, *Isis* 57(4), 442–54.

Needham, J., 1959. *Science and Civilisation in China, vol. 3: Mathematics and the Sciences of the Heavens and the Earth.* Cambridge: Cambridge University Press.

Needham, J., 1965. *Time and Eastern Man*. London: Royal Anthropological Institute of Great Britain & Ireland.

Needham, J., 1986. *The Hall of Heavenly Records: Korean Astronomical Instruments and Clocks, 1380–1780*. Cambridge: Cambridge University Press.

Needham, J., Wang Ling & J. de S.P. Derek, 1986. *Heavenly Clockwork: The Great Astronomical Clocks of Medieval China*, 2nd ed. (Antiquarian Horological Society monograph 1). Cambridge: Cambridge University Press.

Neugebauer, O., 1942. The origin of the Egyptian calendar, *Journal of Near Eastern Studies* 1(4), 396–403.

Neurath, O., 1973. *Empiricism and Sociology* (Vienna Circle collection). Dordrecht: D. Reidel Pub. Co.

Newton-Smith, W.H., 1981. *The Rationality of Science*. Boston: Routledge & Kegan Paul.

Nielsen, B., 2003. *A Companion to* Yi Jing *Numerology and Cosmology: Chinese Studies of Images and Numbers from Han (202* BCE*–220* CE*) to Song (960–1279* CE*)*. London: RoutledgeCurzon.

Nielsen, B., 2009–10. Calculating the fall of a dynasty: divination based on the *Qian zuo du, Zhouyi Studies* 6(1), 65–107.

Niu Weixing 鈕衛星, 2004. Han Tang zhi ji lifa gaige zhong de zuoyong yinsu zhi fenxi 漢唐之際曆法改革中各作用因素之分析, *Shanghai jiaotong daxue xuebao (zhexue shehui kexue ban)* 上海交通大學學報（哲學社會科學版) 12(5), 33–8, 54.

Ōhashi Yukio 大橋由紀夫, 1982. Kōkan Sibunreki no seiritsu katei 後漢四分曆の成立過程, *Sūgakushi kenkyū* 數學史研究 93, 1–27.
Ōhashi Yukio 大橋由紀夫, 1994. Astronomical instruments in classical siddhāntas, *Indian Journal of History of Science* 29(2), 155–313.
Ōhashi Yukio 大橋由紀夫, 1997. Ka Ki no temmon teisūkan ni tsuite 賈逵の天文定數觀について, *Sūgakushi kenkyū* 數學史研究 153, 1–17.
Olberding, G.P.S., 2012. *Dubious Facts: The Evidence of Early Chinese Historiography*. Albany: State University of New York Press.
Osabe Kazuo 長部和雄, 1963. *Ichigyō zenji no kenkyū* 一行禪師の研究. Kobe: Shōka Daigaku Keizai Kenkyūsho.
Pan Nai 潘鼐, 1989. *Zhongguo hengxing guance shi* 中國恆星觀測史. Shanghai: Xuelin chubanshe.
Pan Nai 潘鼐 (ed.), 2005. *Zhongguo gu tianwen yiqi shi* 中國古天文儀器史, colour print ed. Taiyuan: Shanxi jiaoyu chubanshe.
Pankenier, D.W., 2013. *Astrology and Cosmology in Early China: Conforming Earth to Heaven*. Cambridge: Cambridge University Press.
Pankenier, D.W., 2015. Weaving metaphors and cosmo-political thought in early China, *T'oung Pao*, 2nd ser. 101(1–3), 1–34.
Payne, R., 2006. Introduction, in *Tantric Buddhism in East Asia*, ed. R. Payne. Boston: Wisdom Publications, 1–31.
Pingree, D.E., 1967–8. The *Paitāmaha-siddhānta* of the *Viṣṇudharmottarapurāṇa*, *Bhramavidyā* 31–2, 472–510.
Pingree, D.E., 1978. History of mathematical astronomy in India, *Dictionary of Scientific Biography* 15, 533–633.
Plofker, K., 2009. *Mathematics in India*. Princeton: Princeton University Press.
Puett, M.J., 2004. *To Become a God: Cosmology, Sacrifice, and Self-Divinization in Early China*. Cambridge, MA: Harvard University Press.
Puett, M.J., 2007. Humans, spirits, and sages in Chinese late antiquity: Ge Hong's 葛洪 *Master Who Embraces Simplicity* (*Baopuzi*) 抱朴子, *Extrême-Orient, Extrême-Occident* 29(29), 95–119.
Putnam, H., 1974. The 'corroboration' of theories, in *The Philosophy of Karl Popper*, ed. P.A. Schlipp. La Salle, IL: Open Court, 221–40.
Qian Baocong 錢寶琮, 1983. *Qian Baocong kexueshi lunwen xuanji* 錢寶琮科學史論文選集. Beijing: Kexue chubanshe.
Qin Jianming 秦建明, 2008. Gaitian tuyi kao 蓋天圖儀考, *Wenbo* 文博 2008.1, 4–10.
Qiu Guangming 丘光明, 1992. *Zhongguo lidai du liang heng kao* 中國歷代度量衡考. Beijing: Kexue chubanshe.
Qu Anjing 曲安京, 1993. Li Chunfeng deng ren gaitianshuo rigao gongshi xiuzheng an yanjiu 李淳風等人蓋天說日高公式修正案研究, *Ziran kexue shi yanjiu* 自然科學史研究 12(1), 42–51.
Qu Anjing 曲安京, 1994. Zhongguo gudai lifa zhong de jishi zhidu 中國古代曆法中的計時制度, *Hanxue yanjiu* 漢學研究 12(2), 157–72.
Qu Anjing 曲安京, 2008. *Zhongguo shuli tianwenxue* 中國數理天文學 (Shuxue yu kexue shi congshu 數學與科學史叢書 4). Beijing: Kexue chubanshe.
Ran Zhaode 冉昭德, 2009. Guanyu Jin shi de zhuanshu yu Tang xiu *Jin shu* zhuanren wenti 關於晉史的撰述與唐修『晉書』撰人問題, in *Jin shu,* Ba shu, Er shi

yanjiu 晉書、『八書』、『二史』研究, ed. Zhou Wenjiu 周文玖, rpt. Beijing: Zhongguo da baikequanshu chubanshe, 14–27 (first published 1957).

Raz, G., 2005. Time manipulation in early Daoist ritual: the East Well Chart and the Eight Archivists, *Asia Major*, 3rd ser. 18(1), 67–102.

Ren Song 任松, 2008. Lun Tao Hongjing 'ziran louke' de gouxiang 論陶弘景「自然漏刻」的構想, *Hunan gongye daxue xuebao (shehui kexue ban)* 湖南工業大學學報（社會科學版) 13(2), 73–4.

Richter, M.L., 2013. *The Embodied Text: Establishing Textual Identity in Early Chinese Manuscripts*. Leiden: Brill.

Robson, E., 2007. Gendered literacy and numeracy in the Sumerian literary corpus, in *Analysing Literary Sumerian: Corpus-Based Approaches*, ed. J. Ebeling & G. Cunningham. London and Oakville, CT: Equinox, 215–49.

Rochberg, F., 2004. *The Heavenly Writing: Divination, Horoscopy, and Astronomy in Mesopotamian Culture*. Cambridge: Cambridge University Press.

Rogers, M.C., 1968. The myth of the Battle of the Fei River (A.D. 383), *T'oung Pao*, 2nd ser. 54(1–3), 50–72.

Rufus, W.C. & H. Tien, 1945. *The Soochow Astronomical Chart*. Ann Arbor: University of Michigan Press.

Rutt, R., 1996. *The Book of Changes (*Zhouyi*): A Bronze Age Document Translated with Introduction and Notes* (Durham East Asia series 1). Richmond: Curzon.

Said, E.W., 1979. *Orientalism*. New York: Vintage Books.

Saitō Kuniji 齊藤國治 & Ozawa Kenji 小澤賢二, 1992. *Chūgoku kodai no tenmon kiroku no kenshō* 中國古代の天文記錄の檢證. Tōkyō: Yūzankaku.

Salmon, W.C., 1981. Rational prediction, *British Journal for the Philosophy of Science* 32(2), 115–25.

Sanft, C., 2008–9. Edict of monthly ordinances for the four seasons in fifty articles from 5 C.E.: introduction to the wall inscription discovered at Xuanquanzhi, with annotated translation, *Early China* 32, 125–208.

Sarma, S.R., 2009. *Sanskrit Astronomical Instruments in the Maharaja Sayajirao University of Baroda* (M.S. University Oriental Series 24). Vadodara: Oriental Institute.

Sayili, A., 1960. *The Observatory in Islam and Its Place in the General History of the Observatory.* Ankara: Türk Tarih Kurumu Basımevi.

Schaberg, D., 2001. *A Patterned Past: Form and Thought in Early Chinese Historiography.* Cambridge, MA: Harvard University Asia Center.

Schafer, E.H., 1977. *Pacing the Void: T'ang Approaches to the Stars*. Berkeley: University of California Press.

Schipper, K.M. & Wang Hsiu-huei 王秀惠, 1986. Progressive and regressive time cycles in Taoist ritual, in *Time, Science, and Society in China and the West*, ed. J.T. Fraser, N.M. Lawrence & F.C. Haber. Amherst: University of Massachusetts Press, 185–205.

Schlick, M., 1932. The future of philosophy, *College of the Pacific Publications in Philosophy* I, 45–62.

Schuessler, A., 2007. *ABC Etymological Dictionary of Old Chinese*. Honolulu: University of Hawai'i Press.

Shapin, S., 1994. *A Social History of Truth: Civility and Science in Seventeenth-Century England*. Chicago: The University of Chicago Press.

Shapin, S. & S. Schaffer, 1985. *Leviathan and the Air-Pump: Hobbes, Boyle, and the Experimental Life*. Princeton: Princeton University Press.

Shaw, M.J., 2011. *Time and the French Revolution: The Republican Calendar, 1789–Year XIV* (Royal Historical Society studies in history). Woodbridge: Royal Historical Society/Boydell Press.

Shi Yunli 石云里, Fang Lin 方林 & Han Chao 韓朝, 2012. Xihan Xiahou Zao mu chutu tianwen yiqi xintan 西漢夏侯竈墓出土天文儀器新談, *Ziran kexue shi yanjiu* 自然科學史研究 31(1), 1–13.

Shukla, K.S. (ed.), 1976. *Āryabhaṭīya with the Commentary of Bhāskara and Someśvara* (Āryabhaṭīya Critical Edition Series: Published on the occasion of the celebration of the 1500th birth anniversary of Āryabhaṭa 2). New Delhi: Indian National Science Academy.

Sivin, N., 1969. Cosmos and computation in early Chinese mathematical astronomy, *T'oung Pao*, 2nd ser. 55(1–3), 1–73.

Sivin, N., 1982. Why the Scientific Revolution did not take place in China—or didn't it?, *Chinese Science* 5, 45–66.

Sivin, N., 1986. On the limits of empirical knowledge in the traditional Chinese sciences, in *Time, Science, and Society in China and the West*, ed. J.T. Fraser, N.M. Lawrence & F.C. Haber. Amherst: University of Massachusetts Press, 151–69.

Sivin, N., 2009. *Granting the Seasons: The Chinese Astronomical Reform of 1280, with a Study of Its Many Dimensions and a Translation of Its Records*. New York: Springer.

Smith, A., 2011. The Chinese sexagenary cycle and the ritual foundations of the calendar, in *Calendars and Years II: Astronomy and Time in the Ancient and Medieval World*, ed. J.M. Steele. Oxford: Oxbow Books, 1–37.

Song Huiqun 宋會羣, 1999. *Zhongguo shushu wenhua shi* 中國術數文化史. Kaifeng: Henan daxue chubanshe.

Soothill, W.E., 1951. *The Hall of Light: A Study of Early Chinese Kingship*. London: Lutterworth Press.

Soper, A.C., 1959. *Literary Evidence for Early Buddhist Art in China* (Artibus Asiae., supplementum 19). Ascona: Artibus Asiae.

Steele, J.C., 1917. *The 'I Li', or Book of Etiquette and Ceremonial* (Probsthain's oriental series 8). London: Probsthain.

Steele, J.M., 2000. *Observations and Predictions of Eclipse Times by Early Astronomers* (Archimedes: new studies in the history and philosophy of science and technology). Dordrecht: Kluwer Academic Publishers.

Stephenson, F.R., 1997. *Historical Eclipses and Earth's Rotation*. Cambridge: Cambridge University Press.

Su Junlin 蘇俊林, 2010. Guanyu 'zhiri' jian de mingcheng yu xingzhi 關於「質日」簡的名稱與性質, *Hunan daxue xuebao (shehui kexue ban)* 湖南大學學報（社會科學版） 24(4), 17–22.

Sugimoto Kenji 杉本憲司, 1973. Kandai no taishō ni tsuite 漢代の待詔について, *Shakai kagaku ronshū* 社會科學論集 4(5), 85–99.

Sukhu, G., 2005–6. Yao, Shun, and prefiguration: the origins and ideology of the Han imperial genealogy, *Early China* 30, 91–151.

Sun Xiaochun 孫小淳, 2015. Chinese armillary spheres, in *Handbook of Archaeoastronomy and Ethnoastronomy*, ed. C.L.N. Ruggles. New York: Springer, 2127–32.

Sun Xiaochun 孫小淳 & J. Kistemaker, 1997. *The Chinese Sky during the Han: Constellating Stars and Society*. Leiden: Brill.

Takeda Tokimasa 武田時昌, 1989. Isho rekihō kō: Zenkan matsu no keigaku to kagaku no kōryū 緯書暦法考：前漢末の經學と科學の交流, in *Chūgoku kodai kagakushi ron* 中國古代科學史論, ed. Yamada Keiji 山田慶兒. Kyōto: Kyōto daigaku jinbun kagaku kenkyūjo, 55–120.

Tambiah, S.J., 1990. *Magic, Science, Religion, and the Scope of Rationality*. Cambridge: Cambridge University Press.

Tao Lei 陶磊, 2003. *'Huainanzi Tianwen' yanjiu: cong shushu de jiaodu* 『淮南子·天文』研究——從數術的角度. Jinan: Qi Lu shushe.

Tavernor, R., 2007. *Smoot's Ear: The Measure of Humanity*. New Haven: Yale University Press.

Teboul, M., 1983. *Les premières théories planétaires chinoises*. Paris: Collège de France.

Ting Pang-hsin 丁邦新, 1975. *Chinese Phonology of the Wei-Chin Period: Reconstruction of the Finals as Reflected in Poetry* (Special publications (Institute of History and Philology, Academia Sinica) 65). Taipei: Institute of History and Philology, Academia Sinica.

Toomer, G.J., 1998. *Ptolemy's Almagest*, 2nd ed. Princeton: Princeton University Press.

Tseng, L.L., 2011. *Picturing Heaven in Early China*. Cambridge, MA: Harvard University Asia Center.

Twitchett, D., 2002. *The Writing of Official History under the T'ang*. Cambridge: Cambridge University Press.

Vogel, H.U., 1994. Aspects of metrosophy and metrology during the Han period, *Extrême-Orient, Extrême-Occident* 16, 135–52.

Wang Shengli 王勝利, 1982. 'Jiudao' gaishuo 「九道」該說, *Lishi yanjiu* 歷史研究 2, 91–101.

Wang Su 王素, 1989. Qushi Gaochang lifa chutan 麴氏高昌曆法初探, in *Chutu wenxian yanjiu xuji* 出土文獻研究續集, ed. Zhang Qingling 張慶玲 & Guojia wenwu ju gu wenxian yanjiu shi 國家文物局古文獻研究室. Beijing: Wenwu chubanshe, 148–80.

Wang Yumin 王玉民, 2008. *Yi chi liang tian: Zhongguo gudai mushi chidu tianxiang jilu de lianghua yu guisuan* 以尺量天—中國古代目視尺度天象記錄的量化與歸算. Jinan: Shandong jiaoyu chubanshe.

Wang Yumin 王玉民, 2015. 'Huntian-yi' kao 「渾天儀」考, *Zhongguo keji shuyu* 中國科技術語 2015.3, 39–42.

Wang Zhiping 王志平, 1998. Chuboshu yueming xintan 楚帛書月名新探, *Huaxue* 華學 3, 181–8.

Ware, J.R., 1966. *Alchemy, Medicine and Religion in the China of* A.D. *320: The Nei P'ien of Ko Hung*. Cambridge, MA: MIT.

White, W.C. & P.M. Millman, 1938. An ancient Chinese sun-dial, *Journal of the Royal Astronomical Society of Canada* 32, 417–30.

Whitfield, S., 1998. Under the censor's eye: printed almanacs and censorship in ninth-century China, *British Library Journal* 24(1), 4–22.

Wilhelm, R., 1967. *The I Ching Or, Book of Changes*, 3d ed. (Bollingen series XIX). Princeton: Princeton University Press.

Wilson, T.A. (ed.), 2002. *On Sacred Grounds: Culture, Society, Politics, and the Formation of the Cult of Confucius*. Cambridge, MA: Harvard University Asia Center.

Włodarczyk, J., 1987. Observing with the armillary astrolabe, *Journal for the History of Astronomy* 18, 173–95.

Wu Hung 巫鴻, 1989. *The Wu Liang Shrine: The Ideology of Early Chinese Pictorial Art*. Stanford: Stanford University Press.

Wu Jiabi 武家璧, 2007. *Yiwei: Tonggua yan* zhong de guiying shuju 『易緯·通卦驗』中的晷影數據, *Zhouyi yanjiu* 周易研究 2007.3, 89–94.

Wu Shouxian 吳守賢 & Quan Hejun 全和鈞, 2008. *Zhongguo gudai tianti celiangxue ji tianwen yiqi* 中國古代天體測量學及天文儀器 (Zhongguo tianwenxueshi daxi 中國天文學史大系). Beijing: Zhongguo kexue jishu chubanshe.

Xia De'an 夏德安 (Donald Harper), 2007. Zhoujiatai de shushu jian 周家臺的數術簡, *Jianbo* 簡帛 2, 397–407.

Xu Jie 許結, 1999. *Zhang Heng pingzhuan* 張衡評傳 (Zhongguo sixiangjia pingzhuan congshu 中國思想家評傳叢書 26). Nanjing: Nanjing daxue chubanshe.

Xu Xingwu 徐興無, 2003. *Chenwei wenxian yu Handai wenhua goujian* 讖緯文獻與漢代文化構建. Beijing: Zhonghua shuju.

Xu Xingwu 徐興無, 2005. *Liu Xiang pingzhuan: fu Liu Xin pingzhuan* 劉向評傳：附劉歆評傳 (Zhongguo sixiangjia pingzhuan congshu 中國思想家評傳叢書 21). Nanjing: Nanjing daxue chubanshe.

Yabuuti Kiyosi 藪内清, 1963. Astronomical tables in China, from the Han to the T'ang dynasties, in *Chūgoku chūsei kagaku gijutsushi no kenkyū* 中國中世科學技術史の研究, ed. Yabuuti Kiyosi 藪内清. Tōkyō: Kadokawa shoten, 445–92.

Yabuuti Kiyosi 藪内清, 1969. *Chūgoku no temmon rekihō* 中國の天文曆法. Tōkyō: Heibonsha.

Yang Hongnian 楊鴻年, 1985. *Han–Wei zhidu congkao* 漢魏制度叢考. Wuhan: Wuhan daxue chubanshe.

Yang Xiaoliang 楊小亮, 2015. Xihan *Jushe yuan nian liri* zhuihe fuyuan yanjiu 西漢『居攝元年曆日』綴合復原研究, *Wenwu* 文物 2015.3, 70–7.

Yoshimura Masayuki 吉村昌之, 2003. Shutsudo kandoku shiryō ni mirareru rekihu no shūsei 出土簡牘資料にみれる曆譜の集成, in *Henkyō shutsudo mokkan no kenkyū* 邊疆出土木簡の研究, ed. Tomiya Itaru 冨谷至. Kyōto: Hōyū shoten, 459–516.

Yu Shulin 余書麟, 1966. Liang Han sixue yanjiu 兩漢私學研究, *Shida xuebao* 師大學報 11, 109–48.

Yu Yunhao 吳蘊豪 & Li Geng 黎耕, 2008. 'Yi gai' shuo yu *Zhoubi suanjing* yuzhou moxing de zai sikao 「倚蓋」說與『周髀算經』宇宙模型的再思考, *Zhongguo kejishi zazhi* 中國科技史雜誌 29(4), 358–63.

Yu Zhongxin 俞忠鑫, 1994. *Han jian kao li* 漢簡考曆. Taipei: Wenjin chubanshe.

Yuan Min 遠敏 & Qu Anjing 曲安京, 2008. Liang Wudi de gaitianshuo moxing 梁武帝的蓋天說模型, *Kexue jishu yu bianzhengfa* 科學技術與辯證法 25(2), 85–9, 104.

Zhang Peiyu 張培瑜, 1997. *Sanqianwubai nian liri tianxiang* 三千五百年曆日天象. Zhengzhou: Daxiang chubanshe.

Zhang Peiyu 張培瑜, 2007. Genju xinchu liri jiandu shilun Qin he Han chu de lifa 根据新出歷日簡牘試論秦和漢初的曆法, *Zhongyuan wenwu* 中原文物 2007.5, 62–77.

Zhang Peiyu 張培瑜, Chen Meidong 陳美東, Bo Shuren 薄樹人 & Hu Tiezhu 胡鐵珠, 2008. *Zhongguo gudai lifa* 中國古代曆法 (Zhongguo tianwenxueshi daxi 中國天文學史大系). Beijing: Zhongguo kexue jishu chubanshe.

Zhang Peiyu 張培瑜 & Zhang Chunlong 張春龍, 2006. Qindai lifa he Zhuanxu li 秦代曆法和顓頊曆, in *Liye fa jue bao gao* 里耶發掘報告, ed. Hunan Sheng wenwu kaogu yanjiusuo 湖南省文物考古研究所. Changsha: Yuelu shushe, 735–47.

Zhang Wenyu 張聞玉, 2008. *Gudai tianwen lifa jiangzuo* 古代天文曆法講座. Guilin: Guangxi shifan daxue chubanshe.

Zhmud, L., 2006. *The Origin of the History of Science in Classical Antiquity* (Peripatoi 19). Berlin: de Gruyter.

Zhu Wenxin 朱文鑫, 1933. *Tianwen kaogu lu* 天文考古錄. Shanghai: Shangwu yinshuguan.

Zhuang Weifeng 莊威鳳, 2009. *Zhongguo gudai tianxiang jilu de yanjiu yu yingyong* 中國古代天象記錄的研究與應用 (Zhongguo tianwenxueshi daxi 中國天文學史大系). Beijing: Zhongguo kexue jishu chubanshe.

Index

Printed in the United States
by Baker & Taylor Publisher Services